CELLULAR MICROBIOLOGY

CELLULAR MICROBIOLOGY

EDITORS

Pascale Cossart
Unité des Interactions Bactéries-Cellules, Institut Pasteur,
Paris, France

Patrice Boquet
Unité INSERM 452, Faculté de Médecine, Nice, France

Staffan Normark
Microbiology and Tumor Biology Center, Karolinska Institute,
Stockholm, Sweden

Rino Rappuoli
IRIS, Chiron SpA, Siena, Italy

ASM
PRESS
Washington, D.C.

Copyright © 2000 ASM Press
American Society for Microbiology
1752 N Street NW
Washington, DC 20036

Library of Congress Cataloging-in-Publication Data

Cellular microbiology / editors, Pascale Cossart . . . [et al.].
    p.   cm.
  Includes bibliographical references and index.
  ISBN 1-55581-157-4 (hc.)
  1. Virulence. 2. Infection. 3. Host-bacteria relationships. 4. Pathology, Cellular. I.
Cossart, Pascale.

QR175.C43 2000
571.6'29 21—dc21

99-044931

10   9   8   7   6   5   4   3   2   1

Cover and interior design: Susan Brown Schmidler

Cover photo: Induction of virulence gene expression after contact between *Yersinia* and target cells. *Yersinia pseudotuberculosis* harboring a transcriptional fusion between the *yopE* promoter (encoding the cytotoxin) and the *luxAB* operon (encoding luciferase from *Vibrio harveyi*) was used to infect HeLa cell monolayers. Only bacteria in contact with the target cell increased their rate of *yopE* transcription as monitored by direct light emission, using a combination of differential interference contrast microscopy and a Photometrics CCD camera. The pseudo-color image highlights bacteria in yellow, the target HeLa cell in blue, and transcriptional activity from cell-associated bacteria in white (highest intensity) to magenta (lowest intensity). Transcriptional activity was in response to a rapid decrease in cytoplasmic LcrQ, a repressive element of *yop* expression (see chapter 13, Box 13.3, and *Science* **273**:1231–1233, 1996).

Contents

CHAPTER 5

Molecular Basis for Cell Adhesion and Adhesion-Mediated Signaling   81

BENJAMIN GEIGER, AVRI BEN-ZE'EV, ELI ZAMIR, AND ALEXANDER D. BERSHADSKY

CHAPTER 6

Cell Adhesion Molecules and Bacterial Pathogens   97

GUY TRAN VAN NHIEU AND PHILIPPE J. SANSONETTI

CHAPTER 18

Promising New Tools for Virulence Gene Discovery 333

TIMOTHY K. MCDANIEL AND RAPHAEL H. VALDIVIA

Contributors

KLAUS AKTORIES
Institut für Pharmakologie und Toxikologie der Albert-
Ludwigs-Universität Freiburg, Hermann-Herder-Str. 5, D-
79104 Freiburg, Germany

AVRI BEN-ZE'EV
Department of Molecular Cell Biology, Weizmann Institute
of Science, Rehovot 76100, Israel

ALEXANDER D. BERSHADSKY
Department of Molecular Cell Biology, Weizmann Institute
of Science, Rehovot 76100, Israel

PATRICE BOQUET
Unité INSERM 452, Faculté de Médecine, 28 Avenue de
Valombrose, 06102 Nice, France

MICHAEL CAPARON
Department of Molecular Microbiology, Washington
University School of Medicine, St. Louis, MO 63110

G. SINGH CHHATWAL
Department of Pathogenicity and Vaccine Research,
Biozentrum der Technischen Universität, Spielmannstrasse
7, D-38106 Braunschweig, Germany

PETER J. CHRISTIE
Department of Microbiology and Molecular Genetics, The
University of Texas Health Science Center at Houston, 6431
Fannin, Houston, TX 77030

PASCALE COSSART
Unité des Interactions Bactéries-Cellules, Institut Pasteur,
28 rue du Dr. Roux, F-75015 Paris, France

ANTONELLO COVACCI
Chiron SpA, Via Fiorentina, 53100 Siena, Italy

CHANTAL DE CHASTELLIER
INSERM U 411, UFR de Médecine Necker, Paris, France

B. BRETT FINLAY
Biotechnology Laboratory and Departments of
Biochemistry & Molecular Biology and Microbiology &
Immunology, University of British Columbia, Vancouver,
British Columbia, Canada V6T 1Z3

ÅKE FORSBERG
Department of Microbiology, Defence Research
Establishment, S-901 82 Umeå, Sweden

MATTHEW S. FRANCIS
Department of Cell and Molecular Biology, Umeå
University, S-901 87 Umeå, Sweden

RALUCA GAGESCU
Department of Biochemistry, University of Geneva, 1211
Geneva 4, Switzerland

BENJAMIN GEIGER
Department of Molecular Cell Biology, Weizmann Institute
of Science, Rehovot 76100, Israel

JEAN GRUENBERG
Department of Biochemistry, University of Geneva, 1211
Geneva 4, Switzerland

ILONA IDANPAAN-HEIKKILA
Skirball Institute, Department of Microbiology and Kaplan
Cancer Center, New York University School of Medicine,
540 First Avenue, New York, NY 10016

MARC LECUIT
Unité des Interactions Bactéries-Cellules, Institut Pasteur,
28 rue du Dr. Roux, F-75015 Paris, France

SANDRA J. MCCALLUM
Departments of Biochemistry and Microbiology &
Immunology, Stanford University Medical School, Stanford,
CA 94035-5307

TIMOTHY K. MCDANIEL
Department of Microbiology & Immunology, Stanford
University School of Medicine, 299 Campus Drive,
Stanford, CA 94305-5124

JEREMY E. MOSS
Skirball Institute, Department of Microbiology and Kaplan
Cancer Center, New York University School of Medicine,
540 First Avenue, New York, NY 10016

MARIAGRAZIA PIZZA
IRIS, Chiron SpA, Via Fiorentina 1, 53100 Siena, Italy

KLAUS T. PREISSNER
Institute for Biochemistry, Medical Faculty, Justus-Liebig-
Universität, Friedrichstrasse 24, D-35392 Giessen, Germany

DANIEL L. PURICH
Department of Biochemistry & Molecular Biology,
University of Florida College of Medicine Health Science
Center, Gainesville, FL 32610-0277

RINO RAPPUOLI
IRIS, Chiron SpA, Via Fiorentina 1, 53100 Siena, Italy

DAVID G. RUSSELL
Department of Molecular Microbiology, Washington
University School of Medicine, 660 South Euclid Avenue,
St. Louis, MO 63110

PHILIPPE J. SANSONETTI
Unité de Pathogénie Microbienne Moléculaire, INSERM
U389, Institut Pasteur, 28 rue du Dr. Roux, 75724 Paris
Cedex 15, France

KURT SCHESSER
Department of Cell and Molecular Biology, Umeå
University, S-901 87 Umeå, Sweden

FREDERICK S. SOUTHWICK
Departments of Medicine and Biochemistry & Molecular
Biology, University of Florida College of Medicine Health
Science Center, Gainesville, FL 32610-0277

JULIE A. THERIOT
Departments of Biochemistry and Microbiology &
Immunology, Stanford University Medical School, Stanford,
CA 94035-5307

GUY TRAN VAN NHIEU
Unité de Pathogénie Microbienne Moléculaire, INSERM
U389, Institut Pasteur, 28 rue du Dr. Roux, 75724 Paris
Cedex 15, France

EMIL R. UNANUE
Department of Pathology and Center for Immunology,
Washington University School of Medicine, St. Louis, MO
63110

RAPHAEL H. VALDIVIA
Department of Microbiology & Immunology, Stanford
University School of Medicine, 299 Campus Drive,
Stanford, CA 94305-5124

HANS WOLF-WATZ
Department of Cell and Molecular Biology, Umeå
University, S-901 87 Umeå, Sweden

ELI ZAMIR
Department of Molecular Cell Biology, Weizmann Institute
of Science, Rehovot 76100, Israel

ARTURO ZYCHLINSKY
Skirball Institute, Department of Microbiology and Kaplan
Cancer Center, New York University School of Medicine,
540 First Avenue, New York, NY 10016

Foreword

The term "cellular microbiology" was coined in 1996 by P. Cossart, P. Boquet, S. Normark, and R. Rappuoli (1) to describe an emerging scientific discipline that bridged the disciplines of cell biology and microbiology. The idea began to blossom at a scientific meeting held in the summer of 1991 in Arolla, Switzerland. The meeting brought together scientists who were exploring cell biological tools to answer questions of bacterial pathogenesis and cell biologists who were discovering that the fundamental questions of cell biology could be productively addressed by using microbial pathogens and bacterial toxins as cellular probes. It was an exciting scientific meeting where there was a good deal of discussion and a search for a common language and common ideas. On a personal note, I experienced difficulty in addressing cell biologists some years earlier with our first submission to the *Journal of Cell Biology*. The topic of the paper was the entry of *Salmonella* into polarized epithelial cells. One of the reviewers made the comment that the paper was "difficult to understand because the particles added by the authors were alive!" The book *Cellular Microbiology* is a continuation of the mutual education process by cell biologists and microbiologists. The chapters that follow contain information about fundamental cell biology and fundamental microbiology, as well as articles that show the marriage of the two—cellular microbiology.

Purified bacterial toxins and viruses were perhaps the first useful microbial reagents used by cell biologists. It was not necessary to know much in the way of microbiology to use them as biochemical probes of cell biological function. In particular, the dinucleotide-ribosylating enzymes secreted by the cholera-producing *Vibrio*, the diphtheria bacillus, and *Bordetella pertussis* (the agent of whooping cough) were used to probe the function of the large heterotrimeric G proteins. As described in this book, there is now a feast of toxin reagents ranging from those that directly interact with the microfilament network, including the small GTP-binding proteins like Rac, Rho, and Cdc42, to those that induce apoptosis. Similarly, it was exciting to discover that the classical tetanus and botulinum neurotoxins acted identically to each other as proteases for the SNAP25, synaptobrevin, and syntaxin target. This was a key factor in establishing the models for vesicles docking to membranes. At the same time, the medical microbiologist must examine the pathogenesis of infection to understand how one (tetanus) leads to muscle contraction and the other (botulism) leads to flaccid paralysis. For the cellular microbiologist, it is not sufficient to just understand the precise mode of action of a bacterial toxin in terms of the host effect, but it is also crucial to understand the role of the toxins in the pathogenesis of infection. What do these lethal toxins do for the microbe? How did these specific eukaryotic poisons evolve during prokaryote evolution? The tendency of the classical cell biologist is to look at the products of toxin virulence genes as purified reagents. The classical microbiologist may view these proteins as protective antigens from the standpoint of vaccine development. The clinician views the toxins as the causative factors of disease. The cellular microbiologist may be interested in all of these facets but must put the toxin into the perspective of the pathogenesis of infection and the utility of the toxin for bacterial survival, persistence, and transmission.

The first time a cell biologist sees a video of a bacterium like *Salmonella* or *Shigella* entering an animal cell, with the attendant cataclysmic eruption of cytoskeletal elements, there is a sense of awe that so tiny an organism can precipitate such a rapid and dramatic host cell response. However, the rapid assembly and disassembly of cytoskeletal elements are the same as those which occur in such essential host cell functions as phagocytosis, cell division, and adhesion to the extracellular matrix and substratum. How does the bacterium take over the control of these basic, essential eukaryotic cellular functions? We now know that pathogens like *Salmonella, Shigella,* and *Yersinia* inject specific effector proteins directly into the host cell cytoplasm via a proteinaceous appendage assembled on the bacterial surface that looks like a hypodermic needle. The genes encoding this secretory apparatus and the effector proteins have led us to view the evolution of pathogenicity in an entirely new light.

We now understand that pathogens like *Salmonella, Yersinia, Shigella,* and some enteropathogenic *Escherichia coli* strains have blocks of genes that distinguish them from their related commensal brethren. These blocks of genes are ordinarily found as contiguous large DNA chromosomal islands, called pathogenicity islands, or as part of an extrachromosomal element like a plasmid or bacteriophage. In many cases, these blocks of genes encode a specialized secretory pathway, which is configured by evolution to dispense specific effector proteins to the bacterial surface or to effect the transfer of specific effector proteins through a protein needle-like structure that acts as a conduit into the host cell membrane and cytoplasm. The effector proteins delivered in this way are extraordinarily keyed to interact with cytoskeletal elements, to enzymatically catalyze tyrosine phosphorylation or dephosphorylation of host proteins, or even to induce apoptosis. We still have not worked out all the properties of these specialized virulence proteins. We now understand that the genes found on chromosomal islands and plasmid genes associated with virulence have significant sequence and functional homology in bacteria ranging from pathogens of plants to dedicated pathogens of humans only. We can trace the evolution of these blocks. They appear to have been transmitted by horizontal gene transfer and possess a DNA base composition that is overall strikingly different from that of the bulk of the chromosomal genes. It is as if the island DNA once resided in a microbe at best distantly related to the pathogenic bacteria which plague us and in which it is now found.

Thus, the study of the cellular microbiology of bacterial invasion has provided us with new ways to examine the fundamental aspects of the eukaryotic cytoskeleton. It has led, as well, to a deeper understanding of the evolution of pathogenicity and host-parasite relationships. As the bacterial effector proteins become purified and better studied, they will provide important tools to probe the nature of animal cells. To the clinician, these studies provide new clues to prepare measures to thwart infectious agents of disease.

It is not only the cell biology of the bacterial entry process that provides us with new insights into the secrets of the host cell. After the microbe breaches the membrane barrier and enters the host cell, it becomes totally immersed within a eukaryotic environment. The bacteria initially are surrounded by a vesicular membrane. Some bacteria modify this compartment to prevent acidification, whereas others require the normal acidification to progress further with the infectious cycle. Some bacteria, like *Shigella, Listeria,* and the spotted-fever *Rickettsia,* dissolve the vesicular membrane to begin life in the host cell cytoplasm by using a fascinating interaction between synthesized bacterial proteins and host cell cytoskeletal components, as well as a highly developed mechanism for intracellular travel that in some cases undercuts the host cell's cell-tight junctions. Other bacteria, like *Salmonella, Mycobacterium,* and *Legionella,* stay encased in host cell membranes but have developed ways to obtain the nutrients required to persist and replicate, as well as to modify the trafficking pattern of the host to avoid destruction by lysosomal fusion. It seems that no aspect of the biology of the eukaryotic cell has not been fathomed by some microbe that has learned to change subtly and not so subtly to aid microbial domination. It is the job of the cellular microbiologist to discover these interactions and to put them into a broad biological perspective. This focus on the intricacies of the host-parasite interaction has given us a wealth of information for the classical cell biologist and classical microbiologist alike.

Indeed, this volume is written for the classical cell biologist and the classical microbiologist. Of course, the editors hope that it will be useful and appeal to students of all biological disciplines. The chapters are organized to provide both the fundamental "classical" background of each discipline and the cellular microbiology aspect. We are in an era of bacterial genomics. Soon the full DNA sequence of many important microorganisms will be known. In parallel, the full chromosomal DNA sequences of bacterial hosts, from worms to humans, are in the offing. While this volume stresses the interactions between pathogenic microbes and their animal hosts, it also branches out to recognize that the animal body is the home for countless other species. As scientists, each of us can study our favorite organism, but just as the DNA code provided us with the "Rosetta stone" which permitted all biologists to speak the same language, so cellular microbiology will be a wellspring of information to permit us to see more clearly the impact of organismal interactions in shaping the evolution of both microbes and hosts. The basis for understanding bacterial pathogenicity traces its roots back to the first time that amoebas learned to prey on bacteria and vice versa. As my students like to say, "All life, including us, is part of the food chain." This book provides the first organized report of an exciting new aspect of biology. I hope the reader will find it as exciting as the practitioners who wrote the following chapters do.

STANLEY FALKOW
Stanford University
Stanford, California

Reference
1. **Cossart, P., P. Boquet, S. Normark, and R. Rappuoli.** 1996. Cellular microbiology emerging. *Science* **271:**315–316.

Preface

Since it was initially published in 1996 (1), the term "cellular microbiology" has attracted a lot of attention. It has been the title of a scientific meeting (3, 4, 6) that is likely to be the first in a series of meetings with this title, it is the name of a new scientific journal (2, 5), and it is the title of this textbook. The question is whether "cellular microbiology" is just a trendy name that increases the visibility of a popular aspect of today's science or whether there is a real need for a new name, a new discipline, a new journal, or a new textbook. As discussed below, we believe that there is a real need for a new discipline to live up to the expectations of modern science.

To better study the complex phenomena that surround us, science has artificially split the study of natural events into many disciplines, each addressing a limited number of issues. The compartmentalizing into disciplines has been and continues to be a necessary simplification to reduce the complexity of the phenomena studied and to cope with the limits of the technologies available. However, oversimplification in biology is often a dangerous route that analyzes phenomena under artificial conditions far from real life. Our generation of microbiologists did not escape this trend. We have learned to isolate and grow bacteria and viruses under laboratory conditions that these organisms would never encounter in vivo. Similarly, cell biologists have learned to grow cells in perfectly aseptic tissue cultures, forgetting that in real life these cells, when part of functional tissues and organs, may interact with a variety of microbes.

Today's technology has exploded and improved, allowing analysis of increasingly complex systems such as the interactions between microbes and cells and investigations closer to real life than conventional microbiology and cell biology. However, while technology has breached the frontiers between the two disciplines, scientists are not yet able to fully exploit it, because they have been trained as either microbiologists or cell biologists, and their mentality is still compartmentalized. In other words, modern scientists are not well prepared (or could be better prepared!) to fully exploit what technology can offer. Therefore, science should greatly benefit from a new generation of scientists who are neither cell biologists nor microbiologists but, instead, who are equally familiar with both systems and able to study microbes and cells while they interact. We therefore need "cellular microbiologists," "cellular microbiology" courses, workshops, and maybe even "cellular microbiology" departments!

This textbook is a first attempt to form a new generation of scientists who learn from the very beginning that microbes and cells are part of the same world and how they live together. Since it approaches a new discipline, this textbook is not perfect. For instance, when the concept of "cellular microbiology" emerged, efforts were made by us and others to put together bacteriologists and cell biologists, two groups of scientists that had never interacted before, excluding virologists, who had traditionally been using tissue culture cells to grow viruses and therefore were already trained in cell biology. This was a loss for us, although virologists interested in understanding the complex phenomena deriving from the interactions of viruses and eukaryotic cells were rare. This textbook reflects this bias and focuses almost entirely on interactions between bacteria and eukaryotic cells. There is no doubt that if we had to do this book again, we would add some chapters focusing on the beautiful interactions between cells and viruses. We hope that this will be possible in a second edition.

PASCALE COSSART
PATRICE BOQUET
STAFFAN NORMARK
RINO RAPPUOLI

References
1. **Cossart, P., P. Boquet, S. Normark, and R. Rappuoli.** 1996. Cellular microbiology emerging. *Science* **271:**315–317.
2. **Falkow, S.** 1999. Cellular microbiology is launched. *Cell. Microbiol.* **1:**3–6.
3. **Niebuhr K., and S. Dramsi.** 1999. EMBO-EBNIC Workshop on Cellular Microbiology "Host cell-pathogen interactions in infectious disease." *Cell. Microbiol.* **1:**79–84.
4. **Reyrat, J. M., and J. Telford.** 1999. When microbes and cell meet. *Trends Microbiol.* **7:**187–188.
5. **Stephens, R. S., P. J. Sansonetti, and D. Sibley.** 1999 Cellular microbiology—a research agenda and an emerging discipline. *Cell. Microbiol.* **1:**1–2.
6. **Sweet, D.** 1999. Microbial fusion. *Trends Cell Biol.* **9:**239–240.

1

Microbial Pathogens: an Overview

Pascale Cossart and Marc Lecuit

This chapter—which does not claim to be exhaustive—is an introduction to the main human microbial pathogens, with a brief description of the clinical features of the disease and emphasis on the cell biology of the infectious process. For each pathogen, we have also tried to indicate if genetic tools are available. Thus, each section is the "identity card" of the major human pathogens, as seen by cellular microbiologists, and contains references to the chapters where these organisms are mentioned or discussed. Some systems are now paradigms in the field and are thus described in greater detail than those that are only starting to be analyzed. Figure 1.1 gives a schematic description of some of the most striking examples of molecular interactions between microbial and cellular components, and Figure 1.2 shows electron micrographs of the structures and modes of action of some of the pathogens discussed in this chapter. The status of genome-sequencing work for these pathogens is given in Table 1.1.

Bacteria

Strict Intracellular Bacteria
Strict intracellular pathogens cannot be cultivated in broth medium and can replicate only in vivo or in tissue-cultured cells in vitro.

Chlamydia
Chlamydia trachomatis is responsible for genital infections (serovars D to K) and ophthalmic infections (serovars A to C) in humans. It is a leading cause of sexually transmitted disease, causing infertility and extrauterine pregnancy (chronic salpingitis) and also infectious blindness in developing countries. *Chlamydia pneumoniae* causes a community-acquired pneumonia. There is increasing evidence for association of *C. pneumoniae* with atherosclerosis. *Chlamydia psittaci* is primarily an animal pathogen, and only in rare cases is it responsible for human respiratory tract infections, oropharyngitis, and atypical pneumonia.

    Chlamydia has a biphasic developmental cycle with two morphologically different forms: the elementary body (EB), which is the infectious form and is metabolically inactive, and the reticulate body (RB), which results

1

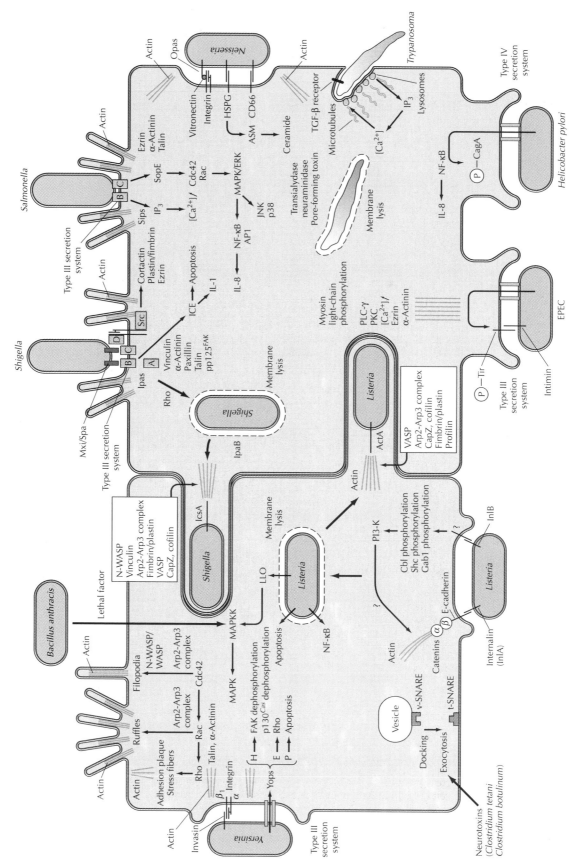

Figure 1.1 Schematic drawing of microbial factors and cellular targets.

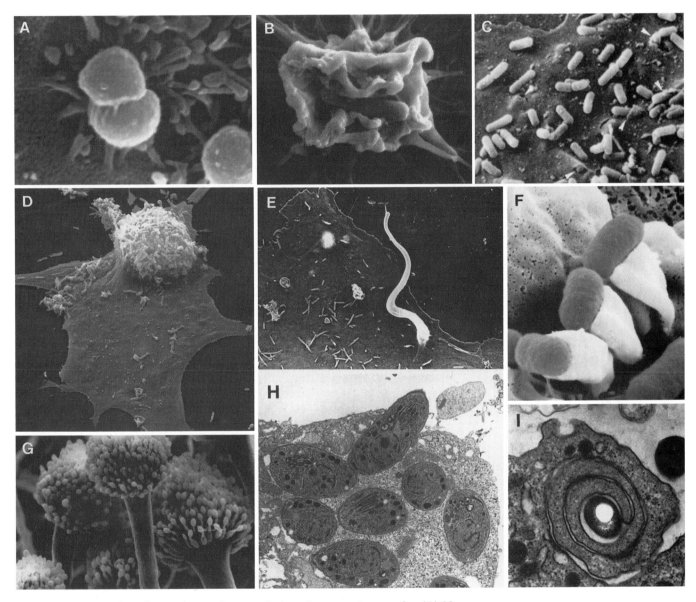

Figure 1.2   Examples of scanning and transmission electron micrographs. **(A)** *N. meningitidis* adhering to a cultured epithelial cell (courtesy of X. Nassif). **(B)** *S. flexneri* entering via macropinocytosis in a cultured epithelial cell by inducing membrane ruffling (courtesy of P. Sansonetti). **(C)** *L. monocytogenes* adherent to and invading a cultured cell (from our laboratory). **(D)** *B. henselae* adherent to and invading a cultured cell (courtesy of C. Dehio). **(E)** *T. cruzi* penetrating a cultured epithelial cell (courtesy of N. Andrews and E. Robbins). **(F)** EPEC on the top of pedestals induced on a cultured cell (courtesy of B. Finlay). **(G)** Conidial heads of *A. fumigatus* (courtesy of J. P. Latgé). **(H)** *T. gondii* tachyzoites free in the cytoplasm of a cultured macrophage, one escaping the cell (reproduced with permission from J. P. Dubey, D. S. Lindsay, and C. A. Speer, *Clin. Microbiol. Rev.* **11:**267–299, 1998). **(I)** *L. pneumophila* invading via coiling phagocytosis a cultured macrophage (reproduced with permission from M. Horvitz, *Cell* **36:**27–33, 1984).

Table 1.1 Genomes that have been sequenced

Organisms	Length (bp) (date sequenced)
Completed	
Borrelia burgdorferi	910,724[a] (12/17/97)
Chlamydia pneumoniae	1,230,230 (3/10/99)
Chlamydia trachomatis	1,042,519 (5/20/98)
Haemophilus influenzae	1,830,138 (7/25/95)
Helicobacter pylori J99	1,643,831 (1/12/99)
Mycobacterium tuberculosis	4,411,529 (6/11/98)
Mycoplasma genitalium	580,073 (10/30/95)
Mycoplasma pneumoniae	816,394 (11/15/96)
Plasmodium falciparum Chr 2	1,000,000
Rickettsia prowazekii	1,111,523 (11/12/98)
Treponema pallidum	1,138,011 (3/6/98)
In progress	
Bartonella henselae	2,000,000
Bordetella pertussis	3,880,000
Campylobacter jejuni	1,700,000
Candida albicans	16,000,000
Clostridium difficile	4,400,000
Enterococcus faecalis	3,000,000
Giardia lamblia	12,000,000
Francisella tularensis	2,000,000
Legionella pneumophila	4,100,000
Leishmania major Chr 1	270,000
Listeria monocytogenes	3,200,000
Mycobacterium avium	4,700,000
Mycobacterium leprae	2,800,000
Neisseria gonorrhoeae	2,200,000
Neisseria meningitidis	2,300,000
Plasmodium falciparum	
Chr 1	800,000
Chr 3	1,200,000
Chr 4	1,500,000
Chr 9	1,800,000
Chr 10	2,100,000
Chr 12	2,400,000
Chr 14	3,400,000
Porphyromonas gingivalis	2,200,000
Pseudomonas aeruginosa	5,900,000
Rhodobacter capsulatus	3,700,000
Salmonella typhimurium	4,500,000
Staphylococcus aureus	2,800,000
Streptococcus mutans	2,200,000
Streptococcus pneumoniae	2,200,000
Streptococcus pyogenes	1,980,000
Treponema denticola	3,000,000
Trypanosoma brucei rhodesiense	35,000,000
Ureaplasma urealyticum	750,000
Vibrio cholerae	2,500,000
Yersinia pestis	4,380,000

[a]One chromosome, two circular plasmids, and nine linear plasmids.

from the differentiation of the EB in the parasitophorous vacuole (also termed inclusion). The RB is a vegetative cell type that is able to divide up to the point where it changes to infectious bodies, which are liberated from the cells and are able to infect other cells.

The *Chlamydia* vacuole is unique in that it is devoid of known host protein markers, in particular those present in vesicles of the endocytic pathway. The lipid composition of the vacuole includes host sphingolipids acquired during the transit of exocytic vesicles from the Golgi apparatus to the plasma membrane.

The sequence of the genome of *C. trachomatis* has recently been determined. Potential virulence genes have been characterized. Several eukaryotic chromatin-associated domain proteins were identified, suggesting a eukaryotic-like mechanism for chlamydial nucleoid condensation and decondensation. The presence of a number of genes encoding proteins normally found in eukaryotes, such as serine/threonine protein kinases or phosphatases, suggests a complex evolution for this obligate intracellular parasite.

Coxiella burnetii

Coxiella burnetii is the etiological agent of Q (query) fever, an aerosol-borne disease. The reservoir is mainly farm animals. *C. burnetii* is responsible for a febrile illness complicated by endocarditis and is believed to be one of the most infectious bacteria. Like *Chlamydia*, *C. burnetii* exhibits a developmental cycle consisting of two morphologically distinct cell types. *C. burnetii* occupies a mature phagosome, whose low pH is required to stimulate the growth and replication of the bacterium. The lysosomal glycoproteins LAMP1, LAMP2, and cathepsin D and the vacuolar H^+ATPase are present on the vacuolar membrane. The *Coxiella* vacuole has the capacity to fuse with fluid-phase markers containing vesicles or intracellular vacuoles containing *Leishmania amazonensis*, *Trypanosoma cruzi*, or *Mycobacterium avium*, indicating that it has the capacity to interact with multiple compartments of the endocytic pathway. Stable genetic transformation has been reported.

Ehrlichia chaffeensis

Ehrlichia chaffeensis is one of the two members of a genus newly recognized as responsible for human diseases. Ehrlichiae are responsible for arthopod-borne disease characterized by headache, fever, and chills. *Ehrlichia chaffeensis* is an obligatory intracellular bacterium of monocytes or macrophages. It replicates in nonlysosomal intracellular vesicles called morulae, where different morphological forms of Ehrlichiae seem to coexist. Morulae in *E. chaffeensis*-infected cells are stained by antibodies against transferrin receptor, a marker for early endosomes, but are not labeled with lysosomal markers.

Mycobacterium leprae

Mycobacterium leprae is discussed in the section Mycobacteria, below.

Rickettsias

The genus *Rickettsia* contains two main subgroups, the typhus group, whose main member is *Rickettsiae prowazekii*, responsible for epidemic typhus (a hemorrhagic fever), and the spotted-fever group, which includes *R. rickettsii*, the agent of Rocky Mountain spotted fever, and *R. conorii*, the etiological agent of Mediterranean spotted fever. Both groups are transmitted to humans by arthropod vectors; typhus is transmitted by fleas, whereas spotted-fever rickettsias are transmitted by ticks.

These bacteria are able to lyse the phagocytic vacuole and to replicate in the cytosol. The role of phospholipase A_2 in the escape from the vacuole has been reported for *R. prowazekii*, but it remains unclear whether this activity is of rickettsial origin or if it is a host latent phospholipase activated upon infection. Some rickettsias like *R. conorii* and *R. rickettsii* are able to polymerize actin and to move intra- and intercellularly, by a process very similar to that used by *Shigella* and *Listeria*. Analysis of the factors involved in virulence has been hampered by the absence of genetic tools. However, an electroporation technique for *R. prowazekii* has recently been reported and the sequence of *R. prowazekii* has recently been determined.

Facultative Intracellular Bacteria (Entry and Multiplication in Phagocytic Cells)

Legionella pneumophila

Legionella pneumophila was relatively recently (1976) discovered as the etiologic agent of Legionnaires' disease, a disease characterized by infiltrative pneumonia, particularly in immunocompromised patients. One particular feature of this gram-negative bacterium is that in the environment, it can survive within amoebae. It is capable of growing within human pulmonary alveolar macrophages, and uptake into these cells may occur by an intriguing coiling phagocytosis. In infected cells, *L. pneumophila* is found in membrane-bound phagosomes. This compartment is significantly less acidic than are phagosomes containing nonpathogens: *L. pneumophila* prevents fusion of its phagosome with lysosomes. For the 2 h after uptake, the *L. pneumophila*-containing phagosome can be observed associated successively with smooth vesicles, mitochondria, and, finally, endoplasmic reticulum. The phagosome is eventually routed to a perinuclear position and appears as a ribosome-studded vacuole in electron micrographs. Bacterial replication is initiated after the formation of this ribosome-decorated compartment, which is therefore referred to as the "replicative phagosome." The ribosome-studded appearance results from an intimate association between the phagosome and the rough endoplasmic reticulum. The morphology of this compartment is similar to that of an autophagic compartment.

Genetic analysis has started to identify the genes and proteins involved in phagosome trafficking, intracellular multiplication, or survival. Among those, the *dot/icm* genes seem critical for the establishment of a replicative phagosome and thus intracellular life. They appear to be transcribed as nine operons, located in two 20-kb regions of the chromosome. Some of the encoded proteins may form a multiprotein secretion apparatus related to the recently described type IV secretion systems, which are capable of secreting a wide range of components including plasmids, linear DNA, and proteic toxins. This type IV secretion system may help the bacterium to transfer protein or other components to the parasitophorous vacuole.

Mycobacteria

Mycobacterium tuberculosis and *Mycobacterium leprae* are the agents of human tuberculosis and leprosy, respectively. *M. tuberculosis* causes chronic pulmonary infections, but also focalized infections (tuberculoma), such as vertebral Pott's disease or central nervous system infections (meningitis and brain tuberculoma). Inadequate treatment and immunodeficiency (AIDS patients) have contributed to the emergence of strains resistant to multiple antibiotics and to the consequent resurgence of tuberculosis.

M. leprae is the etiological agent of leprosy, a long-lasting infection still prevalent in some developing countries and characterized by cutaneous and nervous system lesions. The final host target tissue of *M. leprae* is the peripheral nervous system. Two forms of the disease are known: the tuberculoid and the lepromatous forms, which reflect the effects of active cellular immunity and anergy, respectively. *M. leprae* cannot be grown in vitro but can be grown only in mouse footpads and in the cold-blooded armadillo. The neural tropism of *M. leprae* is attributable to the specific binding of *M. leprae* to the G domain of the laminin-α_2 chain, which in turn binds to α-dystroglycan (itself in an interaction with β-dystroglycan) and consequently to dystrophin and actin.

M. tuberculosis can be grown in vitro and used to infect tissue-cultured phagocytic cells. *M. tuberculosis* remains within vacuoles in infected cells. It prevents its acidification by excluding the proton pump ATPase from the vacuolar membrane. *M. tuberculosis* has a fibronectin-binding protein. For *M. tuberculosis* and a number of nonpathogenic mycobacteria, powerful genetic tools including allele exchange and transposon mutagenesis are becoming available. These tools have led to the identification of a first virulence factor called Erp, a surface protein with proline-rich repeats. Inactivation of the *erp* gene abrogates intramacrophage multiplication and bacterial persistence in a mouse model.

The complete sequence of the *M. tuberculosis* genome has been recently determined. This genome has revealed the presence of two novel families of glycine-rich proteins, called the PE and PPE families due to the presence of repeated Pro-Gln or Pro-Pro-Gln sequences in their N terminus, that may represent a source of antigenic variation. The sequence of *M. leprae* is currently being determined.

Nocardia

Nocardia species belong to the actinomycete group (high-G + C-content gram-positive bacteria). These soil organisms are responsible for two opportunistic infections in humans: nocardiosis, caused essentially by *Nocardia asteroides* and more rarely by *N. brasiliensis* and *N. caviae*, and mycetoma, caused by *N. madurae* and *N. pelletieri*. Nocardiosis is characterized by a chronic pulmonary infection which can, via the lymph and the blood, disseminate to other tissues, in particular to the brain. Mycetomas are severe cutaneous infections which can invade the underlying tissues including the bones. Nocardiosis is sensitive to antibiotic treatment. Surgery is usually required for mycetomas. *N. asteroides* has been reported to reside in non-acidifed vacuoles of macrophages.

Facultative Intracellular Bacteria (Entry and Multiplication in Nonphagocytic Cells)

Bartonella

The genus *Bartonella* contains the two human pathogens *Bartonella henselae* and *B. quintana*. *B. henselae* is thought to be responsible for cat scratch disease, a chronic, necrotizing swelling of the lymph nodes, and for endocarditis in immunocompetent individuals. Immunocompromised patients develop severe clinical manifestations such as bacillary angiomatosis and bacillary peliosis, characterized by angioproliferative lesions of the skin and liver, respectively. *B. quintana* is responsible for trench fever and cases of bacillary angiomatosis.

Invasion of human umbilical vein endothelial cells by *B. henselae* has been analyzed in detail. Bacteria seem to be internalized by groups of hundreds of bacteria, in a structure called the invasome. Internalization is actin dependent. The genes involved in entry have not been identified. This identification should be facilitated by the recent possibility of manipulating *B. henselae* with plasmids and transposons (Tn*5*).

Brucella

The genus *Brucella* contains six species named *Brucella melitensis, B. abortus, B. suis, B. canis, B. ovis,* and *B. neotomae.* All are pathogenic for animals, but only the first four are known to be pathogenic for humans, inducing a polymorphic illness characterized by both generalized and localized infections (septicemia and osteoarticular and neurological disorders). During the course of infection in their primary hosts (goats for *B. melitensis* and cattle for *B. abortus*), professional phagocytes are the first targets of the bacteria, but extensive replication occurs in epithelial cells of the genitourinary tract. In vitro, *B. abortus* replicates in a broad range of epithelioid and fibroblastic cell lines (HeLa, NIH 3T3, Vero, BHK, HEp-2, etc.). Bacterial entry is inhibited by both cytochalasin D and nocodazole, suggesting a role for both actin and microtubules. After a transient passage through early endosomal compartments, *B. abortus* is distributed in multimembrane autophagosome-like vacuoles characterized by the presence of the lysosome-associated membrane proteins LAMP1 and LAMP2. Later during infection, *B. abortus* replicates in the endoplasmic reticulum of host cells.

Francisella tularensis

Francisella tularensis is the causative agent of tularemia, a disease affecting primarily wild animals and transmitted to humans by tick bites or consumption of contaminated game. It is responsible for chronic ulceroglandular lesions, which can be complicated by dissemination to the bloodstream and endotoxemia.

In laboratory animals, *F. tularensis* has the capacity to invade and grow in nonphagocytic cells, particularly hepatocytes. In vitro, infection of macrophage cell lines has been the most extensively analyzed mechanism. In these cells, *F. tularensis* resides in an acidic compartment, whose low pH facilitates the availability of the iron essential for *Francisella* growth. *F. tularensis* has the capacity to vary its lipopolysaccharide (LPS) when present in the vacuole of a macrophage. The normal LPS fails to stimulate the production of significant levels of nitric oxide, thereby allowing intracellular multiplication. The bacterium can spontaneously vary its LPS to induce macrophages to produce increased levels of NO, thereby suppressing intramacrophagic growth. This is the first report of a phase variation phenomenon which modulates intracellular growth and the innate immune response.

Listeria monocytogenes

The gram-positive bacterium *Listeria monocytogenes* is responsible for severe food-borne infections in humans and animals. Listeriosis is characterized by central nervous system infections and maternofetal infections, and it affects mainly immunocompromised individuals, the elderly, neonates, and pregnant women. This bacterium, first a model for the study of induction of T-cell-mediated immunity in mice, is now one of the best known invasive bacteria. The genus *Listeria* contains five other species, *L. seeligeri*, *L. innocua*, *L. welshimeri*, *L. murrayi*, and *L. ivanovii*, an animal pathogen.

In vitro, *L. monocytogenes* infects a wide variety of phagocytic or nonphagocytic cell types. Entry occurs by a zipper-type mechanism, which requires an active actin cytoskeleton and at least one tyrosine kinase. It is mediated by at least two surface proteins, internalin (InlA) and InlB, which by themselves are sufficient to allow entry into mammalian cells. The receptor for internalin is the cell adhesion molecule E-cadherin. InlB, whose receptor is unknown, activates phosphoinositide 3-kinase (PI3-K). This stimulation is required for entry. Upon entry, PI3-K is recruited to the plasma membrane by interacting with at least three phosphorylated proteins (Gab1, Shc, and Cbl). How the production of 3'-phosphoinositides leads to cytoskeletal rearrangements required for entry is unknown.

After lysis of the internalization vacuole, by the bacterial pore-forming toxin listeriolysin O (LLO) and in some cells by the phosphoinositide-phospholipase C or the phosphatidylcholine phospholipase C, bacteria move intra- and intercellularly via an actin-based motility mediated by the bacterial surface protein ActA. This protein, which is sufficient to induce actin polymerization and movement, has become a paradigm to study the mechanism underlying actin-based motility (Figure 1.3). The ActA protein is a composite protein. The N-terminal part was recently shown to act in concert with the Arp2-Arp3 complex to stimulate actin polymerization in vitro. The central proline-rich region and the C-terminal part are homologous to the actin-binding protein zyxin. ActA as zyxin binds VASP (vasodilator phosphoprotein), which would recruit profilin and actin to stimulate the actin polymerization process in the N-terminal part of the protein.

▶ *For Figure 1.3, see color insert.*

Intracellular movement leads, as with *Shigella*, to direct cell-to-cell spread. This event probably occurs in the liver, when bacteria move from macrophages to hepatocytes and induce apoptosis by a process which has not been deciphered at the molecular level. In contrast, in an endothelial cell line, apoptosis is mediated by LLO. During infection of HeLa cells, mitogen-activated protein kinases (MAPK) are activated, with the MEK1/ERK2 pathway being required for entry while LLO stimulates both ERK1 and ERK2. Infection of cultured endothelial cells or macrophages activates NF-κB. Cell wall and virulence factors are involved in this process.

Most of the genes involved in virulence are clustered in a pathogenicity islet of 15 kb which is also present in *L. ivanovii* and *L. seeligeri*. In *L. seeligeri*, the virulence genes are not expressed.

Salmonella

The genus *Salmonella* contains several species. Among them, the species *S. enterica* has received the most attention. It contains the two widely studied serovars, *S. enterica* serovar typhi (*S. typhi*) and *S. enterica* serovar typhimurium (*S. typhimurium*).

S. typhi is a human enteroinvasive bacterium that causes septicemia after dissemination to the liver and spleen. It is responsible for a variety of indirect systemic symptoms that comprise typhoid fever. *S. typhimurium* is an enteric pathogen causing gastroenteritis in humans. In mice, it is the etiologic agent of a systemic infection similar to typhoid fever and thus is used as a model to study the pathophysiology of typhoid fever.

S. typhimurium enters nonphagocytic epithelial cells by a process of macropinocytosis, in which the bacterium becomes entrapped in a spacious phagosome. The morphological events that allow entry are characterized by large membrane ruffles and important cytoskeletal rearrangements. For entry into epithelial cells, bacteria use a battery of genes which encode proteins secreted and translocated by a type III secretion system located on a pathogenicity island (PAI) named SPI1. This system encodes surface proteins which form a supramolecular structure recently visualized by electronic microscopy and similar (albeit not identical) to flagellar bodies. The role of the small G protein Cdc42 and, to a lesser extent, Rac in the entry process is now established. The bacterial protein involved in the interaction with the small GTpases in SopE, a protein translocated by the type III secretion system acting as an exchange factor.

S. typhimurium can also be internalized by macrophages. Intramacrophage survival is mediated by genes located on a second pathogenicity island, SPI2, which encodes a second type III secretion system. *Salmonella* induces apoptosis of macrophages once it is internalized. There is evidence that replication in the phagocytic vacuole is favored by the low pH of the mature vacuole.

During infection, *Salmonella* activates the mitogen-activated protein kinases (MAPKs) ERK, JNK, and p38. Stimulation of the MAPK pathways results in activation of the transcription activators NF-κB and AP1, which regulate the expression of interleukin-8 and other cytokine gene promoters. These nuclear responses are also mediated by Cdc42 and Rac.

Shigella

The genus *Shigella* contains four species, *Shigella dysenteriae, S. flexneri, S. boydii,* and *S. sonnei,* which are responsible for bacillary dysentery, a bloody and purulent diarrhea reflecting the ability of this bacterium not only to invade and extensively damage the intestinal epithelium but also to cause local inflammation by triggering polymorphonuclear transmigration.

Both in vivo and in vitro studies have led to a detailed description of the successive steps of the infectious process, which probably starts at the level of the M cells, which underlie the Peyer's patches. Then bacteria reach the underlying lamina propria and invade either the enterocytes or the resident macrophages, where they induce apoptosis with concomitant release of interleukin-1, a proinflammatory cytokine which contributes to the attraction of neutrophils at the site of infection. In epithelial cells, bacteria enter by a process morphologically similar to that of *Salmonella* (membrane ruffles and intense rearrangements of the cytoskeleton). Then the bacteria are trapped in a membrane-bound vacuole, which they lyse to reach the cytosolic compartment, where they start moving. When the bacteria reach the plasma membrane, they induce the formation of protuberances that allow direct cell-to-cell spread of the bacteria. Most *Shigella* virulence genes are carried by the virulence plasmid, which on two divergent operons, encodes a type III secretion system responsible for the secretion of four invasion proteins, IpaA, IpaB, IpaC, and IpaD. In the bacterium, IpaB and IpaD prevent secretion. When secretion is induced upon cell contact, IpaB and IpaC associate and might insert in the mammalian cell membrane. β_1-integrins seem to play a role in the early contact of IpaB, IpaC, and IpaD with the cell. IpaB also appears to interact with CD44, the acid hyaluronic receptor. This receptor interacts with ezrin after its activation by rho. This event would then allow IpaA translocation within the eukaryotic cell and its interaction with vinculin, which in turn would recruit α-actinin and somehow down-regulate the actin cytoskeletal rearrangements just after the internalization step. Entry also requires the small GTPases Cdc42 and Rac and the Src kinase, which would phosphorylate cortactin. The actin-bundling protein fimbrin (plastin) would play a role in organizing the actin filaments. Taken together these data suggest that the entry process mimics the formation of a pseudo-adhesion plaque.

Nonsporulating Extracellular Bacteria

Acinetobacter

A single species, *Acinetobacter baumannii*, is responsible for human noso-comial infections, especially in patients with catheters in intensive care units. It is also responsible for respiratory and urinary tract infections and septicemia.

Aeromonas

In humans, *Aeromonas* infections are mostly opportunistic. They usually follow contact with contaminated water and are due to the four species of the group *A. hydrophila-punctata*, *A. hydrophila*, *A. sobria*, and *A. caviae*. They are usually limited to the digestive tract but can also appear as septicemias, meningitidis, endocarditis, and osteomyelitis. *Aeromonas* produces many virulence factors, including adhesins, hemolysin, enterotoxins, and prote-ases. Aerolysin is a secreted 48-kDa channel-forming protein which binds to specific receptors and heptamerizes to form a 1.5-nm channel, leading to cell permeabilization and death.

Bordetella pertussis

Bordetella pertussis is the etiologic agent of whooping cough, which was a very common disease before the large immunization programs. Clinical manifestations are usually divided into three stages: the catarrhal and highly communicable stage, mimicking a common cold; the paroxysmal stage, marked by severe and repetitive paroxysmal cough, with excess mu-cus production, vomiting, and convulsions (neurological damage and bron-chopneumonia are possible complications); and, finally, a recovery stage marked by a progressive diminution of the coughing episodes. The vaccine used for many years was whole killed bacteria, which caused numerous side effects, the most serious of which was irreversible brain damage. Con-cern about side effects has led to the development of a novel acellular vac-cine.

B. pertussis has many different virulence factors, which are under the control of a pivotal activator, the BvgA-BvgS two-component regulator sys-tem. Virulence factors include adhesins, namely, filamentous hemaggluti-nin and pertussis toxin (Ptx), both of which mediate binding to ciliated cells and macrophages and are important for colonization of the airway. Pertac-tin and pili also play a role in adherence. *Bordetella* produces several toxins. Ptx is an AB-type toxin acting on large G proteins. Ptx catalyzes the ADP-ribosylation of the α_1 subunit of G protein and prevents the G_i complex from interacting with receptors. The complex thus remains in a GTP-bound form, unable to inhibit adenyl cyclase, thus leading to constitutive produc-tion of cyclic AMP (cAMP). *Bordetella* produces adenylate cyclase (ACase), which enters cells and also increases the host cell cAMP level. ACase is an invasive cyclase that also increases the host cell cAMP level. It is active only in the presence of calmodulin. Ptx and ACase may contribute to symptoms of the disease such as death of ciliated cells and mucin secretion. Another possible activity is to lower the oxidative burst in PMNs and macrophages, thus allowing bacterial survival. There is evidence that ACase induces apoptosis in macrophages. Other toxins are dermonecrotic toxin and tracheal cytotoxin, a peptidoglycan fragment that kills ciliated cells. Tracheal cytotoxin and LPS contribute to eliciting an inflammatory response.

Recently, a type III secretion system under the control of Bvg has been discovered.

Borrelia and the Other Spirochete, *Treponema pallidum*

Borrelia burgdorferi is the etiological agent of Lyme disease, a recently described tick (*Ixodes scapularis*)-borne disease endemic in certain temperate-forest areas where the vector lives. It is characterized by three successive phases if untreated: (i) a cutaneous phase (spreading annular rash), in which the inoculation region becomes inflamed and erythematous (erythema migrans); (ii) an acute phase with lesions in the joints, heart, nervous and musculoskeletal systems, and skin; and (iii) a chronic phase with chronic arthritis of large joints, chronic fatigue, paralysis, and dementia.

In addition to its chromosome, *Borrelia* has two circular plasmids and nine linear plasmids, all of which have been sequenced. These bacteria are fastidious to cultivate, they have a long generation time, and there is no genetic system available yet. Multiple integrins mediate the attachment of Lyme disease spirochetes to different mammalian host cells, including platelets. Different classes of proteoglycans contribute to the attachment of *B. burgdorferi* to cultured endothelial cells and brain cells. Decorin, a proteoglycan which decorates collagen fibers, binds to two *Borrelia* proteins and mediates the adherence of bacteria to collagen in skin and other tissues. Finally, *B. burgdorferi* may invade endothelial cells, and OspA and OspB, the two major outer membrane lipoproteins, may play a role in this process.

Treponema pallidum is the etiologic agent of syphilis, a sexually transmitted disease characterized by three clinical phases: (i) a phase characterized by a lesion (chancre); (ii) a phase in which bacteria penetrate the mucosa and enter the bloodstream, leading to fever, rash, and mucocutaneous lesions; and (iii) finally, a phase in which bacteria can invade the heart, the musculoskeletal system, and the central nervous system. Like *B. burgdorferi*, *T. pallidum* can attach to endothelial cells and stimulate the expression of intracellular cell adhesion molecule 1 (ICAM-1), in turn attracting polymorphonuclear neutrophils, which may disrupt the endothelial barrier. These cells are responsible for the strong inflammation characterizing syphilis. *T. pallidum* cannot be cultivated in laboratory medium. Its genome has been sequenced. Potential virulence factors include a family of 12 membrane proteins and several putative hemolysins.

Campylobacter

Campylobacter jejuni is a leading cause of bacterial diarrhea ranging from mild abdominal symptoms to severe invasive enteritis. Extradigestive manifestations such as arthritis and endocarditis can occur. *C. jejuni* adheres and invades tissue-cultured cells. Like several other enteropathogens (various *Escherichia coli* strains and a few *Shigella* species), *C. jejuni* produces a cytolethal distending toxin, which induces actin assembly and arrest of cell division at G_2 in infected Chinese hamster ovary cells. *C. jejuni* binds to fibronectin through a 37-kDa outer membrane protein. It seems to have several proteins acting as adhesins, including CBF1 and the structural sub-unit of flagella, flagellin (FlaA). Some *C. jejuni* strains have an LPS with a terminal tetrasaccharide similar to that in GM_1 ganglioside. Patients with Guillain-Barré syndrome have autoantibodies to GM_1 ganglioside, a membrane component of peripheral nerves. This is the first example of a molecular mimicry between nerve tissues and the infectious agent (that elicits Guillain-Barré syndrome). *C. jejuni*, like *Citrobacter freundii*, invades epithelial cells by a process involving microtubules but not microfilaments, as shown by its insensitivity to cytochalasin D.

Corynebacterium diphtheriae

Corynebacterium diphtheriae is the etiological agent of diphtheria, a disease with direct local effects of bacterial infection on the upper respiratory tract (formation of membranes and inflammation due to neutrophil attraction at the site of bacterial multiplication) and systemic effects (cardiac and neurological) due to the diffusible toxin diphtheria toxin. This toxin, whose gene is iron regulated and carried on a phage, belongs to the AB toxin family which enter sensitive cells by receptor-mediated endocytosis. The enzymatically active A fragment subsequently enters the cytosol, where it inhibits protein synthesis by inactivating the elongation factor 2. The diphtheria toxin receptor is the heparin-binding epidermal growth factor (EGF)-like precursor (HB-EGF), which forms a complex with membrane protein DRAP27/CD9. The structure of diphtheria toxin has been solved.

Escherichia coli

The species *Escherichia coli* includes the nonpathogenic *E. coli* K-12, which has been widely used as a model in bacterial physiology and genetics. In humans, *E. coli* is the major constituent of the digestive commensal flora. Some *E. coli* strains have acquired specific virulence factors and are thus responsible for mild to severe digestive or urinary tract infections or for septicemia and meningitis. Depending on the symptoms that they are associated with, they are classified in four main categories: enteropathogenic *E. coli* (EPEC), enterohemorrhagic *E. coli* (EHEC), enteroinvasive *E. coli* (EIEC), and enterotoxinogenic *E. coli* (ETEC).

One of the two categories which have received most attention recently is EPEC. EPEC strains are a major cause of diarrhea in the developing world. EPEC infection is characterized by an intimate attachment to host epithelia. This attachment leads to the formation of attaching and effacing (A/E) lesions characterized by the degeneration of the epithelial brush border and the formation of actin-rich pedestals. The first step of these events is loose attachment via the bundle-forming pilus, resulting in localized adherence. Protein secretion, by a type III secretion system, induces a series of signals. Among the secreted proteins are EspA, which forms filamentous organelles on the bacterial surface, EspB, and a protein previously named Hp90 and now renamed Tir, which becomes tyrosine-phosphorylated and acts as a receptor for intimin. Intimin is similar to the invasin of *Yersinia pseudotuberculosis*. The Tir-intimin interaction results in intimate attachment and recruitment of a number of cytoskeletal elements such as α-actinin and ezrin, which induce the formation of actin-rich pedestals on which bacteria reside, ready to send more signals into the cell. Pedestal formation is accompanied by an increase in intracellular Ca^{2+} levels, activation of protein kinase C (PKC) and phospholipase C-γ, and phosphorylation of myosin light chain. This last event plays a key role in altering intestinal epithelium permeability.

The second category which has received considerable attention recently is EHEC, which is responsible for large food-borne outbreaks of bloody diarrhea and hemolytic-uremic syndrome. The most notorious EHEC serotype is O157:H7. EHEC shares a number of virulence factors with EPEC, in addition to the most important virulence determinant, the potent cytotoxin known as Shiga toxin or Vero toxin, which is phage encoded. This toxin consists of one A subunit and five identical B subunits. The B subunit binds to a glycolipid receptor, enters cells via clathrin-coated pits, and is transported to the endoplasmic reticulum. The A subunit inhibits host cell protein synthesis by an *N*-glycosidase activity which removes an adenine residue from the 28S rRNA. In addition to production of Shiga toxin, EHEC produces the A/E lesion. It also produces an intimin similar to that of EPEC, albeit different in the C-terminal part; the receptor of this intimin has not been precisely identified. EHEC also has a type III secretion system, and the secreted proteins are similar to Esp proteins. Genes for intimin, the type III secretion system, and the EspA, EspB, and EspD secreted proteins are, as in EPEC, located on a 35-kb pathogenicity island called the locus of enterocyte effacement. There might be other virulence factors responsible for disease, including a 104-kDa protein with homology to a family of proteins called autotransporters, which includes immunoglobulin A1 (IgA1) of *Neisseria gonorrhoeae*, the vacuolating cytotoxin (VacA) of *Helicobacter pylori*, and pertactin of *Bordetella pertussis*. The sequence of the genomes of *E. coli* K-12 and O157:H7 interestingly differ in many more loci than previously anticipated.

Haemophilus influenzae and *H. ducreyi*

The gram-negative bacterium *Haemophilus influenzae* is one of the leading causes of upper respiratory tract infections and otitis in children. It is also responsible for meningitis. Invasive disease (septicemias and meningitis) is caused by encapsulated bacteria, whereas nonencapsulated *H. influenzae* strains cause localized infections (e.g., middle ear infections). Colonization is facilitated by pili. The erythrocyte receptor for one of the pili has been identified; it is the Anton receptor (An^{Wj}), a blood group antigen. A very efficient polyosidic capsular vaccine is available.

H. influenzae is capable of entering cultured human cells. One factor involved in this process, Hap, appears to be a serine type IgA1 protease. Phase variation seems to control LPS expression, and a cell envelope protein, OapA, responsible for colony opacity, may help in colonization. Gangliosides GM_1, GM_2, GM_3, and CD19 inhibit pilus-mediated hemagglutination and binding to buccal epithelial cells.

The genome of *H. influenzae* was the first bacterial genome sequenced.

H. ducreyi, which is not closely related to *H. influenzae*, is responsible for chancroid, a sexually transmitted disease characterized by genital ulceration and inguinal lymphadenopathy.

Helicobacter pylori

Helicobacter pylori is a newly described bacterial species, previously classified in the genus *Campylobacter*. *H. pylori* infection is the most common bacterial human infection. It is particularly prevalent in developing countries (90%, compared to 30% in industrialized countries). Infection is acquired mainly during childhood, and, to date, there has been no conclusive evidence for the existence of an animal reservoir. Long-term infection is associated with chronic gastritis and peptic ulceration, and an epidemiological link has been established with mucosa-associated lymphoid tissue lymphoma and gastric adenocarcinoma. However, it remains unclear why infection with *H. pylori* is associated with such a wide diversity of pathologic findings extending from mild to very severe disease. It is most likely that a combination of host, bacterial, and environmental factors determines the eventual clinical outcome.

H. pylori possesses factors essential for colonization of the gastric mucosa, such as motility and urease activity. Less is known about how *H. pylori* is able to persist in the stomach and avoid elimination by the host immune response. It is likely that expression of variable and multiple adhesins with different binding specificities contributes to the persistence of the organism. Among the genetically characterized adhesins, the Lewis b binding adhesin (BabA) is now recognized as one of the members of a large family of 32 outer membrane proteins capable of phenotypic phase variation. Other bacterial factors which participate in the induction of proinflammatory responses and cell toxicity have been identified; these factors contribute either directly (e.g., VacA) or indirectly (via autoimmunity, inflammation, and alteration of acid secretion) to the genesis of gastric lesions. Among the proinflammatory factors is the *cag* PAI. The *cag* PAI consists of 40 genes encoding several proteins, some of which induce interleukin-8 secretion by epithelial cells via an NF-κB-mediated mechanism. The *cag* PAI is not present in all clinical isolates, and this variation is likely to reflect ethnic tropism of certain *H. pylori* strains for different populations.

Klebsiella pneumoniae
Klebsiella pneumoniae is the causative agent of nosocomial and community-acquired pneumonia. *Klebsiella* strains are usually resistant to a wide spectrum of antibiotic agents. Virulence is associated with a capsule which confers resistance to phagocytosis by polymorphonuclear neutrophils and macrophages. It is also mediated by fimbrial adhesins, which mediates binding to collagen.

Leptospira interrogans
Leptospira interrogans is the etiological agent of leptospirosis, which is essentially a zoonosis. Humans can be infected through animal contact (mainly with cattle, swine, and rats). After entry into the body directly through the skin or via the respiratory tract or the conjunctiva, the microorganism is responsible for a severe illness characterized by fever and chills associated with myalgia. In rare cases, severe hepatic and renal complications appear with hemorrhagic manifestations. *L. interrogans* induces apoptosis in cultured macrophages. It can invade Vero cells. A sphingomyelinase, which is similar to *Staphylococcus aureus* beta-hemolysin and *Bacillus cereus* sphingomyelinase C, has been identified.

Mycoplasmas
Mycoplasma organisms are a leading cause of sexually transmitted diseases. They belong to a special class of small bacteria (length, 0.3 μm) called mollicutes. They have no cell wall and have a special plasticity. They are extracellular but live tightly associated with cells. The main pathogen is *M. pneumoniae,* which is responsible for severe pneumopathies frequently associated with cold agglutinins. The others are commensals which can give

rise to opportunistic infections and include *Ureaplasma urealyticum*, *M. hominis*, and *M. genitalium*, which are responsible for sexually transmitted genital infections in men and women.

The genomes of both *M. genitalium* and *M. pneumoniae* have been sequenced. The genome of *M. genitalum* (468 genes) is the smallest among the known cellular life-forms. Mycoplasmas can be manipulated genetically, albeit not easily (via transposon mutagenesis with plasmid delivery). For *M. pneumoniae* and *M. genitalum*, adhesins are among the putative important virulence factors. For other species, the adhesive properties are probably associated with lipoproteins. Most mycoplasmas are able to stimulate cytokine production by macrophages. Signaling events include the MAPK pathway, NF-κB and AP-1.

Neisseriae

Neisseria gonorrhoeae is the etiologic agent of the most common sexually transmitted disease, gonorrhea, which is a highly contagious genital infection of males and females and occasionally becomes an invasive infection, leading, by hematogenous seeding, to localized infections such as septic arthritis. *N. meningitidis* is a commensal in 10 to 30% of healthy individuals, yet it is one of the most harmful pathogenic bacteria in susceptible humans. In young people, it causes purulent meningitis, with a very high mortality rate and a high incidence of sequelae. It also occasionally causes purpura fulminans, a lethal septicemic syndrome associated with septic shock.

One of the most striking features of *Neisseria* species is their capacity to vary their surface proteins (pili or opacity proteins, Opa and Opc) by phase variation and/or antigenic variation. These bacteria can invade tissue cultured cells but do not seem to multiply intracellularly. Both *N. gonorrhoeae* and *N. meningitidis* are internalized via interaction between one of the bacterial outer membrane Opa proteins and a member of the CD66 family of carcinoembryonic antigens on the host cell. CD66 proteins are members of the Ig superfamily of receptors and are expressed in many cell types including activated neutrophils, epithelial cells, and endothelial cells. Invasion can also be mediated by interactions between some Opa proteins and heparan sulfate proteoglycans on some epithelial cells. Moreover, heparan sulfate proteoglycan-mediated entry into some epithelial cells is enhanced by vitronectin, which binds to Opas on the bacterial surface. The GTPase Rac, a rho family member, but not Cdc42 is involved in the CD66-mediated uptake of *N. gonorrhoeae* by neutrophils, which is activated upon entry. This activation is needed for efficient uptake and appears to be controlled by Src family kinases, possibly Hck and Fgr. Entry into epithelial cells involves lipid metabolism, activation of a phosphatidylcholine-phospholipase C, and the stimulation of acidic sphingomyelinase. An important factor mediating adhesion to host cells is PilC, a minor component of pili. A pilus receptor, probably the PilC receptor, was identified as CD46.

For *N. meningitidis*, the surface opacity proteins Opa and Opc (also called class 5 proteins) are able to mediate meningococcus-cell interactions if the bacteria are nonencapsulated. Opc has the capacity to interact with polarized endothelial cells by a sandwich mechanism that involves vitronectin and endothelial integrins, especially αvβ3 (VnR). For encapsulated bacteria, class 5 proteins do not seem to affect bacterial interactions with host cells. Recent evidence indicates that neisserial components other than Opa and Opc are involved in transducing signals. The IgA1 protease may be involved in intracellular survival by cleaving the LAMP1 protein and preventing phagolysosomal fusion.

Pasteurella multocida
Pasteurella multocida is responsible mainly for infections in animals. However, it can lead to infections in humans after animal bite wounds. It produces an RTX toxin.

Pseudomonas aeruginosa
Pseudomonas aeruginosa bacterium is an opportunistic pathogen frequently responsible for nosocomial infections with a variety of clinical features and locations (skin, urinary tract, lungs, digestive tract, eyes, ears, septicemias, and endocarditis). *P. aeruginosa* is also responsible for deleterious chronic infections in cystic fibrosis patients. Indeed, it chronically infects the lungs of over 85% of cystic fibrosis patients. It is ineradicable by antibiotics and is responsible for airway mucus overproduction, which contributes to airway obstruction and death. *P. aeruginosa* activates a c-Src–Ras–MEK1–MAPK–pp90^{rsk} signaling pathway that leads to activation of NF-κB (p65/p50). Activated NF-κB binds to the 5′-flanking region of the *MUC2* gene and activates *MUC2* mucin transcription. *Pseudomonas* also produces a number of exocellular enzymes and toxins which appear to require cell-to-cell communication for full expression. This communication is mediated by a diffusible molecule termed *Pseudomonas* autoinducer.

Staphylococci
Staphylococci are very common in the environment and are highly resistant to hostile conditions such as heat, desiccation, or salinity. They are often commensals on the skin or mucosa of humans and animals.

The most pathogenic species is *Staphylococcus aureus*, which is responsible for cutaneomucosal polymorphic infections (ranging from simple local erythemas to abscesses which can disseminate to secondary sites and cause generalized infections), toxic shock syndromes (mainly affecting young women using vaginal tampons), and food-borne toxic infections which occur very rapidly after ingestion of contaminated food. If *S. aureus* is also often the cause of nosocomial infections, resulting in the symptoms described above, nosocomial infections are also often due to coagulase-negative staphylococci such as *S. epidermidis* and, in rare cases, *S. haemolyticus*. *S. epidermidis* has the capacity to colonize plastic catheters. Urinary tract infections due to another coagulase-negative species, *S. saprophyticus*, appear to be the second most common urinary tract infections among young women in the United States.

The virulence of *S. aureus* is due mostly to the secretion of several toxins and enzymes. Some toxins, such as alpha, beta, gamma, and delta toxins, have a dermonecrotic effect; others, such as leucocidin, are cytotoxic for and destroy polymorphonuclear cells. Some toxins, such as exfoliatin and toxic shock syndrome toxin type 1 (TSST-1), have an exceptional capacity to rapidly diffuse into tissues and produce disease while the bacteria remain confined to the initial site of infection. The roles of the many extracellular enzymes secreted by *S. aureus* are not understood. The coagulase would be responsible for the formation of a clot. Those of other enzymes (hyaluronidase, deoxyribonuclease, lipase, esterase, and catalase) have not been clearly established.

In addition to secreted proteins, *S. aureus* has a number of surface proteins which may play a role in infections such as protein A, which binds the Fc fragment of Igs and abrogates the action of specific Igs, and several proteins which bind extracellular matrix proteins such as fibronectin (FnbA, FnbB), collagen (Cna), fibrinogen (ClfA, Fib), vitronectin (60-kDa protein), elastin (EbpS), and probably others.

There is a major concern about the increasing proportion of methicillin-resistant strains as well as the appearance of vancomycin-resistant isolates.

Some toxins, and in particular TSST-1, are superantigens. Superantigens link antigen receptor of T cells with major histocompatibility complex (MHC) class II by interacting with some Vβ subunits of T cells and the MHC class II molecule of antigen-presenting cells. This leads to the activation (cytokine secretion and proliferation) or deletion of a large subpopulation of Vβ T cells in vivo and in vitro. The consequences in vivo are shock or immunosuppression, i.e., reduced antibody levels against some staphylococcal products and persistence of *Staphylococcus* cells in the host. Another explanation for this persistence derives from the recent discovery that staphylococci which were assumed to be exclusively extracellular can enter cultured epithelial cells. One report even describes escape from the internalization vacuole, probably due to the action of some of the toxins described above and induction of apoptosis.

Streptococci

Streptococci are divided into many subgroups. The most important clinically are group A streptococci (GAS), whose prototype is *Streptococcus pyogenes*, group B streptococci (GBS), whose prototype is *S. agalactiae*, and *S. pneumoniae*. Streptococci have long been considered extracellular pathogens, and their capacity to bind to extracellular matrix proteins or to produce a wide range of toxic molecules was considered the major features for their virulence. It has been recently recognized that the three groups described here have the capacity to invade tissue-cultured cells.

S. pyogenes. *S. pyogenes* is responsible for cutaneomucosal infections (the classic tonsilitis, scarlet fever, erysipelas), skin infections (impetigo), and occasionally endocarditis and postinfectious manifestations such as immune complex glomerulonephritis and rheumatic fever. The initial encounter of GAS with the host is with the pharynx or the skin.

S. pyogenes produces a wide variety of surface-associated and secreted components which are important for their virulence, including the M protein, fibronectin-binding proteins, hyaluronic acid capsule, extracellular enzymes, and toxins which allow streptococci to colonize and/or cause invasive diseases. It has been suggested that expression of different proteins constitutes the basis for the cellular tropism of different *S. pyogenes* isolates; the fibronectin-binding proteins have been implicated in bacterial attachment to epithelial cells of the upper respiratory tract, whereas the M protein mediates binding to skin cells. Recent studies show that the fibronectin-binding protein (Sfb1 or protein F1) and also protein M mediate internalization. The surface protein SDH, a surface glyceraldehyde-3-phosphate dehydrogenase, can trigger signal transduction events. Other factors may be involved in internalization. Invasion of deeper tissues requires the expression of several toxins or enzymes; *S. pyogenes* produces well-known toxins: streptolysin O, a pore-forming toxin which binds to cholesterol, oligomerizes and thus permeabilizes cells (hence its use in cell biology), and the pyrogenic exotoxins SPEA and SPEB. These molecules, like TSST-1 produced by *S. aureus* and the staphylococcal enterotoxin SE, display superantigen activity.

S. agalactiae.   *S. agalactiae* causes neonatal pneumonia, sepsis, and meningitis. This pathogen can translocate within brain microvascular endothelial cells, suggesting that direct invasion and injury of the blood-brain barrier allow bacteria to gain access to the central nervous system, leading to meningitis.

S. pneumoniae.   *S. pneumoniae* is a major cause of meningitis and septicemia in children and of pneumonia and septicemia in adults. Pneumococci undergo spontaneous phase variation marked by switching from opaque to transparent colony morphotypes. Transparent strains are adapted to nasopharyngeal adherence and display less capsule, more surface choline, and more of the adhesin CbpA than do opaque strains. Opaque strains show improved survival in the bloodstream and bear more capsule, less choline, and more of the protective antigen PspA. Pneumococci were the first bacteria shown to have surface proteins attached to the bacteria via choline residues present in the cell wall. The presence of choline in the cell wall has also been described for *Haemophilus, Pseudomonas, Neisseria,* and *Mycoplasma* species. Choline-binding proteins include several important virulence factors: PspA, a well-described antigen; LytA, the autolytic enzyme; and CbpA, a major adhesin. The Cbp proteins contain 3 to 10 repeats which bind choline. CbpA participates in binding of pneumococci to activated human cells and in nasopharyngeal colonization.

The major pneumococcal toxin pneumolysin acts as a protective antigen and also activates complement.

CbpA is absolutely required for pneumococcal traffic across the blood-brain barrier; this entry involves cytokine activation of the cerebral endothelial cells, which then passively let the pneumococci follow the platelet-activating factor receptor-recycling pathway, which leads them into the cell. How pneumococci drive the endocytosed vesicle across the cell to release bacteria into the cerebrospinal fluid remains to be understood.

The genome sequence has been completely determined.

Tropheryma whippelii

Tropheryma whippelii was recently identified as the etiological agent of Whipple's disease, a chronic digestive and systemic disorder. It is the first bacterial agent which has been indirectly identified by PCR and sequencing. It is a gram-positive actinomycete that is not closely related to any known genus.

Vibrio cholerae

Cholera is a serious epidemic disease that is acquired by drinking water contaminated with human feces or by eating food washed in contaminated water. The disease is characterized by an aqueous diarrhea, essentially due to cholera toxin (CT), a potent enterotoxin that is an AB-type ADP-ribosylating toxin containing one A enzymatic subunit and five identical B subunits. The X-ray structure of the toxin has been established. The toxin is secreted when *Vibrio cholerae* is on the intestinal mucosa; it binds to a host cell ganglioside GM_1, which is a sialic acid-containing oligosaccharide covalently attached to a ceramide lipid and found on many cell types. The

A1 subunit of CT ADP-ribosylates the α subunit of a G_s protein, leading to constitutive activation of adenylate cyclase. The high levels of cyclic AMP alter the activities of sodium and chloride transporters and are responsible for the aqueous diarrhea.

The mechanism by which CT migrates within cells is now well known. In addition to CT, *V. cholerae* virulence requires the expression of TCP (toxin-coregulated pili). Both CT and TCP are regulated by ToxR, a special membrane-associated response regulator. It has recently been shown that the CT gene is encoded by a filamentous bacteriophage, called CTXφ. CTXφ uses TCP as receptor and infects *V. cholerae* cells more efficiently within gastrointestinal tracts of mice than in broth media. TCP and a gene regulator called *toxT* were found in a 39.5-kb PAI, absent in nontoxigenic strains of *V. cholerae*.

Yersiniae

The genus *Yersinia* has three main species, *Yersinia pestis*, *Y. pseudotuberculosis*, and *Y. enterocolitica*. *Y. pestis* is the etiologic agent of bubonic and respiratory plague and is one of the most virulent bacteria. It is transmitted by fleas, in contrast to *Y. pseudotuberculosis* and *Y. enterocolitica*, which are present in contaminated food products. *Y. pseudotuberculosis* causes adenitis and septicemias. *Y. enterocolitica* is responsible for a broad range of gastrointestinal syndromes in humans. In spite of differences of infection routes, the three species have a common tropism for lymphoid tissues.

Y. pseudotuberculosis, after initial gastrointestinal colonization, translocates across the ileum into deeper host tissues. In animal infection models, the organism is internalized by M cells, which are intercalated into the epithelium overlaying ileal lymphoid follicles called Peyer's patches. Efficient translocation across Peyer's patches requires the 986-amino-acid bacterial outer membrane protein invasin. Invasin has become a paradigm for a bacterial protein mediating bacterial entry into mammalian cells via a zipper-type mechanism. Its receptor was identified as β_1-integrin. The invasin-integrin interaction is the most extensively described bacterial ligand-cellular receptor interaction. However, the downstream events resulting from the initial interaction and leading to actin rearrangements and bacterial uptake are far from being understood. The role of invasin seems to be critical for the initial step of infection, i.e., entry into M cells. Later during infection, bacteria remain extracellular.

Y. enterocolitica has been widely used to demonstrate the existence of a class of proteins which are secreted by an unconventional type of secretion, now known as the type III secretion system. In *Yersinia*, these proteins are called Yops (*Yersinia* outer membrane proteins). They are mainly secreted after host cell contact and are directly injected into the mammalian cells. These Yops have various functions. Several of them are involved, in macrophages, in a process of antiphagocytosis affecting either the host phosphorylation machinery (YopH) or the cytoskeleton (YopE and YopT), or other functions, e.g., apoptosis (YopP).

Sporulating Extracellular Bacteria

Bacillus anthracis

The gram-positive bacterium *Bacillus anthracis* is responsible for a disease called anthrax, which affects mostly animals but also humans, mainly in

Africa and Asia. The human disease starts by inoculation of bacilli or spores in a small wound, which leads to a growing necrotic lesion that can lead to fatal edema. It is a potential weapon in biological warfare.

The virulence of *B. anthracis* is due mainly to a polyglutamate capsule present over an S-layer which prevents phagocytosis and to a tripartite toxin consisting of three polypeptides, PA (protective antigen), EF (edema factor), and LF (lethal factor), which associate either as PA plus LF or as PA plus EF. The so-called anthrax lethal toxin (Letx) is the toxin made of PA and LF, both of which are required for toxicity. PA binds to cellular receptors. After proteolytic activation on the host surface, it forms a membrane-inserting heptamer that translocates the LF, the catalytic moiety, across the plasma membrane into the cytosol. Sequence analysis suggested that LF might be a protease. Its substrate has been recently identified as MAPKK1 and MAPKK2. The proteolytic cleavage inactivates MAPKK1 and inhibits the MAPK signal transduction pathway. PA can also translocate EF, a calmodulin-dependent adenylate cyclase which inhibits phagocytosis.

Clostridia

Clostridia are anaerobic bacteria which are widespread in the environment and belong to the normal flora of the digestive tract. They can enter organisms either by the oral route or through the skin via wounds. Clostridia produce various toxins and hydrolytic enzymes that are responsible for the local lesions and general symptoms of infection. Most of the clostridial toxins damage cell membranes or disrupt the cytoskeleton. Two *Clostridium* groups (*C. tetani* and *C. botulinum*) produce potent neurotoxins.

C. botulinum. *C. botulinum* is a ubiquitous bacterium that contaminates food products, in particular preserved foods. It produces the botulinum neurotoxin, which is responsible for flaccid paralysis and death. Botulinum neurotoxins are a group of closely related protein toxins (seven different serotypes) which show absolute tropism for the neuromuscular junction, where they bind to unidentified receptors. They enter into the cytoplasm of motoneurons and cleave their intracellular targets, which are VAMP/synaptobrevin, a protein of synaptic vesicles, or syntaxin for serotype C or SNAP25 for serotypes C, A, and E. This cleavage results in the blockade of transmitter release. *C. botulinum* also produces the C3 toxin, which has been widely used in cell biology due to its capacity to inhibit rho function. This species can now be transformed.

C. difficile. *C. difficile* is responsible for a highly contagious nosocomial antibiotic-associated diarrhea called pseudomembranous colitis. It produces many toxins, encoded by chromosomal genes, including two large cytotoxins, ToxA and ToxB, which both act on small G proteins and are, like C3, very useful tools in cell biology. It also produces one enterotoxin. *C. difficile* has never been transformed.

C. perfringens. *C. perfringens* is responsible for food poisoning and wound contaminations leading to clostridial myonecrosis (gas gangrene), cellulitis, intra-abdominal sepsis, and postabortion and postpartum infections (puerperal fever). It produces the largest number of potential virulence factors of any bacterium, including alpha, beta, epsilon, and iota toxins. However, all strains do not produce all of these toxins. The best-characterized toxin is alpha toxin, which is a phospholipase C, and the pore-forming toxin perfringolysin.

C. tetani.   *C. tetani*, a soil bacterium, is the etiologic agent of tetanus, a fatal neurological disease that is due to tetanus toxin. This neurotoxin is responsible for spastic paralysis and death. Its receptor has not been identified. The toxin binds to neuronal cells and enters the cytosol, where it specifically cleaves vesicle-associated membrane protein/synaptobrevin at a single peptide bond. This selective proteolysis prevents the assembly of the neuroexocytosis apparatus and consequently the release of neurotransmitter. *C. tetani* has recently been transformed.

Parasites

Entamoeba histolytica

Entamoeba histolytica is the agent of amoebic dysenteria and visceral amoebiasis, the third leading cause of death due to parasitic disease. It is transmitted to humans primarily by ingestion of contaminated water and food. *E. histolytica* is an anaerobic amoeba, and the trophozoites lack mitochondria. Cyst germination occurs, and cells grow both on and in mucosal cells. Continued growth leads to ulceration of the intestinal mucosa, causing diarrhea and severe intestinal cramps. The diarrhea is then replaced by a condition referred to as dysentery, characterized by intestinal bloody and mucoid exudates. If the condition is not treated, trophozoites of *E. histolytica* can migrate to the liver, lungs, bones, and brain. Growth in these tissues can cause large abscesses.

E. histolytica induces cell death, albeit without using the two common pathways of apoptosis (i.e., the Fas pathway and the tumor necrosis factor alpha pathway). *E. histolytica* can transfer a galatose-specific lectin to the lateral surface of enterocytes in culture. This transfer precedes the killing of target cells. The potent cytolytic activity of *E. histolytica* is due to a 77-amino-acid polypeptide similar to eukaryotic saposins and surfactant-associated protein B. *E. histolytica* contains cysteine proteinase, which is able to cleave laminin, an activity that seems to correlate with the capacity to produce liver abscess. The 170-kDa galactose-inhibitable lectin, which is homologous to components of the complement cascade and CD59, an inhibitor of complement component C5b9, is responsible for amoebic resistance to complement. This lectin is also implicated in parasite adhesion to host cells. Interestingly, its cytoplasmic domain has homology to the region of the β_2-integrin cytoplasmic tail implicated in the regulation of integrin-mediated adhesion and may regulate parasite adhesion to the host cell. The availability of a transfection system should facilitate the identification of genes involved in virulence.

Leishmania Species

Leishmania spp. are the etiological agents of leishmaniasis, which range from local (cutaneous) to systemic (visceral) disease and affect 15 million people, mainly in areas where the hematophagous sand fly circulates. The broad spectrum of clinical features depends on the *Leishmania* species and on the genetic and immunological status of the host. Immunocompromised hosts such as AIDS patients are at high risk for visceral leishmaniasis.

In mammals, *Leishmania* spp. multiply exclusively in phagocytic cells as amastigotes (without flagella). They reside in a PV, where they can divide by binary fission. For all *Leishmania* spp. examined, the PVs exhibit the properties of a late endocytic compartment containing lysosome-associated membrane proteins and hydrolases.

Dermal mononuclear phagocytes harboring transmissible parasites can be ingested by the sand fly during a blood meal. Within the sand fly midgut, amastigotes are released from the phagocytes and start to multiply as extracellular flagellated parasites named promastigotes. Then they migrate toward the anterior region of the digestive tract, where they differentiate in metacyclics, the infective stage delivered in the skin during the second blood meal of the sand fly.

Several genetic tools are now available, including the *Drosophila* transposable-element mariner and some mariner derivatives allowing gene fusion. These tools have already been useful for identifying genes implicated in the infectious cycle of the parasite, such as those necessary for the synthesis of the key surface glycoconjugate, lipophosphoglycan, but genetic studies are hampered by the fact that *Leishmania* spp. are diploid organisms.

Plasmodium falciparum

The protozoan parasite *Plasmodium falciparum*, which is transmitted by a mosquito (anophele), is the agent of malaria. The clinical symptoms (chills, fever, and headaches) and the more severe forms such as cerebral malaria are mainly due to the intraerythrocyte development of the parasite. The virulence of *P. falciparum* is mediated in part by adhesins expressed on the infected erythrocyte, mediating adhesion to noninfected erythrocytes or endothelial cells of cerebral, pulmonary, renal, or placental capillaries. Many different endothelial receptors for infected erythrocytes have been identified, such as CD36 and thrombospondin, ICAM-1, vascular cell adhesion molecule (VCAM), and chondroitin sulfate A (CSA). An association between the degree of binding to ICAM-1 and clinical illness in nonanemic patients has been suggested. Plasmodial adhesins are encoded by 50 to 100 genes, which undergo antigenic variation by a mechanism currently under investigation. Recent work has analyzed some aspects of the process of invasion into the erythrocyte and the intraerythrocyte life-style, highlighting a role for hemoglobin degradation and pyrimidine and phospholipid biosynthesis. The role of these processes in virulence will soon be testable by genetic means, since techniques to inactivate genes first set up in *P. berghei*, a rodent *Plasmodium*, are now available for *P. falciparum*. These techniques have demonstrated the role of the circumsporozoite in parasite development in the mosquito and that of TRAP, a protein which contains a functional integrin-like domain involved in the gliding motility and infectivity of *Plasmodium* sporozoites. However, the precise molecular events controlling sporozoite invasion and exoerythrocytic development within hepatocytes remain largely unknown.

Toxoplasma gondii

Toxoplasma gondii is one of the most successful protozoan parasites, infecting 10 to 25% of the world's human population. Clinical toxoplasmosis is most frequent in newborns whose mothers contracted infection during pregnancy; this condition results in abortions or congenital malformations. Postnatally acquired toxoplasmosis is usually asymptomatic but results in lifelong persistence of slowly replicating parasites. In immunocompromised patients (AIDS patients or transplant recipients), reactivation of latent parasites leads to intracerebral toxoplasmic abscesses and sometimes diffuse encephalitis and generalized infection. *T. gondii* is an obligate intracellular parasite, which can invade a broad range of host cells including phagocytic and nonphagocytic cells. In the infected cell, the parasite occupies a specialized vacuole in the cytoplasm (the parasitophorous vacuole [PV]), where it rapidly divides by binary fission, ultimately leading to lysis of the host cell and liberation of an infectious progeny.

Invasion is an active process driven by the parasite, which usually takes about 5 to 10 s; the host cell is a passive spectator (no membrane ruffling, no participation of host actin, no tyrosine phosphorylation of host proteins). Extracellular parasites display gliding motility on the surface of host cells. This twisting motion is dependent on the parasite cytoskeleton (actin and myosin) and is necessary for cell invasion. Entry is initiated by reorientation of the parasite to create contact between its apical end and the host cell plasma membrane. Passage in the host cell involves the formation of a region of very close apposition of the two plasmalemmas located at the site of entry, called the moving junction. The moving junction has the morphological characteristics of a tight junction, although it has not been characterized at the molecular level. When formed, the PV does not fuse with lysosomal compartments and is rapidly modified by the parasite secretory proteins in a compartment suitable for intracellular multiplication. Note that when parasites are primed by the immune response and phagocytosed (opsonized), they are killed; only active invasion leads to parasite development. The PV formed by live parasites has nonselective pores that allow the diffusion of molecules of up to 1,200 Da. The vacuole is surrounded by a layer of endoplasmic reticulum and mitochondria, which may be used for the metabolic requirements of the parasite.

The cell infection process involves the sequential exocytosis of three different apical secretory organelles called micronemes, rhoptries, and dense granules. Microneme proteins are apparently used for host cell recognition, binding, and possibly gliding motility; rhoptry proteins are used for PV formation; and dense-granule proteins are used for remodelling the vacuole into a metabolically active compartment. The stimuli and molecular mechanisms which trigger exocytosis of apical organelles are unknown. Unfortunately, genetic tools are not very efficient.

Trypanosoma cruzi

Trypanosoma cruzi is the etiological agent of Chagas' disease in humans. The vector for *T. cruzi* is an insect of the Reduviidae family. Transfusion can also lead to contamination in areas of endemic infection (South America). In the digestive lumen of these insects, trypomastigotes differentiate in epimastigotes, which replicate extensively by binary fission. When liberated in insect feces, newly differentiated trypomastigotes actively penetrate the vertebrate mucosa or cutaneous wounds.

T. cruzi is intracellular in vertebrates but extracellular in insects. It enters cells by a mechanism characterized by a smooth diving into the cell. This process is coupled to the recruitment of lysosomes at the site of entry. While inhibitors of microtubules such as nocodazole inhibit entry, the actin cytoskeleton does not seem to be involved in the entry process. Disruption of the actin cytoskeleton by cytochalasin D even increases the rate of entry. Availability of intracellular Ca^{2+} is a prerequisite for internalization. Ca^{2+} is probably released from intracellular stores upon inositol triphosphate production, suggesting the action of a phospholipase C. Ca^{2+} transients are thought to be required for rearrangements of actin microfilaments, which might facilitate lysosome access to the plasma membrane. The Ca^{2+} signaling activity expressed by tissue culture trypomastigotes is generated through the action of a cytosolic enzyme, *T. cruzi* oligopeptidase B, which probably acts as a processing enzyme that generates an active Ca^{2+} agonist. The role of this peptidase has recently been confirmed by using oligopeptidase B gene knockout mutants. Signaling through transforming growth factor β receptors is also required for *T. cruzi* trypomastigote invasion.

T. cruzi does not stay in the parasitophorous vacuole. It is released in the cytosol less than 2 h through the action of a 65-kDa pore-forming toxin active at pH 5 and of a transialydase/neuraminidase that is thought to disorganize the vacuolar membrane by desialylation of the vacuole constituents and to facilitate the action of the toxin. Once released into the cytosol, the trypomastigotes are ready to differentiate in amastigotes and to replicate. After lysis of the first infected cells, the parasites migrate into deeper tissues and/or circulate in the blood before invading other cells. Recent advances in genetic manipulation (targeted gene replacement) should help identify the molecular mechanisms underlying trypanosome infections.

Yeasts and Molds

Aspergillus fumigatus

Aspergillus fumigatus is one the most common of the airborne saprophytic fungi. Humans and animals constantly inhale numerous conidia of this fungus. The conidia are normally eliminated in immunocompetent hosts by innate immune mechanisms, and aspergilloma and allergic bronchopulmonary aspergillosis, both uncommon clinical syndromes, are the only infections observed in such hosts. With the increasing number of immunosuppressed patients, there has been a dramatic increase in severe and fatal invasive aspergillosis, now the most common mold infection worldwide. The lung is the site of infection by *Aspergillus* spp., and since alveolar macrophages are the major resident cells of the lung alveoli, along with neutrophils, which are actively recruited during inflammation, they are the two main cells involved in the phagocytosis of *A. fumigatus*. Several lines of evidence suggest that nonoxidative mechanisms are essential for the killing of conidia by the alveolar macrophages of the immunocompetent host. The second line of phagocytic cells, the neutrophils, plays a role in containing the conidia that resist intracellular killing and germinate. Contact between neutrophils and hyphae triggers a respiratory burst, secretion of reactive oxygen intermediates, and degranulation. Killing of hyphae required oxidants, but oxidant release by polymorphonuclear neutrophils could not mediate hyphal killing without concomitant fungal damage by granule constituents. The cell biology events involved in the phagocytosis of *A. fumigatus* conidia and hyphae and the down-regulation of the immune system by immunosuppressive drugs are poorly understood. Several strategies are available to produce single or multiple mutants of *A. fumigatus*, which is a haploid organism. The classical method involves the disruption of the gene of interest by the insertion of an antibiotic resistance gene. A *PYRG* blaster, very similar to the *URA* blaster developed for *Candida albicans*, has been developed for *A. fumigatus*. Screening randomly STM-REMI-generated mutants for loss of virulence has failed in *A. fumigatus*.

Efforts to sequence the genome of *A. fumigatus* are under way.

Candida Species

Candida is the most frequent cause of human mycosis. Several species of *Candida* cause human infections, and the major etiologic agent of candidiasis is *C. albicans*. *C. albicans* is a commensal of the gastrointestinal and genitourinary tracts. It causes infections in immunocompromised patients or in patients with local predisposing factors. A wide range of infection types have been reported. The most common diseases are superficial oropharyngeal and vulvovaginal candidiasis. The most severe forms of the disease are hematogenously disseminated candidiasis, which occurs in immunocompromised patients. Natural resistance to infection is believed to rely on polymorphonuclear cells and on mononuclear phagocytes. The alternative complement pathway plays an important role in enhancing the effect of phagocytic cells via opsonization of the organism. Polymorphonuclear neutrophils can ingest *C. albicans*. Once in the phagosome, *Candida* yeasts are killed in the immunocompetent host. Typical morphological features of *Candida* include budding yeasts and pseudohyphae with branching and production of oval blastospores. With the exception of *C. glabrata*, all *Candida* species produce hyphae and pseudohyphae in tissue. Most of the efforts aimed at understanding the virulence of this pathogen at the molecular level have been centered on the identification of genes regulating the dimorphism of *C. albicans*. *C. albicans* is naturally diploid. A *URA* blaster method has been developed, allowing successive rounds of gene disruption.

The *C. albicans* genome sequence will soon be available.

Cryptococcus neoformans

Infections caused by the encapsulated basidiomycetous yeast *Cryptococcus neoformans* are initiated by inhalation of the yeast into the lungs and show a remarkable propensity to spread to the brain and meninges (causing cryptococcal meningitis). Immunosuppression, especially AIDS, is the leading predisposing factor for cryptococcosis. Known virulence factors are the polysaccharide capsule and the phenol oxidase responsible for melanin formation from phenolic substrates. Human polymorphonuclear neutrophils and macrophages are able to ingest and kill yeasts in vitro. Antibody- and complement-mediated opsonization is essential during phagocytosis. Evidence that encapsulated and acapsular yeasts may differently trigger certain macrophage functions also provides an important model to study the cell biology of receptor signaling at the macrophage cell surface. Gene inactivation strategies have been developed. They vary for the different serotypes, and different DNA delivery systems are used. In all cases, the rate of homologous integration is very low (1 in 1,000 or less). Therefore, strategies for selecting these rare events have been developed.

Histoplasma capsulatum

Histoplasmosis is caused by *Histoplasma capsulatum*, which is found in temperate and tropical climates. *H. capsulatum* grows as a saprophyte and is acquired by inhalation of airborne conidia. Most infections are mild or subclinical, except in immunosuppressed individuals, especially AIDS patients, who are at high risk for disseminated histoplasmosis. *H. capsulatum* is typically dimorphic; the pathogenic phase is the yeast form, which is obtained at 37°C. Although biochemical and molecular differences between the yeast and mycelial forms have been observed, the mechanism of dimorphism, as for *Candida*, remains poorly understood. The primary host cell is the macrophage, in which *H. capsulatum* proliferates in a phagoly-sosome. The yeast form is able to modulate the pH of this intracellular compartment, avoiding damage by the degradative lysosomal enzyme, thus allowing intramacrophage survival and eventually leading to macro-phage death. A calcium-binding protein and a cell wall α1-3-glucan con-tribute to intracellular survival. Homologous transformation is a rare event in *H. capsulatum*, and transforming DNA mostly integrates randomly.

Pneumocystis carinii

Pneumocystis carinii is a frequent cause of pneumonia in immunocompro-mised patients, especially AIDS patients. Historically, this extracellular pathogen was thought to be a protozoan, but recent molecular data sug-gested that it is a fungus. The reservoir and the mode of acquisition of the infection remain undetermined, although molecular analysis has shown the presence of *P. carinii* in the air. In infected lungs, *P. carinii* is intimately associated with alveolar cells, suggesting that it may use cell-based nutri-ents or stimuli to proliferate. This attachment involves the major surface glycoprotein and fibronectin. One of the striking features of *P. carinii* is its ability to generate surface variation via programmed gene rearrangements. The genes involved in this process encode various isoforms of the major surface glycoprotein. Data from many studies indicate that infection is host species specific. One of the major obstacles in *Pneumocystis* research is the lack of a method of sustained in vitro culture of this organism. The absence of axenic growth has limited the extent of molecular studies.

A sequencing project of the entire *P. carinii* genome is in progress.

Selected Readings

Alberts, B., D. Bray, J. Lewis, M. Raff, K. Roberts, and J. D. Watson. 1994. *Molecular Biology of the Cell*, 3rd ed. Garland, New York, N.Y.

Cossart, P., P. Boquet, S. Normark, and R. Rappuoli. 1996. Cellular microbiology emerging. *Science* **271:**315–316.

Cossart, P., and J. Miller (ed.). 1999. Host-microbe interactions: bacteria. *Curr. Opin. Microbiol.* **2:**1–106.

Finlay, B., and P. Cossart. 1997. Exploitation of mammalian host cell functions by bacterial pathogens. *Science* **276:**718–725.

Finlay, B., and S. Falkow. 1997. Common themes in microbial pathogenicity re-visited. *Microbiol. Mol. Biol. Rev.* **61:**136–169.

Finlay, B., and P. Sansonetti (ed.). 1998. Host-microbe interactions: bacteria. *Curr. Opin. Microbiol.* **1:**1–129.

Mandell, G. L., J. E. Bennett, and R. Dolin. 1996. *Principles and Practice of Infectious Diseases*, 4th ed. Churchill Livingstone, New York, N.Y.

Miller, V. L., J. B. Kaper, D. A. Portnoy, and R. R. Isberg (ed.). 1994. *Molecular Genetics of Bacterial Pathogenesis*. ASM Press, Washington, D.C.

Mims, C. 1995. *Mims' Pathogenesis of Infectious Disease*, 4th ed. Academic Press, Inc., New York, N.Y.

Salyers, A. A., and D. D. Whitt. 1994. *Bacterial Pathogenesis: a Molecular Approach*. ASM Press, Washington, D.C.

2

Cell Biology: an Overview

Patrice Boquet

Mechanisms controlling basic cellular functions such as cell division, motility, adherence, differentiation, and death are extremely highly conserved throughout the animal kingdom. Many of these mechanisms are parasitized by microbes when they colonize or invade host cells. This chapter aims to introduce briefly some of these basic cell biology mechanisms.

Microbes first encounter the outside of the cell. Many of the proteins on the outside of the cell are used for communication. Extracellular messengers bind to cell surface receptors; this induces modifications in these membrane receptor proteins including dimerization or oligomerization and autophosphorylation. The signals are transduced by nucleotide binding or reversible phosphorylation, and signaling networks are assembled by the interaction of specific modular protein domains. Cascades of kinases are the effector molecules of cell signaling systems in the cytosol and the nucleus.

Once inside the cell, microbes are often caught up in the membrane trafficking system, which is used primarily to move proteins throughout the cell. Intracellular vesicles are formed by budding from a donor membrane, and targeting and fusion to an acceptor compartment are mediated by the assembly and disassembly of protein complexes. Special motor proteins move these vesicles, their cargo, mRNAs, and perhaps regulatory proteins. Movement is along intracellular tracks made by the cytoskeleton. The cytoskeleton is made of building blocks such as actin or tubulin, which polymerize into supramolecular structures (either dynamic or stable) that are also involved in motility, maintenance of cell shape, and resistance to external stresses. The cytoskeleton is also important in the adhesion of cells to their substratum or to neighboring cells, since this involves activation and clustering of transmembrane proteins and their reversible attachment to the cytoskeleton. Loss of this attachment is one of the many reasons why cells self-destruct. This cell suicide is carried out primarily by proteases; other proteases are activated when cells need to terminate the life span of regulatory proteins or are undergoing starvation conditions. Any decision on cell suicide is largely dependent on the status of the cell cycle clock, which is driven by the periodic activation of cyclin-dependent kinases, while associated checkpoint mechanisms preserve genome integrity.

General Cell Organization

Cell organization is outlined in Figure 2.1. Cells are enclosed by a membrane made up of a lipid bilayer (1) (shown under a magnifying glass in

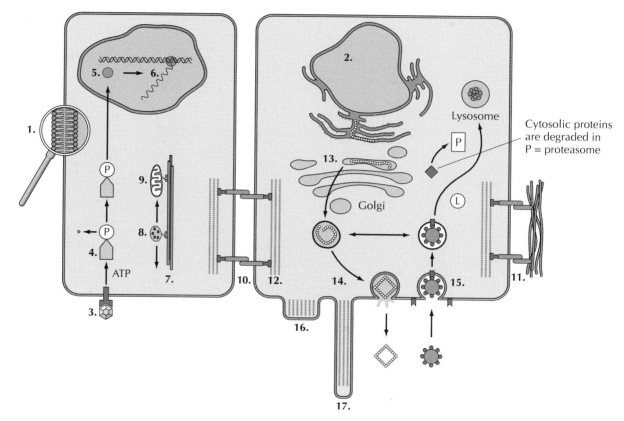

Figure 2.1 General cell organization. See the text for details.

the figure). The cytoplasm contains a nucleus (2) surrounded by a nuclear envelope, which acts as a scaffold for the DNA. Signals from outside the cell bind to membrane receptors (3), triggering phosphorylation of kinase cascades (4), which phosphorylate cytoplasmic proteins or, upon entry into the nucleus (5), phosphorylate nuclear factors regulating DNA transcription (6) and cell division factors (see "Signal transduction and cell regulation" and "Cell cycle and organelle inheritance" below). Microtubules (7) are used as tracks for vesicles carrying cargo (8) (see "The cytoskeleton" below) or mitochondria (9), which provide energy to the cell. Cells are attached to each other (10) or to their substratum (11) by adhesion molecules that are linked to the actin cytoskeleton (12) (see "Cell adhesion and morphogenesis" below). The nuclear envelope is continuous with the endoplasmic reticulum (ER) (13), into which secreted proteins are translocated during their synthesis (see "Organelle biogenesis" below). Secreted proteins are glycosylated and transported via the Golgi apparatus to the cell membrane into vesicles (14) (see "Membrane traffic" below). Cells take up molecules from the external medium by the endocytic pathway (15). These molecules are degraded in lysosomes (L in Figure 2.1). Cytosolic proteins can be proteolyzed by the proteasome machinery (P in Figure 2.1; see Box 2.2). Actin filaments (12) can form wide membrane extensions called lamellipodia (16) or long protrusions called filopodia (17), both of which are involved in cell locomotion (see "The cytoskeleton" below). All of these cellular processes can be studied by a variety of techniques, including the application of the activator and inhibitor chemicals listed in Table 2.1. The

Table 2.1 Main inhibitors and activators used in cell biology

Cellular target	Inhibitor or activator
Targets of inhibitors	
Tyrosine kinases	Genistein/tyrphostin/herbimycin
Ser/Thr kinases	Staurosporine
Protein kinase A	8-Bromo-cyclic AMP
Protein kinase C	Calphostin C
Myosin kinase	Butanedione monoxime
PI3-K	Wortmannin
Phosphatases (general)	Okadaic acid/vanadate
Ca^{2+}/calmodulin-activated phosphatase 2B	Cyclosporin/FK506
Phospholipase C	Neomycin sulfate
ATP synthesis	Apyrase
Sodium/potassium membrane ATPases	Ouabain
Vesicular ATPases	Bafilomycin A1/concanamycin A
Sodium/proton antiporter	Amiloride
Calcium ions	BAPTA
Calmodulin	Trifluoperazine
ER calcium entry pumps	Thapsigargin
IP_3 calcium release	D-*myo*-Inositol-1,4,5,-P_3
Farnesylation and geranylation	Lovastatin
Sterol and fatty acid biosynthesis	Cerulenin
Actin polymerization	Cytochalasin D
Tubulin polymerization	Nocodazole
MAPK	PD 98059
Proteasome degradation	Lactacystin
Cysteine proteases	E-64
Serine/threonine proteases	Dichlorocoumarin
Zinc-dependent proteases	Phosphoramidon
Glycoprotein processing	Castanospermine
Lysosomal alpha mannosidases	Swainsonine
Heterotrimeric G proteins G_0/G_i	Pertussis toxin
Rho	*C. botulinum* exoenzyme C3
Ras	Ras-Raf-binding peptide
v-SNAREs/cellubrevin	Tetanus toxin/botulinum toxin B
t-SNAREs	Botulinum toxin C
SNAP25	Botulinum toxin A
Caveolae endocytic pathway	Filipin/cyclodextrin
Coated pit-coated vesicle endocytic pathway	Chlorpromazine
COP1 binding to ARF-GTP	Ilimaquinone
ARF1 GEF (Golgi transport)	Brefeldin A
Targets of activators	
Actin polymerization	Jasplakinolide
Actin filament stabilization	Phalloidin
Protein kinase C	Phorbol esters
Protein kinase A	Dibutyryl cyclic AMP
Tyrosine kinases	D-*erythro*-Sphingosine
Adenylate kinase	Forskolin
Phospholipase A_2	Melittin
GTP-binding protein (general)	GTPγS
Heterotrimeric G proteins (general)	Mastoparan
Heterotrimeric αG_s proteins	Cholera toxin
Rho	*Escherichia coli* CNF1 toxin
MT stabilization	Taxol
Calcium cytosolic entry	Ionomycin/A23187

processes can also be visualized in real time by video microscopy (Table 2.2).

Signal Transduction and Cell Regulation

Detection and Initial Transduction

Molecular messengers that alter cell behavior must be detected, and their message must be transmitted and translated. Messengers are detected by membrane receptors, and a primary means of transmission is by phosphorylation or dephosphorylation of target proteins by specific kinases or phosphatases. Phosphorylation or dephosphorylation allows activation or deactivation of proteins involved in the regulation of either structural molecules or DNA transcription. Coordination between different pathways is achieved when the signaling proteins bind to each other via a limited number of recognition motifs to create a signaling network. Deactivation occurs by GTP hydrolysis, dephosphorylation, proteolysis, or endocytosis followed by degradation of membrane receptors.

Membrane receptors can be divided into three main classes according to their response to ligand binding: the ligand-gated ion channels open a selective pore (these channels are, however, usually restricted to the nervous system, and so they are not discussed here further); the seven transmembrane receptors are linked to heterotrimeric GTP-binding proteins; and various receptors are linked to enzymes. Many of the receptors in the last group have enzymes as part of their cytoplasmic extension, such as the large group of tyrosine kinase receptors (TKRs). The TKRs dimerize upon ligand binding, and they link to other kinases and small GTP-binding proteins of the $p21^{ras}$ superfamily. Other receptors have no intrinsic enzymatic activity but can recruit cytosolic kinases. The following sections first address the activation of the receptors and the proteins directly attached to them and then discuss the methods by which these signals are amplified and transmitted throughout the cell. Some of the same transmission mechanisms are shared by many different receptors, even those of different types.

The members of the family of seven transmembrane (or serpentine) receptors, such as the receptor for the growth factor thrombin, are linked by their cytoplasmic tails to heterotrimeric GTP-binding proteins (also designated as large GTPases due to their molecular mass of 45 kDa). A large GTPase is a complex of three proteins—α, β, and γ—in which α is the GTP-binding protein. When the α subunit is linked to GDP, the $\alpha\beta\gamma$ complex is

---

Table 2.2   Videos

Videos of cellular processes described in this chapter have recently been collated by the journal *Molecular Biology of the Cell*. The relevant references are as follows:

Inoué S., and R. Oldenbourg. 1998. Microtubule dynamics in mitotic spindle displayed by polarized light microscopy. *Mol. Biol. Cell* **9**:1603–1607. (Three videos.)

Presley, J. F., C. Smith, K. Hirschberg, C. Miller, N. B. Cole, K. J. Zaal, and J. Lippincott-Schwartz. 1998. Golgi membrane dynamics. *Mol. Biol. Cell* **8**:1617–1626. (Nine videos.)

The videos can be viewed on the journal Web site at http://www.molbiolcell.org.

formed and is associated with the cytoplasmic tail of the serpentine receptor (Figure 2.2A). A serpentine receptor activated by its ligand will provoke, through a transmembrane modification in its shape, the removal of GDP bound on α. The α subunit, now free of GDP, binds GTP, since there is a large excess of this nucleotide in the cell. Binding of GTP to the α subunit separates α from βγ. Activated α subunit (free to diffuse in the cytosol) and the βγ complex (associated with the membrane) then leave the serpentine receptor to bind and regulate different proteins (or protein complexes) implicated in signaling pathways. Once the α subunit has activated its downstream target protein, GTP is hydrolyzed into GDP, α-GDP reassociates with βγ, and the complex returns to the serpentine receptor.

TKRs, such as the receptor for epidermal growth factor, have a single transmembrane domain. Their cytoplasmic tails have tyrosine kinase activity. Ligand binding induces TKR dimerization; the cytoplasmic tails of the

Figure 2.2 Signal transduction and modular protein domains. **(A)** Signal transduction by TKRs and serpentine receptors converge on the MAPK pathway, which controls proliferation and differentiation through activation and repression of gene transcription. Phosphatases (PPs) negatively regulate this cascade. TKRs also activate the Rho family of GTP-binding proteins, which control cell shape by modulating actin cytoskeleton organization. **(B)** Signal transduction is carried out by molecular complexes that are held together by modular protein domains. The modular protein domain SH2 recognizes and binds phosphorylated tyrosine; the SH3 domain recognizes and binds polyproline residues; and the PH domain recognizes and binds phosphorylated lipids in the membrane (such as PIP₂). **(C)** Activation of the MAPK pathway by TKRs requires association of different proteins by modular protein domains. See the text for details.

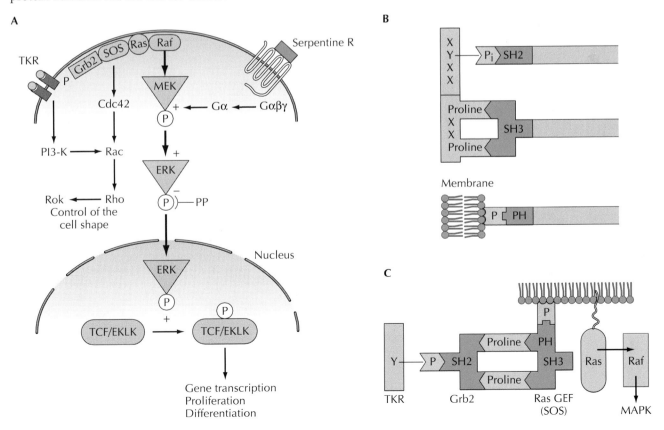

TKRs then cross-phosphorylate each other on specific tyrosine residues. The phosphorylated tyrosines are bound by adapter proteins (such as Grb2) using a specific motif called SH2 (src homology 2) on the adapter protein (Figure 2.2). The adapter proteins have other binding motifs such as the SH3 domain, which binds to polyproline residues (Figure 2.2C). This links the receptors to activators (called guanine exchange factors [GEFs]) of small GTPases of the p21^{ras} superfamily. SH2 domains also direct the binding of other enzymatic activities, such as phosphoinositide 3-kinase (PI3-K), the Src tyrosine kinase or phospholipase C-γ (PLC-γ).

Other types of receptors that span the membrane only once turn on different signaling pathways. For example, receptors that bind the immune system signaling molecules called interleukins are usually linked to the JAK (Janus kinase)/STAT (signal transducers and activators of transcription) pathway. As with TKRs, initial activation is by dimerization. The cytosolic kinase JAK associates with the cytosolic portion of the activated membrane receptor and phosphorylates the transcription factor STAT. Phosphorylated STAT proteins leave the receptor, dimerize, and migrate into the nucleus, where they bind DNA and activate the transcription of certain genes.

Some receptors are activated by signals—such as bacterial lipopolysaccharide or tumor necrosis factor (TNF)—that give acute-phase responses. These receptors oligomerize upon ligand binding. Proteins, including the kinases IKKα and IKKβ, are added to an 800-kDa complex that binds to the cytoplasmic tail of the oligomerized receptors. These kinases phosphorylate IκB, which is then rapidly degraded by the ubiquitin proteasome pathway (see Box 2.2 and Figure 2.3). This has the effect of releasing the NF-κB that was bound to IκB, so that the NF-κB can enter the nucleus and activate transcription.

Secondary Signal Transduction

This section addresses the events occurring after ligand binding and receptor activation. These events include signaling by GTPases, kinases, phosphoinositide kinases, and calcium.

As mentioned above, G proteins have varied targets, depending on the identity of the particular G protein. Some G-protein α subunits attach to the enzyme adenyl cyclase, either increasing or decreasing its production of the signaling molecule cyclic AMP. An increase in the amount of cyclic AMP turns on protein kinase A (PKA), which has many targets. Other α subunits attach to and turn on phospholipase C-β (PLC-β), which breaks apart phosphatidylinositol-4,5-bisphosphate [PtdIns(4,5)P$_2$, or PIP$_2$] into inositol-1,4,5-triphosphate (InsP$_3$) and diacylglycerol (DAG). Both molecules eventually turn on kinases. InsP$_3$ triggers the release of calcium, which binds to a protein called calmodulin. The calcium-calmodulin then binds and activates a kinase called calcium/calmodulin-dependent protein kinase. DAG directly turns on certain isoforms of the Ser/Thr-specific protein kinase C (PKC), which sets in motion a cascade of protein kinase activation, thus turning on the mitogen-activated protein kinases (MAPK).

Responses to the Ras superfamily of GTPases are also diverse, depending on which of the five branches—Ras, Rho, Rab, Ran, and ADP-ribosylating factor (ARF)—is activated. (ARF was first discovered as a factor that was indispensable for cholera toxin to exert its ADP-ribosyltransferase activity.) Members of each of these branches control a different set of cellular mechanisms. Ras controls cell division and cell differentiation, Rho controls actin cytoskeleton organization, Rab controls in-

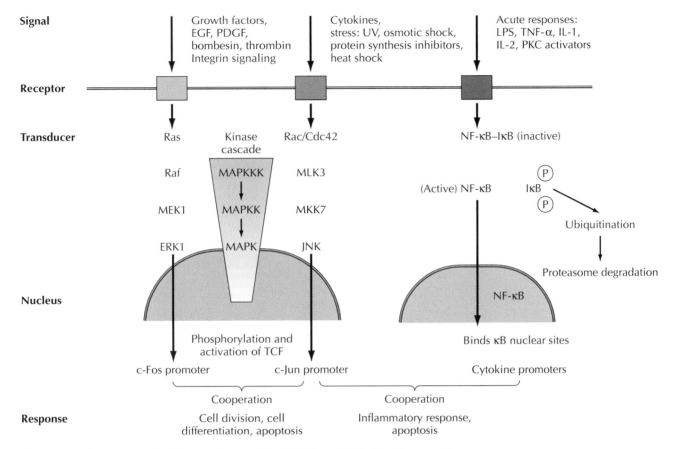

Figure 2.3 Important signaling pathways: the MAPK and NF-κB pathways. The MAPK pathways are kinase cascades that end in activation of a terminal kinase (either ERK or JNK). These terminal kinases phosphorylate and activate proteins that control the transcription of either the Fos or Jun genes. Other stimuli lead to phosphorylation of the IκB protein so that it dissociates from NF-κB. The phosphorylated IκB is polyubiquitinated and degraded by the proteasome machinery (Box 2.2), and the free NF-κB enters the nucleus and activates transcription. The ultimate response of the cell to activation of these pathways is indicated at the bottom of the figure.

tracellular vesicular traffic, Ran controls the transport of proteins through nuclear pores, and ARF controls the formation or budding of intracellular vesicles.

These small GTPases (named for their low molecular mass of ~21 kDa) are timer molecules. They are defined as "on" when bound to GTP and "off" when the GTP is hydrolyzed into GDP. Small GTPases are activated by stimulation of a GEF, which promotes the exchange of the bound GDP for GTP as described for large G proteins. Small GTPases are deactivated by GTP hydrolysis. GTP affects the shape of two GTPase domains called switch 1 and 2. In the GTP-bound form of the small G protein, the switch 1 region (a polypeptide of 10 residues) moves so that it binds to a protein or protein complex termed the effector, which often has Ser/Thr kinase activity. This activates the kinase, which turns on other kinases such as the members of the MAPK family (Figure 2.2A). The switch 2 domain contains amino acids crucial for GTP hydrolysis. These residues are not, however, properly oriented for active GTP hydrolysis until a GTPase-activating protein binds. This protein introduces an arginine residue called the arginine finger into the GTP-bound form of the small G protein, allowing fast hydrolysis of GTP.

The Rho subfamily of GTPases encompasses Cdc42, Rac, and Rho. These proteins induce the formation of various actin-rich structures including protrusions called filopodia (by Cdc42-GTP), a dense actin mesh on the cell periphery that makes up lamellipodia (by Rac-GTP), and stress fibers and focal adhesion complexes (by Rho-GTP). The mechanism by which Cdc42 (and probably Rac) induces actin polymerization at the plasma membrane requires its binding to a protein named N-WASP (neuronal Wiskott-Aldrich syndrome protein). This protein, activated by Cdc42-GTP, harbors a recognition motif at its amino terminus which binds polyproline stretches contained within proteins (such as zyxin or vinculin) that associate with receptors such as integrins involved in the attachment of cells to the extracellular matrix. N-WASP also contains specific motifs that recruit the actin-nucleating Arp2/3 complex and/or profilin, an accelerator of F-actin polymerization. Rho-GTP activates a Ser/Thr kinase (Rho kinase [Rok]). Rok phosphorylates and inactivates myosin light-chain phosphatase (MLCP), so that the activity of the opposing myosin light-chain kinase (MLCK) predominates. Phosphorylation of myosin by MLCK leads to activation of this motor protein so that it binds to actin filaments and causes them to contract.

The autophosphorylation of a TKR, followed by adapter and then Ras GEF binding, leads to the activation of Ras. Ras in turn stimulates the first kinase of the MAPK cascade, Raf (Figure 2.2C). The next kinases in this cascade, the MAPKKs (MEKs), are dual Ser/Thr and Tyr kinases of narrow specificity. MEKs activate MAPKs by phosphorylation (Figure 2.3). (The particular MAPK in the pathway downstream of Raf is also called an ERK [extracellular signal-regulated protein kinase].) MAPKs are Ser/Thr kinases with broader specificity than MEKs. Scaffolding proteins such as JIP-1 or MP1 bind a specific sequence of MAPKs and may channel their activities to certain sites. MAPKs phosphorylate cytosolic proteins such as microtubule-binding proteins and transcription factors required for cell division or differentiation. Parallel and very similar to the MAPK cascade of kinases, the c-Jun kinase and p38 kinase pathways are controlled at the level of the cell membrane by receptors implicated in the detection of stresses such as osmotic shock or temperature stress. Apparently, the small GTP-binding proteins that turn on the c-Jun and p38 kinase pathways are not Ras but Rac and Cdc42 (Figure 2.3).

The TKRs can also bind PLC-γ. As with the serpentine receptors and PLC-β, this produces DAG (activating PKC) and InsP$_3$ (releasing calcium). Other kinases of note include the PI3-K-activated PKB (also known as AKT), and tyrosine kinases such as p125^{fak} (focal adhesion kinase), p60^{src}, or p130^{cas}, which are usually associated with focal contacts and are important in cross talk with TKRs or serpentine receptors and the activation of kinase cascades such as the MAPKs.

Proteins regulated by phosphorylation at Ser/Thr or Tyr residues require dephosphorylation to return to their active or inactive initial states. One example of such regulation is control of the cell cycle kinase Cdc2 by Cdc25 and other phosphatases. Activities of Ser/Thr phosphatases and Tyr phosphatases are regulated by a combination of targeting and regulatory subunits. The time during which regulatory proteins are phosphorylated is pivotal for a cell to decide between division and differentiation. The mechanism determining the balance between kinase and phosphatase activities is therefore a basic binary system by which cell fate can be determined.

As noted briefly above, ions such as calcium are important intermediates in signaling cascades. Cells tightly regulate their intracellular level of calcium, via calcium-binding proteins and extrusion mechanisms. This allows calcium to be used in localized cellular regulation such as muscle contraction, actin filament organization, membrane hyperpolarization, and regulated exocytosis, but it also means that calcium released outside the cell or from intracellular stores can be used as a rapidly mobilizable messenger. To maintain a low cytosolic calcium concentration, cytosolic calcium is actively pumped into the ER. Calcium is sequestered in the ER by binding to specialized proteins (such as calsequestrin or calreticulin). Several mechanisms may introduce small bursts of calcium into the cytosol, either from the ER or across the plasma membrane. In nonexcitable cells, the $InsP_3$ pathway is the major system used to induce a burst of calcium in the cytosol. Membrane receptors coupled to either heterotrimeric GTPases or TKRs activate PLC-β (heterotrimeric G protein) or PLC-γ (TKRs) that convert PIP_2 into $InsP_3$ and DAG. $InsP_3$ binds the ryanodine receptor on the ER membrane, triggering the release of calcium. Receptors that induce $InsP_3$ formation cause a slow release of calcium into the cytosol. Conversely, stimulation of certain membrane receptors induces a fast release of calcium into the cytosol through the activation of plasma membrane Ca^{2+} channels. These channels are activated when the ER calcium store is depleted, thus rapidly replenishing the calcium in the cell. Excitable cells also contain voltage-dependent calcium channels that enable them to increase their calcium concentration very rapidly.

The second group of small molecule signals is the family of 3,4,5- and 4,5-phosphorylated inositol lipids (PIP_3 and PIP_2). Two kinases, termed 3-kinase and 5-kinase (the numbers refer to the position of phosphorylation on the inositol ring), dominate this regulation by producing PIP_3 and PIP_2, respectively. The PI3-K 85-kDa regulatory subunit binds to activated TKRs (such as the epidermal growth factor receptor) on phosphorylated tyrosines and activates the 110-kDa catalytic subunit. Its product, PIP_3, is involved in the formation of lamellipodia (membrane ruffling) and in the mechanism that closes the phagocytic cup during macrophage phagocytosis. In contrast to PI3-K, PI5-K is associated with the cytoskeleton. PI5-K phosphorylates phosphatidylinositol-4-phosphate to produce PIP_2. PIP_2 modulates the activity of many actin-binding proteins such as profilin, gelsolin, vinculin, CapZ, and cofilin, thereby regulating the assembly and disassembly of the actin cytoskeleton.

Membrane Traffic

Cells contain many distinct membrane-bound compartments or organelles. Each organelle has specialized functions and therefore needs a unique combination of lipids and proteins. Vesicles transport proteins and lipids between compartments. Donor compartments such as the ER and Golgi complex have specific mechanisms that allow selective packaging of proteins (or cargo) into vesicles. The vesicles can then recognize the membranes of acceptor compartments. Cargo molecules transported to the new compartment can either stay there, be transported to another organelle, or be recycled to the donor compartment.

Proteins that leave a donor compartment associate with vesicle proteins. The p23/24 family of transmembrane proteins are probably the major molecules that, using their luminal extensions, bind cargo molecules in the ER and the *cis* part of the Golgi. The cytoplasmic tail of p23/24 also allows a direct or indirect interaction with the coat protein complex COP1, which is involved in the formation or budding of the vesicle. Budding from the Golgi apparatus involves three sets of proteins: ARF, COP1, and p23/24. The GTP-bound form of ARF is favored by ARNO, a GEF for ARF, and associates with the vesicle donor membrane through a myristic acid modification of the amino terminus. This allows binding of the COP1 coat proteins to both ARF and the cytosolic extension of p23/24. Binding of COP1 induces budding. Binding of COP1 also seems to activate the hydrolysis of GTP into GDP on ARF. After GTP hydrolysis, ARF-GDP is then released from the vesicle, inducing the loss of the coat protein complex COP1.

Vesicles migrate to the acceptor compartment, guided by the proteins GM130 and p115 (see "Cell cycle and organelle inheritance" below). A family of highly conserved proteins called SNAP receptors (SNAREs) perform targeting and fusion of vesicles to the acceptor compartment. Vesicles display v-SNAREs (vesicle-associated SNAREs) on their surface. Each v-SNARE can interact with two t-SNAREs (target for v-SNARE) localized on the acceptor compartment. This tight complex, stable even after heating at 90°C or treatment with ionic detergents such as sodium dodecyl sulfate, is formed by coiled-coiled associations and glutamine-arginine interactions. The binding of a v-SNARE to t-SNAREs is used not only for recognition and docking of vesicles to membranes but also for fusion of the lipid bilayers. Upon vesicle fusion, the v-SNARE–t-SNAREs complex is dissociated by *N*-ethylmaleimide-sensitive factor (NSF) in the presence of an adapter protein (soluble NSF attachment protein [SNAP]) and ATP. NSF may interfere with the glutamine-arginine ionic interactions between v-SNARE and t-SNAREs, and thus prime the dissociation of the complex.

Rab GTPases are also important in vesicle docking and fusion. More than 30 different Rabs have been isolated to date. Rab function appears to involve shuttling between membranes (a location favored by the GTP-bound form) and the cytosol (in the GDP-bound form). One important function of Rabs might be to control the final docking and fusion of vesicles. Ypt1, a yeast protein that binds Rab-GTP, activates the v-SNARE–t-SNARE association by removing Sec1p, a yeast protein bound to the t-SNARE. This allows the SNARE proteins to associate. Rab5 seems to control early endosome fusion by allowing the binding of a specific protein called EEA1 (early endosome-associated protein). Rab5-GTP associates with early endosomes and binds the amino terminus of EEA1. The carboxy terminus of EEA1 contains a specific domain that binds to the membrane lipid phosphatidylinositol 3-phosphate, which is generated by PI3-K. Thus, EEA1 can cross-link early endosomes so that they are ready for fusion. Other roles for Rab proteins have been described, including control of organelle locomotion on microtubules (by a complex of Rab6 and Rabkinesin) and control of exocytosis (by a complex of Rab3 and Rabphilin).

Organelle Biogenesis

Although vesicles transport many proteins throughout the cell, other proteins arrive at distinct locations directly. These targeting systems allow the ER, mitochondria, and peroxisomes to accumulate their own particular combinations of proteins and lipids so that they can carry out their specialized functions. Each organelle selects its own set of proteins by using specific receptors. Proteins synthesized on ribosomes are targeted to the ER, mitochondria, or peroxisomes by specific sets of amino acids called signal sequences. The proteins are then transferred through membranes by a translocating apparatus so that they end up in the lumen of the organelle or spanning the membrane. There are two main mechanisms to deliver proteins into the ER. The signal recognition particle and its receptor are used to deliver nascent proteins from the ribosome (a cotranslational mechanism). An alternative, posttranslational pathway uses chaperone proteins (such as Hsp70) to unfold proteins that have already been synthesized and deliver them into the ER. Transfer of proteins into mitochondria or peroxisomes uses posttranslational mechanisms. For mitochondria, this involves unfolding the protein followed by recognition of a mitochondrial outer membrane receptor. This receptor is associated with a Tom protein complex, which forms a transmembrane pore through which the protein is transferred. Depending on a second signal in the sequence of the translocating protein, the protein can either stay in the mitochondrial intermembrane space or be transferred to the mitochondrial matrix by another translocating complex called Tim.

Compared to our knowledge of the delivery of proteins to organelles, we know relatively little about how lipids traffic between the different membranes. One fundamental point is that organelles share a set of identical lipids but the relative abundance of these lipids is unique to each organelle. Lipids are synthesized at different sites in the cell. The ER and mitochondria produce glycerolipids, and the Golgi apparatus produces sphingolipids. Lipids travel between organelles by the secretory and endocytic pathways via vesicle-mediated transport or via lipid-carrier proteins present in the cytosol (protein lipid transporter). Direct contact between membranes, such as those of the ER and mitochondria, may also allow the transfer of lipids.

Certain sphingolipids may cluster in discrete mobile areas of the lipid membrane, creating lipid rafts (Box 2.1). Signaling molecules may be brought together in these rafts by their affinity for certain sphingolipids or for scaffolding proteins specifically present in these rafts (such as caveolin). Signaling proteins in rafts might generate signals controlling intracellular vesicle trafficking or other cellular functions such as adherence to the extracellular matrix.

The Cytoskeleton

The cell cytoplasm is structured by a cytoskeleton encompassing three types of elements: actin filaments (microfilaments), intermediate filaments (IFs), and microtubules (MTs). These are classified according to their diameters of 6 nm (actin), 10 nm (IFs), or 23 nm (MTs). Actin filaments provide the driving force for cells to move and divide. IFs form an intracellular scaffold which allows cells to resist external stresses. MTs segregate chromosomes during mitosis and serve as tracks along which motor proteins transport vesicles loaded with cargo (Figure 2.4).

There are more than 50 different IF genes in humans, but all IFs have the same basic structure—a conserved head and tail with a long α-helical center. The α-helices of two monomers associate and intertwine to form a parallel coiled-coil rod. Association of two dimers produces a tetrameric protofibril, and association of several protofibrils forms a filament 10 nm

BOX 2.1

Rafting on the lipid membrane

The membrane is made of different lipids asymmetrically distributed in the exoplasmic and cytoplasmic leaflets. Clustered sphingolipids and cholesterol in the lipid membrane form microdomains called rafts, which move within the plane of the membrane. Sphingolipid-cholesterol rafts are insoluble in Triton X-100 at 4°C, forming detergent-insoluble glycolipid-enriched complexes (DIGs). Due to their low density, DIGs can be separated from other detergent-insoluble complexes by gradient centrifugation.

Rafts can recruit proteins. Glycosylphosphatidylinositol (GPI)-anchored proteins and doubly acylated tyrosine kinases of the Src family are associated with DIGs on the exoplasmic leaflet and the cytoplasmic leaflet of rafts, respectively. Caveolin, a 21-kDa protein that makes a hairpin loop in the lipid membrane, is also associated with DIGs and binds signaling molecules such as p21^{ras}. Lipids with a high turnover that are involved in signal transduction, such as PIP$_2$, are also localized in DIGs.

Rafts have been implicated in membrane signaling and trafficking. The microdomain organization of cell membranes may allow signal transduction proteins to cluster in different microdomains. GPI-anchored proteins, by clustering into DIGs, may activate specific signaling pathways. This activation might involve the formation of a complex between GPI-anchored proteins and caveolin, allowing possible stimulation of tyrosine kinases of the Src family. Sphingolipid-cholesterol rafts that form invaginations of the membrane called caveolae are also involved in the transport of proteins in endocytic pathways. Caveolae are formed by self-association of caveolin molecules within lipid rafts. In contrast to coated pits, which are constitutively endocytosed, caveolae remain associated with the plasma membrane until a signal (still unknown) induces their release. Caveolae are involved in the transcytosis of proteins and in potocytosis—a mechanism for the internalization of small molecules by the reversible closing and opening of caveolae.

Inclusion in a particular raft may constitute a sorting signal. This would be an alternative to the classical sorting signals located on the cytoplasmic domains of transported proteins, which are used after endocytosis in clathrin-coated vesicles.

in diameter. IFs include the keratins (the major structural protein of hair and epidermal cells), the nuclear lamins (which link chromatin to the nuclear membrane), vimentin (a cytosolic scaffolding molecule found in fibroblasts), and neurofilaments (which form the backbone of neuronal axons and determine the axonal diameter). The main role of IFs is to provide protection against mechanical stresses. For that purpose, IFs are anchored to the cell membrane and to the rest of the cytoskeleton, forming a dense mesh. IFs attach to cell–cell adhesion points called desmosomes via the protein desmoplakin and form a network with actin filaments, MTs, and myosin by binding to the cross-linker protein plectin.

Actin filaments are composed of two twisted linear actin protofilaments made of polymerized actin monomers. An actin filament is a dynamic structure with a functional polarity. The barbed or plus end of the filament is the fast-growing end, where there is rapid association of actin monomers. The pointed or minus end is the depolymerizing end, where ADP-associated actin monomers dissociate. Association and dissociation of actin monomers are controlled by regulatory actin-binding molecules. At the barbed ends, profilin and thymosin β$_4$ regulate polymerization by controlling monomer availability whereas capping proteins such as CapZ, gelsolin, and ADF/cofilin regulate the extension of preexisting filaments. At the pointed end, dissociation of actin monomers may be blocked by the Arp2/3 complex. This complex can also nucleate new microfilaments.

Actin polymerization at the leading edge of the cell drives the formation of membrane protrusions (or filopodia) and lamellipodia and thus drives actin-based motility. The mechanochemical properties of myosins—the actin motor—are also needed for motility. Myosin motors form a large

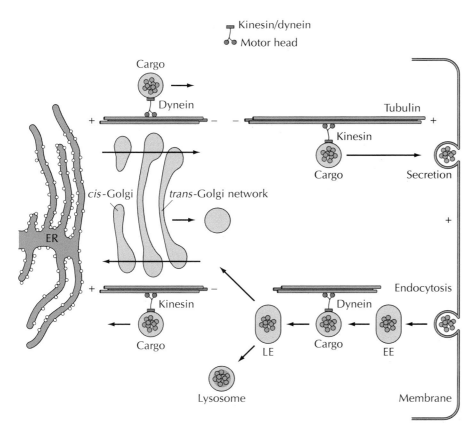

Figure 2.4 Motor proteins and intracellular traffic. The involvement of the motor proteins kinesin and dynein in cellular vesicular trafficking is shown.

family of proteins. The type II myosin proteins have two actin-binding motor heads, which bind ATP and are linked via an elongated coiled-coiled domain. At the hinge region between the motor head and the elongated domain lie the myosin light chains, which activate myosin upon phosphorylation by MLCK. Myosins move rapidly on actin filaments toward the plus end and are involved in actin contractility. Myosins read the actin filament in a linear fashion and bind sites separated by 36 nm (a distance dictated by the architecture of the two twisted linear actin protofilaments). This distance is too great for the myosin head to span, so after binding and hydrolysis of ATP, the motor detaches and jumps from one binding site to the other. The fact that myosin motors detach from the actin filament implies that many myosin motors (usually several hundred) must work together to ensure that movement along the actin filament is efficient.

MTs are produced by polymerization of dimers of α- and β-tubulin. The amino acid sequences of the α- and β-tubulin monomers are 50% homologous to each other, but it is β-tubulin that binds GTP and hydrolyzes it to yield GDP. Dimers of α- and β-tubulin are nucleated (by assembly of two or three dimers) from centrosomes that contain a special type of tubulin (γ-tubulin), which stimulates this step. Thirteen protofilaments associate laterally to form a hollow cylindrical MT 25 nm in diameter.

MTs are dynamic structures which grow or shrink rapidly. The faster-growing plus end exposes β-tubulin, whereas the slower-growing minus end exposes α-tubulin. MT dynamics are thought to be accounted for by two phenomena. MTs are said to treadmill when there is a constant incorporation of GTP-bound tubulin dimers at one end of the filament and a balanced loss of GDP-bound dimers at the opposite end. Dynamic instability, however, describes the abrupt transitions between extended states of MT growth and shrinking at one end of the MT. The transition from shrinking to growing is called rescue, and the transition from growing to shrinking is called catastrophe.

Kinesins and dyneins are motor proteins that drive the motion of vesicles, organelles, and perhaps regulatory complexes along MTs (Figure 2.4). Kinesins move toward either the plus or minus ends of MTs, whereas dyneins move toward the minus ends. The structural organization of dyneins and kinesins is similar to that of type II myosin: there are two motor heads linked by an elongated neck domain through coiled-coil interactions. Kinesin movement has been extensively analyzed, although the exact sequence of events is not yet known. Kinesins bind only to the β-tubulin subunits of MTs. Each β-subunit is separated by 8 nm, which corresponds to the spacing between the two heads of a kinesin. The kinesin is thought to move hand-over-hand. In one theory, one head is thought to remain bound as the second head hydrolyzes its bound ATP, lets go of its β-tubulin subunit, and then swings to the next free β-tubulin. Another theory postulates that ATP binding to the rearward head induces a conformational change, producing a modification in the rearward neck structure that displaces the forward head from its tubulin subunit. The rearward head then binds to this β-tubulin subunit, while the forward head attaches to the next β-tubulin subunit.

Grafting experiments with the neck and motor regions of different kinesins show that the directionality of kinesins is determined not by their heads but, rather, by the adjacent neck regions. It must be stressed that kinesins (and dyneins), unlike myosin motors, never detach from MTs during their movement cycle, since one head is always associated with a β-tubulin subunit. This is why a single kinesin is capable of moving a vesicle or other cargo for considerable distances.

MTs are stabilized by specific proteins called MT-associated proteins, which bind MTs in a nucleotide-insensitive manner. These proteins are regulated by phosphorylation, which decreases their MT-stabilizing activities. MTs can also be destabilized by specific factors. Indeed, certain kinesins can induce dramatic destabilization of MTs.

Cell Adhesion and Morphogenesis

Multiprotein adhesion complexes link cells to the extracellular matrix (ECM) or to one another. The adhesion complexes usually contain transmembrane receptors, signal transduction proteins, linker proteins, and cytoskeletal proteins.

Cells are attached to the ECM through focal adhesions. These structures coordinate cell adhesion and cell motility. Highly motile cells do not contain easily observable focal adhesions, probably because these structures are transient. Focal adhesions are prominent in adherent stationary cells. The assembly of focal adhesions is regulated both by binding to the ECM and by intracellular signaling. Binding to the ECM is mediated mostly by the integrin class of transmembrane receptors. Integrins form a large family of heterodimeric transmembrane proteins with different α and β subunits. The affinity of integrins for the ECM can be modulated: this process of activation is the main mechanism for regulating the binding of cells to their substratum. Binding affinity is increased by the binding of proteins (such as the plaque proteins talin and paxillin or tyrosine kinases such as $p60^{src}$, $p125^{fak}$, or $p130^{cas}$) to the integrin cytoplasmic tail. This binding induces a conformational change, which is propagated across the membrane to cause modification of the extracellular ligand-binding site.

Intracellular signaling networks regulate focal adhesions mainly through the Rho GTPase. Rho-GTP induces the binding of plaque proteins to actin filaments, the subsequent clustering of integrins in focal adhesions by an actin- and myosin-dependent process, and the bridging of non-integrin transmembrane receptors and actin filaments by ezrin, radixin, or moesin (ERM). Rho-GTP appears to act through two mechanisms. It activates PI5-K, which increases the concentration of PIP_2. The PIP_2 binds to a protein called vinculin, unfolding it and thereby unmasking its actin- and integrin-binding domains. Integrin clustering, however, is induced by actin filament contraction that follows Rho-GTP activation of Rok. Rho-GTP activation of ERM proteins may occur either by PIP_2 production and binding or by Rok phosphorylation.

Cell-cell adhesion is mediated at adherens junctions by cadherins, which are transmembrane calcium-dependent homophilic adhesion receptors. The E-cadherins, expressed in epithelial tissues, have been the most extensively studied and are required for epithelial cells to remain tightly associated. To be functional, cadherins must form complexes with both the cytoplasmic catenin proteins and the actin cytoskeleton. There are two types of catenins: α and β. The α-catenin is required for cadherin-cadherin interaction but also links cadherins to actin microfilaments. The β-catenin links α-catenin to the cytoplasmic tail of cadherin. Phosphorylation of β-catenins, in response to growth factors, induces a loss of adhesion. In addition to their role in cadherin adhesion, catenins act as DNA transcription factors. To turn off the activities of catenins as transcription factors, catenins are phosphorylated and then degraded.

Desmosomes are intercellular junctions in epithelia and cardiac muscle that are linked to IFs. The adhesion receptors of desmosomes—the desmogleins and desmocollins—are members of the cadherin superfamily. These receptors bind to IFs via specific proteins (desmoplakin and plakoglobin). Desmosomes and IFs operate together to maintain cell integrity during extracellular mechanical stresses.

Occluding or tight junctions are the most important elements for the formation of a permeability barrier in tissues such as epithelia and endothelia. Tight junctions are used to regulate the permeability of the paracellular space between adjacent cells and to divide the cell membrane into two distinct biochemical and functional areas. The transmembrane protein occludin contributes to the barrier function of tight junctions and interacts with two cytosolic proteins, ZO-1 and ZO-2. Several cytoskeletal proteins (such as cingulin) link actin filaments with ZO-1 and ZO-2.

The Rho family of GTP-binding proteins is involved in the regulation of both tight and adherens junctions. Rho may affect tight junctions by inducing the contraction of actin filaments that are linked to the occludin complex. Rac and Cdc42 may regulate adherens junctions by modulating the adhesive properties of E-cadherins. Rab GTPases are probably another important regulator of adherens junctions. Rabs, by regulating membrane traffic, may control the flux of membrane-associated receptors and effector molecules that tune cadherin-based adhesion.

Adhesion mechanisms are associated with morphogenesis. Indeed, similar molecular components are required for adhesion and for dynamic tissue changes such as the mesenchymal-epithelial transition or blastocyst formation. During the mesenchymal-epithelial transition, mesenchymal cells that are loosely attached progressively form a monolayer made of adherent cells with both tight and adherens junctions, in a process called compaction. Cadherins are the major proteins that induce compaction. Adhesion mechanisms also operate in cell rearrangements such as the formation of elongated tubular cells during blood vessel formation.

Apoptosis

Cells can self-destruct via an intrinsic program of cell death. Apoptosis, or programmed cell death, is characterized by blebbing of the plasma membrane, condensation of the cytoplasm and nucleus, and cellular fragmentation into apoptotic bodies. At the molecular level, apoptosis is characterized by degradation of chromatin, first into large fragments and subsequently into pieces of less than 200 bases. Apoptosis minimizes the leakage of cytoplasmic proteins such as proteases which could damage adjacent cells, since apoptotic cells are immediately engulfed by macrophages. This feature distinguishes apoptosis from necrosis. Cell necrosis, which usually results from physical trauma, causes cells to swell and lyse, releasing cytosolic proteins that may stimulate an inflammatory process.

Apoptosis is divided into three phases. Detection of the apoptotic signal by sensors is followed by conversion of the signal so that it can trigger the execution phase of apoptosis. Finally comes the execution phase itself.

Apoptotic sensor molecules are proteins such as the death ligands Fas and TNF-α, together with their membrane receptors. Stimulation of an apoptotic sensor activates a Ser/Thr kinase (probably PKB), resulting in the phosphorylation of a protein (Bad) belonging to the Bcl-2 family. Unphosphorylated Bad is associated with another protein of the Bcl-2 family, Bcl-x_L. Upon phosphorylation of Bad, Bcl-x_L is free to bind to the mitochondrial outer membrane. It forms a channel, which allows the release of cytochrome c. The cytochrome c binds a protein called Apaf-1 and thus helps activate the cysteine protease caspase 9, which in turn activates caspases 3, 6, and 7—the effectors of apoptosis.

Caspases are cysteine proteases that, upon activation by proteolysis, cleave a large variety of essential cellular proteins such as the nuclear lamins, and proteins implicated in DNA repair or regulation of the actin cytoskeleton.

Apoptotic cells are efficiently recognized by macrophages and destroyed by phagocytosis, thus avoiding the onset of an inflammatory process. This occurs because the membrane composition of cells undergoing apoptosis is modified. Phosphatidylserine, a typical inner-leaflet plasma membrane lipid, is exposed at the cell surface in apoptotic cells. Specific receptors on macrophages are able to bind apoptotic cells that have phosphatidylserine exposed at their surface, allowing rapid engulfment of these cells by the macrophages.

Cell Cycle and Organelle Inheritance

The cell cycle is an orderly progression of duplication (e.g., DNA synthesis during S phase) and division (e.g., chromosome segregation during M phase, or mitosis), separated by two gap phases: G_1 before S phase and G_2 after S phase. When a cell prepares to divide, it must coordinate the duplication and division steps. To ensure good quality control, the transitions between phases must be governed by a decision-making process that assesses such things as the quality of the duplicated DNA and the successful reorganization of the cellular machinery. During G_1, the cell must make sure that it has reached a sufficient size and has sufficient nutrients to complete a cell cycle. Cells that do not meet these criteria may decide to stop division and stay quiescent or differentiate. In G_2, the cell must dissolve its nuclear membrane, duplicate its centrosome, and assemble an array of MTs into a spindle that will pull apart, or segregate, the duplicated chromosomes. Intracellular membrane-bound compartments such as the ER and the Golgi apparatus must also be duplicated. All these events are triggered by Ser/Thr kinases called cyclin-dependent kinases (Cdks). The periodic activation and deactivation of Cdks ensures the progression of the cell division cycle. Cdks are activated by association with regulatory subunits called cyclins and by phosphorylation and dephosphorylation of the Cdks. Cyclins are proteolyzed by the proteasome after ubiquitination (Box 2.2) to deactivate Cdks.

Initiation of the cell cycle (the transition from the resting state of G_0 to the G_1 state) occurs when Ras activates the MAPK pathway, leading to the phosphorylation of the inhibitory factor pRb (retinoblastoma protein). Phosphorylated pRb dissociates from the E2F transcription factor, allowing E2F to bind to DNA and activate cyclin E gene transcription. Cdk2, stimulated by association with cyclin E, induces the initiation of DNA synthesis by triggering the phosphorylation and activation of transcription factors.

In G_2, a Cdk called Cdc2 associates with cyclin B but accumulates in an inactive state due to the phosphorylation of a tyrosine residue located in the Cdc2 catalytic site. Cdc2-cyclin B is dephosphorylated, and thus activated, by the phosphatase Cdc25; it then induces mitosis by phosphorylating various proteins including those involved in chromatin condensation. Cdc2 is deactivated by proteolysis of cyclin B by the proteasome machinery.

BOX 2.2

Cells digest selected meals

Protein degradation is an absolute requirement for the cell. Indeed, cellular structures are constantly reshaped in response to external stimuli. Proteins must be degraded if they are misfolded or are needed to generate immunocompetent peptides or if their life span must be short so that they can act as effective regulatory molecules. Protein degradation in the cytosol is not, however, a random process, since proteins not assigned to proteolysis must be carefully preserved. The basic system used by cells to control protein degradation is to address the selected proteins to a proteolytic compartment by marking them with a sorting signal. Proteins chosen for degradation are first tagged by polyubiquitination: the covalent coupling of a protein called ubiquitin to several lysine residues near their amino-terminal end. An E1-activating enzyme hydrolyzes ATP and forms a thioester bond between the active-site cysteine of the enzyme and the ubiquitin carboxy terminus. A conjugating E2 enzyme, often with the assistance of a targeting E3 protein, then transfers the activated ubiquitin to the target protein. The ubiquitinated protein is recognized by the 19S caps at the ends of the barrel-shaped 20S proteasome degradation machine. The caps unfold the ubiquitinated protein in a process that probably requires ATP, which allows the protein to pass through a narrow pore controlling the entrance to the core proteasome. Proteolysis occurs in a hollow inner cavity several nanometers in diameter. The basic core proteasome is built by the assembly of α and β subunits. The β subunits are serine proteases which are present at the surface of the proteasome cavity and are activated by removal of an amino-terminal propeptide.

Checkpoint pathways check the quality of DNA duplication and segregation and stop the progression of the cell cycle if there is a problem. DNA damage between G_1 and S triggers the synthesis of the p53 protein, a transcriptional activator named the guardian of the genome. The p53 protein is normally produced at a low level and degraded by the proteasome machinery after ubiquitination, but after DNA damage it is phosphorylated and therefore protected from degradation. The p53 protein then increases the transcription of a gene for an inhibitor of Cdk2-cyclin complexes. DNA damage between G_2 and M triggers the inhibition of Cdc2 by phosphorylation of Cdc25. Phosphorylated Cdc25 binds to the 14-3-3 protein and can therefore no longer activate Cdc2.

The total number of division cycles that a cell can undergo is limited, often to around 40 or 50. The maximum number of divisions may be controlled, in part, by the length of telomeres, which are the ends of chromosomes. DNA duplication fails to reproduce the very end of the telomere in each cell cycle, leading to a progressive shortening. When telomeric DNA becomes too short, the cell becomes quiescent or apoptosis is triggered and the cell is eliminated.

During G_2, each organelle must double in size so that later it can be correctly divided between the daughter cells. The strategy used for organelle inheritance is based on having multiple copies of organelles (such as mitochondria, lysosomes, or chloroplasts) randomly dispersed in the cytosol. During cytokinesis, these organelles are separated equally between daughter cells. The system of nuclear and ER membranes and the Golgi apparatus are normally single-copy organelles, but they also use this strategy by fragmenting before division and re-forming after division. Fragmentation is driven by Cdc2-dependent phosphorylation of lamins (which link chromatin to the inner face of the nuclear envelope) and of GM130, a vesicle-docking protein on the acceptor Golgi compartment. Phosphorylation of GM130 prevents normal vesicle fusion from taking place, even as the vesicle formation involved in Golgi transport continues. The result is vesiculation of the Golgi.

After mitosis, the phosphorylations performed by Cdc2 are reversed and vesicles reassociate around chromatin. Golgi-derived vesicles self-associate into stacks, which are actively transported by motor proteins along MTs to the pericentriolar area of the cell. The stacks then fuse with each other to re-form the conventional Golgi structure. Recent observations with tagged Golgi enzymes have shown that the stochastic partitioning of the Golgi apparatus into small vesicles during mitosis may not be the whole story. Indeed, Golgi stacks seem to be fragmented into tubulovesicular structures that are much larger than vesicles. Just as with chromosomes, these tubulovesicular structures may use the MT spindle to segregate into daughter cells during mitosis.

Conclusion

To conclude this chapter, it is worth quoting Georges Palade, a cell biology pioneer: "Although cells appear to have a large variety of mechanisms to fulfill their tasks, when they have found an efficient system they use it repetitively for different purposes." For example, nucleotide binding, reversible phosphorylation, proteolysis, and protein oligomerization are repeatedly used in many pathways. In unraveling how microorganisms interact with cells or how virulence factors damage the eukaryotic host, we should remember this fact.

Selected Readings

Green, D., and G. Kroemer. 1998. The central executioner of apoptosis. *Trends Cell Biol.* **8:**267–270.

Gumbiner, B. M. 1996. Cell adhesion: the molecular basis of tissue architecture and morphogenesis. *Cell* **84:**345–357.

Hall, A. 1998. Rho GTPases and the actin cytoskeleton. *Science* **279:**509–514.

Hirokawa, N. 1998. Kinesin and dynein superfamily proteins and the mechanism of organelle transport. *Science* **279:**519–526.

Koepp, D. M., J. W. Harper, and S. J. Elledge. 1999. How the cyclin became a cyclin: regulated proteolysis in the cell cycle. *Cell* **97:**431–434.

Lowe, M., C. Rabouille, N. Nakamura, R. Watson, M. Jackman, E. Jämsä, D. Rahman, D. J. C. Pappin, and G. Warren. 1998. Cdc2 kinase directly phosphorylates the cis-Golgi matrix protein GM130 and is required for Golgi fragmentation in mitosis. *Cell* **94:**773–782.

Mitchison, T. J., and L. P. Cramer. 1996. Actin-based cell motility and cell locomotion. *Cell* **84:**371–379.

Rohatgi, R., L. Ma, H. Miki, M. Lopez, T. Kirchausen, T. Takenawa, and M. W. Kirschner. 1999. The interaction between N-WASP and the Arp2/3 complex links Cdc42-dependent signals to actin assembly. *Cell* **97:**221–231.

Warren, G., and W. Wickner. 1996. Organelle inheritance. *Cell* **84:**395–400.

Weber, T., B. V. Zemelman, J. A. McNew, B. Westermann, M. Gmachl, F. Parlati, T. H. Söllner, and J. E. Rothman. 1998. SNAREpins: minimal machinery for membrane fusion. *Cell* **92:**759–772.

3

Extracellular Matrix and Host Cell Surfaces: Potential Sites of Pathogen Interaction

Klaus T. Preissner and G. Singh Chhatwal

Tissue cells produce and release a variety of macromolecules which form a complex structural network within the extracellular space. This extracellular matrix (ECM) not only serves as a structural support for resident cells but also provides a support for infiltrating pathogenic bacteria to colonize, particularly in the setting of injury or trauma. Similarly, integral cell surface or cell-associated adhesive components of the host are often recognized by pathogenic bacteria in a tissue- or cell-specific manner, and pathogens come into contact with host tissue fluids that also contain a variety of adhesive components. The initiation of infection requires bacterial adherence to nonphagocytic cells or the ECM, and many invasive microorganisms enter host cells after binding to specific surface structures (Box 3.1). Microbial binding may lead to structural and/or functional alterations of host proteins and to activation of cellular mechanisms that influence tissue and cell invasion by pathogens. Not only do particular mammalian receptors facilitate the entry of pathogens into host cells, but also microbes may escape certain antibiotics. Interactions with soluble or immobilized host components can mask the microbial surface and thereby interfere with antigen presentation and provide an overall immune evasion strategy.

Bacteria express surface-associated adhesion molecules, generally termed adhesins, that recognize the eukaryotic cell surface, the ECM protein, or carbohydrate structures. Nonadherent prokaryotes are rapidly cleared by the local nonspecific host defense mechanisms (peristalsis, ciliary movement, and fluid flow) or by turnover of epithelial cells and the mucus layer. The identification and characterization of host cell and ECM adhesion components and complementary bacterial adhesins are thus central to understanding the molecular aspects of pathogenesis and, given the emergence of multidrug-resistant bacteria, will lead to alternative targets for antimicrobial therapy.

Structural and Functional Components of the Extracellular Matrix

The major structural components of the eukaryotic ECM are collagens that form different types of interstitial or basement membrane networks, including fibril-forming collagens (types I, II, and III) and the two-

BOX 3.1

How do bacteria get into touch with the host tissue?

A number of specific bacterial interactions with host components determine the initial contact of pathogens and the subsequent infection cycle in the host organism. The connective tissue matrix and host cell surfaces provide a support for infiltrating pathogenic bacteria to colonize and invade, particularly under conditions of injury or trauma. Adhesion molecules found in the ECM, such as collagens, fibronectin, and other matrix proteins, or adhesion receptors, such as integrins on host cells, can interact directly with bacteria through their surface adhesins and facilitate the entry of pathogens into tissues. As a consequence, pathogens may also activate cellular mechanisms which influence their invasion. In contrast, nonadherent microbes are cleared by local nonspecific host defense mechanisms.

dimensional collagen type IV network (Table 3.1). For further organization of the ECM, additional collagenous and noncollagenous glycoproteins, proteoglycans, hyaluronan, and many other components, such as growth factors and proteases, become associated with interstitial or basement-membrane ECM, giving rise to their specialized structure and function at different locations in the body. The main portion of interstitial ECM, which determines the specific character of each tissue or organ, is produced and deposited by different embedded connective tissue cells. Consequently, the tropism of pathogenic microorganisms is determined largely by the local composition of this complex network of host-derived factors.

More recently, research in molecular cell biology has indicated that in a variety of cellular systems, cellular shape, orientation, differentiation, and metabolism are intimately linked to and determined by the ECM. Of particular interest are cellular receptors for ECM components, such as integrins and proteoglycans, that serve to make cell-to-cell or cell-to-ECM contacts, thereby providing a physical link between the cellular cytoskeleton (interior) and the ECM (exterior). The time-dependent modification and rebuild-

Table 3.1 Structural and functional components of the ECM

Interstitial ECM
    Fibril-forming collagens (I, II, III)
    FACIT (e.g., collagen IX)
    Fibronectin fibrillar network
    Elastin/fibrillin microfibrils
    Hyaluronan
Basement membranes
    Collagen (IV) network
    Laminin network
    Nidogen

Interstitial ECM and basement membranes
    Proteoglycans
    Adhesive and counter-adhesive glycoproteins
    Growth factors
    Proteases
    Transglutaminase

ing of ECM at different locations in the body is essential for inflammation and wound healing. At these sites, the provisional ECM may provide bacterial entry and colonization, whereas intact tissues or epithelia, which are covered by a variety of adhesive and mucoidal components, are mostly protected against bacterial infection. Furthermore, pathologically disturbed ECM can often be the basis for the initiation and progression of various organ or tissue defects, resulting in the increased susceptibility to microbial pathogens. Conditions such as atherosclerosis, liver cirrhosis, glomerulonephritis and glomerulosclerosis, scleroderma, lung fibrosis, and micro- and macrovascular complications of diabetes are associated with dysfunctional ECM and may provide exposed sites for bacterial adherence. Besides stimulating inflammation, pathogens may directly cause alterations of the ECM, such as in rheumatic diseases.

Collagens

More than 25 different collagen gene products have been defined on a molecular basis, and most (>90%) of the body's collagens are of the fibril-forming types I to III. These collagens are produced and assembled during normal wound-healing processes but also contribute to the formation of granulation tissue or various types of fibrosis. Processing and posttranslational modifications of collagens are essential to achieve their triple-helical conformation, which is needed, for example, for mechanical stability in various fibrillar networks. Tissue transglutaminase-dependent cross-linking of ECM components with each other or with the eukaryotic cell surface adds to the tight linkage of epithelial cells to their underlying basement membrane. Some bacteria, such as *Staphylococcus aureus*, are covalently connected to host connective tissue fibronectin by an analogous mechanism. There are three different mechanisms by which cells (and bacteria) can interact with the various forms of collagen: (i) recognition of both triple-helical and denatured collagen (as in wound-healing areas or at sites of ECM degradation) in a conformation-independent fashion, (ii) a triple-helix-dependent binding mechanism, and (iii) a type of recognition that requires fibrillar structures. In contrast to adhesive glycoproteins, such as fibronectin or laminin, which have defined cell-binding domains (see below), collagens provide multiple interaction sites for cells (and bacteria) along the triple-helical structures within the fibrils or networks. Due to the tight association of other adhesion proteins, such as fibronectin, vitronectin, von Willebrand factor, laminin, nidogen, and proteoglycans with collagens to form supramolecular aggregates of variable structure and composition, the interaction with host cells or bacteria is not determined solely by the collagen component.

  S. aureus is the most frequent cause of bacterial arthritis and osteomyelitis. Infection is initiated by hematogenous spread or by direct inoculation after trauma or surgery, and collagenous proteins present a common target for pathogenic bacteria that recognize the typical triple-helical structural motifs. The expression of a specific collagen-binding adhesin on *S. aureus* was shown to be necessary and sufficient for adhesion to cartilage. Moreover, matrix-associated bone sialoprotein provides another target for these microorganisms in inflamed areas of connective tissue. Contacts with host components often are initiated by different gram-negative bacteria utilizing their pili. As an example, fimbriae of uropathogenic *Escherichia coli* and the type III fimbriae of *Klebsiella pneumoniae* bind specifically to collagen type IV or V, respectively. Other bacterial adhesins, such as polymeric YadA, which confers pathogenic functions of *Yersinia enterocolitica* associated with joint diseases, recognize multiple ECM components, including fibrillar collagens, at sites distinct from the tripeptidic Gly-X-Y repeats. In infectious endocarditis, which is characterized by the formation of septic masses of platelets on the surfaces of heart valves and which is most commonly caused by streptococci, collagen-like "platelet aggregation-associated protein" of *Streptococcus sanguis*, as well as direct interactions with host ECM collagens, enhances platelet accumulation and subsequent bacterial colonization.

Laminins

As a major constituent of basement membranes, laminins are the first ECM protein to be produced during embryogenesis. Through its specific interactions with type IV collagen, proteoglycans, and other ECM components, as well as with several cell types, laminin fulfills a central structural and functional role within basement membranes. This 900-kDa glycoprotein has a cruciform structure and is composed of three different chains, α, β, and γ, linked to each other by disulfide bridges. Due to variations in chain composition, there are at least 10 different isoforms of laminin, which are produced by a large variety of cell types and whose biosynthesis appears to be tightly linked to the mesenchymal-epithelial-cell transition during tissue differentiation. Specific interaction sites exist for self-association and for binding of other basement membrane components or cell surface receptors which promote disparate biological activities including cell attachment, chemotaxis, neurite outgrowth, and enhancement of angiogenesis.

Exposure of laminin to pathogenic bacteria is most frequently seen at damaged or inflamed tissues. For example, following epithelial-cell denudation in the wounded human respiratory tract, *Pseudomonas aeruginosa* cells can attach to exposed basement membrane laminin via nonpilus adhesins. This pathogen is responsible for the most common infection observed in patients with cystic fibrosis, and further damage of the respiratory epithelium is mediated by bacterial as well as host-derived proteinases. Administration of protease inhibitors may dampen the fatal outcome of this genetic disorder. Moreover, both lipopolysaccharide and a protein adhesin of *Helicobacter pylori* mediate a specific interaction of this gastroduodenal pathogen with laminin, thereby interfering with the binding between gastric mucosal cells and laminin receptor, with possible loss of mucosal integrity. While *Streptococcus* strains (such as *S. viridans*) associated with endocarditis frequently express a laminin-binding adhesin(s), this bacterial receptor has a lower expression frequency in oral streptococcal isolates.

Elastin and Fibrillins

The elasticity of lung, skin, and other tissues, particularly blood vessels, is achieved by the recoiling property of elastic fibers embedded into the ECM, forming up to 50% of its dry mass. The main component (90%) of elastic fibers is elastin, which is a highly cross-linked polymer of the nonproteolytically modified, hydrophobic precursor tropoelastin. The other components of elastic fibers are 10- to 12-nm-thick microfibrils, which are composed of different fibrillins and other associated glycoproteins. A number of mutations within the fibrillin genes are known, and the clinical phenotype of Marfan syndrome is characterized by malformations of the skeletal muscle, cardiovascular, and ocular systems and the occurrence of bacterial endocarditis. As with many other ECM components, *S. aureus* expresses an elastin-binding protein that contributes to the colonization of elastin-containing organs, such as the lungs, skin, and blood vessels, by this bacterial species.

Proteoglycans and Hyaluronan

Proteoglycans represent the most abundant, heterogeneous and functionally most versatile nonfibrillar component of ECM, and their properties are determined by the variable length and structures of the glycosaminoglycan side chains, which are covalently attached to a specific core polypeptide. As well as the most prominent secreted extracellular heparan proteoglycan, perlecan, which is found in all basal membranes of the body, decorin and biglycan are two smaller ECM proteoglycans with leucine-rich repeats. The specific binding of *Borrelia burgdorferi* to decorin via two adhesins may increase the adherence of the spirochete to collagen fibers in skin and other tissues.

Hyaluronan is a high-molecular-weight nonsulfated polysaccharide with a high hydration capacity that is found in the ECM of most animal tissues. It binds tightly to the chondroitin sulfate proteoglycan aggrecan or other proteoglycan members and several link proteins. These aggregates determine the viscoelastic properties of cartilage or other tissues. Cell surface hyaluronan receptors, or "hyaladherins," such as the family of CD44 isoforms on a variety of normal and transformed cell types, contribute to ECM assembly and turnover or are crucial for cellular invasion. Recognition and binding of hyaluronan by *Treponema denticola* mediates the adherence of spirochetes to epithelial cells in periodontal tissue. Because gram-positive bacteria, such as group A streptococci, contain capsular hyaluronan, their binding to proteoglycans or CD44 may strengthen their interactions with host tissue. Likewise, capsular polysaccharides of *E. coli* K5 are structurally related to heparan sulfate and may contribute to the interaction of this bacterium with heparin-binding host proteins to gain access to ECM or cell surface sites. Conversely, hyaluronate- or heparin-degrading lyases of other microorganisms contribute to tissue destruction and may allow the penetration of pathogens.

Adhesive Glycoproteins

Upon vessel wall injury, particularly at sites of wound healing, initial adhesion of platelets to the exposed subendothelium and subsequent platelet aggregation are dependent on adhesive glycoproteins present in the subendothelial cell matrix as well as those stored inside platelets and secreted during this initial phase of hemostatic plug formation. In addition to their strong attachment-promoting activity, residing predominantly in the Arg-Gly-Asp (RGD)-containing epitope and being the predominant recognition site for integrins, these multifunctional proteins are of major importance in the initial adherence phase of pathogens. The circulating forms of these adhesive proteins may differ from those in the subendothelium and α-granules of platelets, due to alternative splicing (fibronectin), differences in the state of polymerization (von Willebrand factor), different conformational forms (vitronectin), or the transition into a self-aggregating molecule (fibrinogen-fibrin).

Fibronectin

Fibronectin is a ubiquitious adhesive matrix protein which is essential for the adhesion of almost all types of cells. The 30-kDa amino-terminal fragment contains the major acceptor site for factor XIIIa-mediated cross-linking and also bears the binding sites for heparin, fibrin, and bacteria, including *S. aureus*. Fibronectin serves as the prototype of adhesion proteins that bind specifically to microorganisms. It binds to over 16 bacterial species, and the interaction with gram-positive bacteria has been extensively characterized (Table 3.2). In particular, most *S. aureus* isolates, as well as various streptococcal strains, specifically bind and adhere to fibronectin, and the interaction is mediated by several different bacterial adhesins with highly homologous recognition motifs for the adhesion protein (Figure 3.1). In addition to the amino-terminal fragment of fibronectin, cooperative binding sites for staphylococci and streptococci are located within the fibrin-binding carboxy terminus. Although most characterized interactions with gram-negative bacteria involve recognition of lectins and carbohydrate structures in the host tissue, several types of fimbriae of enterobacteria exhibit specific interactions with fibronectin, laminin, or other adhesion proteins. Certain strains of *E. coli* provide surface-associated "curlin" subunits that mediate fibronectin-binding during wound colonization. In addition to fibronectin-mediated adherence of bacteria at sites of blood clots or damaged tissue, pathogens are able to colonize artificial devices, such as intraocular lenses, prosthetic cardiac valves, vascular grafts, prosthetic joints, and intravascular catheters, via biomaterial-adsorbed matrix proteins, particularly fibronectin and fibrinogen-fibrin. Synthetic peptides representing the fibronectin-binding domain of *S. aureus* adhesins as well as antibodies against the fibrinogen-binding domain of *S. aureus* "clumping factor" (see below) have been successful in blocking bacterial colonization on implanted foreign materials.

Vitronectin

The multifunctional adhesive glycoprotein vitronectin is synthesized predominantly by liver cells. It is found as a single-chain polypeptide with a molecular mass of 78 kDa in the circulation and becomes associated as a multimeric, heparin-binding form with different ECM sites, particularly during tissue or vascular remodeling. Histochemical studies suggest that vitronectin is deposited in a fibrillar pattern in loose connective tissue, in association with dermal elastic fibers in skin and with renal tissue and the vascular wall at sites of arteriosclerotic lesions. Vitronectin functions as an inhibitor of cytolytic reactions of the terminal complement pathway and serves as a major regulator in cellular interactions related to migration and invasion.

Vitronectin has equivalent effects on cellular adhesion and bacterial binding to those demonstrated for fibronectin, yet specific interactions with bacteria are likely to occur in damaged or altered tissues where the protein is deposited. Specific interactions of vitronectin with various strains of staphylococci and group A and G streptococci, as well as with gram-negative bacteria, have been described (Table 3.2). The adhesin(s) of *S. aureus* responsible for vitronectin interaction may also bind to heparan sulfate. The adherence of streptococci to the luminal side of cultured endothelial cells or to epithelial cells is mediated by fibronectin-independent vitronectin-specific interactions. Hemopexin-type repeats in vitronectin, as

Table 3.2 Specific interactions of bacterial species with host ECM components

Microorganism	Collagen types I and II	Collagen type IV	Elastin	Fibronectin	Vitronectin	Fibrinogen-fibrin	Laminin	Thrombospondin	Bone sialoprotein II	Glycosamino-glycans
Aeromonas hydrophila	X									
Candida albicans		X		X	X	X				
Escherichia coli	X			X	X	X				
E. coli fimbriae		X								X
Helicobacter pylori				X	X		X			X
Leishmania spp.				X						X
Mycobacterium tuberculosis				X						
Neisseria gonorrhoeae					X					X
Plasmodium falciparum				X				X		
Pneumocystis carinii				X						
Porphyromonas gingivalis				X	X					
Staphylococcus aureus	X		X	X	X	X	X	X	X	X
Staphylococcus (coagulase negative)			X	X	X				X	X
Streptococcal groups C and G				X	X	X		X		
Streptococcus dysgalactiae					X	X				
Streptococcus equi						X				
Streptococcus sanguis				X						
Streptococcus pyogenes		X		X	X	X	X			X
Streptococcus pneumoniae					X					
Treponema denticola							X			
Treponema pallidum				X						
Trypanosoma cruzi				X						X
Vibrio cholerae			X							
Yersinia spp.				X						

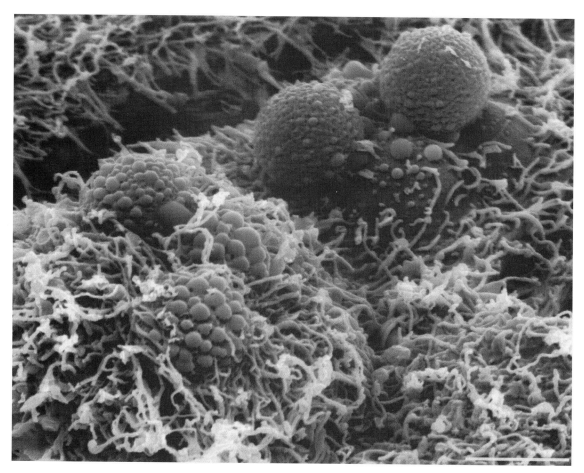

Figure 3.1   Scanning electron micrographs demonstrating the interaction between latex beads, coated with fibronectin-binding streptococcal adhesin SFBI, and HEp-2 cells. Different stages of adhesion, engulfment, and internalization are apparent. Bar, 3 μm. Photograph: M. Rohde (Braunschweig, Germany).

well as in hemopexin itself, have been identified as primary binding sites for group A streptococci. For internalization, gonococci expressing the OpaA adhesin require interaction with heparan sulfate proteoglycans on the host cell surface followed by entry in a vitronectin-dependent fashion.

Fibrinogen-Fibrin

The fibrin precursor of fibrinogen is a major 350-kDa plasma glycoprotein, composed of two identical sets of three polypeptide chains, and serves as predominant macromolecular substrate for thrombin in the blood-clotting cascade. Together with vitronectin and other adhesion proteins, the fibrin clot constitutes the majority of the initial provisional ECM network for sealing a wound site, thereby protecting this area against infiltration by opportunistic, infectious microorganisms. Cell surface receptors for fibrinogen that belong to the family of integrins have been identified on mammalian cells, of which the platelet $\alpha_{IIb}\beta_3$-integrin (GP IIb/IIIa) is principally required for platelet aggregation. $\alpha_M\beta_2$-Integrin (complement receptor 3) on phagocytes may also recognize the ligand in situations of wound healing and defense in which phagocytic clearance of fibrin(ogen)-associated clot

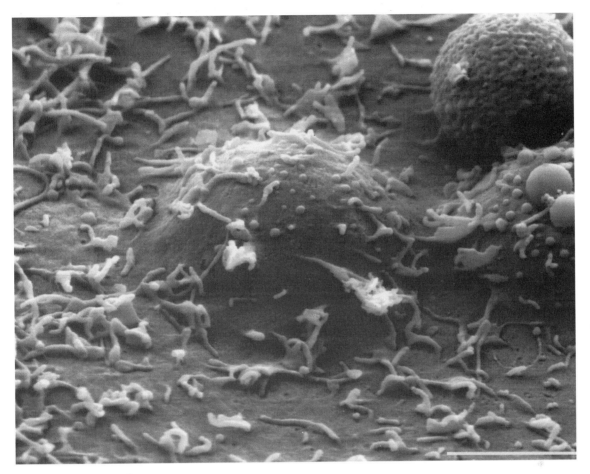

Figure 3.1 *continued*

or (bacterial) cell fragments is required. Due to its dimeric structure, fibrinogen may serve as a bridging component between surface receptors on different cells or other extracellular sites once they become exposed. Likewise, *S. aureus* adherence to endothelial cells is mediated predominantly by fibrinogen as a bridging molecule leading to acute endovascular infections. Although *S. aureus* can induce platelet aggregation via a fibrinogen-dependent mechanism, this process is independent of the aforementioned principal $\alpha_{IIb}\beta_3$-integrin-binding interactions with fibrinogen.

Different bacterial fibrinogen-binding proteins mediate pathogen colonization in wounds or catheters; the proteins expressed by *S. aureus* are the best characterized. In particular, "clumping factor" serves to recognize the carboxy-terminal portion of the fibrinogen α chain in a manner analogous to $\alpha_{IIb}\beta_3$-integrin binding. Moreover, the homology between metal ion-dependent adhesion sites of integrin subunit α_{IIb} or α_M or clumping factor and an integrin-like protein from *Candida albicans* indicates common mechanisms of fibrinogen binding in mammalian cells, lower eukaryotes, and prokaryotes. Staphylocoagulase serves as an additional fibrinogen-binding factor, which is not involved in bacterial clumping but, due to prothrombin-binding and conversion, serves to promote fibrin formation or bacterial attachment onto fibrinogen-coated surfaces. Fibrinogen binding to streptococci of groups A, C, and G leads to inhibition of complement fixation and subsequent phagocytosis, indicating an important role for this interaction. Bacterial colonization is reduced in a mouse mastitis model by vaccination with *S. aureus* fibrinogen-binding proteins, providing a new concept for antimicrobial therapy.

In addition to staphylococcal or streptococcal surface proteins that interact with ECM components, proteins released by these bacterial species can directly influence fibrin formation or dissolution. While staphylocoagulase binds prothrombin and mediates its conversion to thrombin, staphylokinase and streptokinase interact stoichiometrically with plasminogen, resulting in plasmin formation, whereby the former fibrinolytic agent acts in a fibrin-specific manner. These strategies apparently allow effective fixation and subsequent penetration of bacteria into wound areas.

Thrombospondin and Other "Matricellular" Proteins

Members of the thrombospondin family of extracellular proteins are found in different tissues. Together with structurally unrelated members of the tenascin protein family as well as with osteonectin (SPARC, BM40) or osteopontin, they associate with collagen fibrils or basement membranes, promoting divergent cellular functions including counteradhesive activities. While thrombospondin interactions with bacteria have been studied to some extent, bacterial binding to the other matricellular proteins remains to be analyzed. In particular, *S. aureus* adherence to activated platelets, to blood clots, or to ECM in pyogenic infections is mediated by heparin-dissociable interactions with thrombospondin. Importantly, thrombospondin binds *Plasmodium falciparum*-parasitized erythrocytes and, together with its cell surface receptor, CD36, mediates their adherence to endothelial and other cells; it has thus has been implicated as an adhesive mediator of malaria infection.

ECM Degradation

A prerequisite for bacteria to invade normal tissues or wound sites is the degradation of matrix proteins, accomplished by bacterial collagenases or elastases, as produced by various clostridial strains. Host-derived plasmin appears to play a prominent role in pathogen-mediated tissue destruction as well. In situ, urokinase-type and tissue-type plasminogen activators, which are produced and secreted by a variety of eukaryotic cells, particularly under inflammatory conditions, are responsible for plasmin formation by limited proteolysis. A number of bacteria, such as *Yersinia pestis*, produce plasminogen activators that activate host-derived plasminogen. The spirochete *Borrelia burgdorferi,* the causative agent of Lyme disease (characterized by inflammatory manifestations in the skin, joints, heart, and central and peripheral nerve systems), induces monocytes to secrete urokinase-type plasminogen activator. Moreover, due to the availability of plasmin(ogen) receptors on the surface of this microorganism, accelerated plasmin formation and protection against inactivation by host inhibitors is achieved, resulting in efficient pathogen invasion. Likewise, other pathogens, such as *E. coli,* group A, C, and G streptococci, and *Neisseria gonorrhoeae,* express surface receptors for plasmin(ogen) that facilitate pericellular proteolysis and invasion. The structural similarity of neutrophil-derived polypeptide "defensins" to plasminogen kringle motifs suggests that their antimicrobial activity can be related to interference with plasmin formation, thereby preventing the spread of infection. As an acute-phase reactant in host defense, the circulating broad-spectrum proteinase inhibitor α_2-

macroglobulin serves to eliminate complexed proteinases via receptor-mediated endocytosis and may do so with microbial enzymes as well. Interestingly, various strains of streptococci exhibit specific binding to α_2-macroglobulin and may thereby gain access to host tissues, possibly via the α_2-macroglobulin receptor.

Cell Surfaces and Bacterial Interactions

The bacterial tropism to colonize the restricted range of hosts, tissues, and cell types is dependent, among other factors, on the availability of specific host cell receptors for a given bacterial adhesin(s). Although not all bacterium-host cell interactions result in cellular entry, adherence to epithelial mucous membranes, particularly of the respiratory, the gastrointestinal, and urogenital systems, or entry via the eye, the ear, or wound sites is a prerequisite for the infectious process and is considered an important virulence factor. The predominant parts of the accessible host cell surface glycocalyx are membrane-anchored glycoconjugates, including glycoproteins, proteoglycans, and glycolipids. In addition, secreted and epithelial surface-associated mucin glycoproteins are thought to act in a lubricative and protective manner by shielding the gastrointestinal, respiratory, and urogenital tracts against physical damage, dehydration, and bacterial infection. Despite the common oligosaccharide structures in the different groups of glycoconjugates, their high compositional diversity allows variability among species, among cell types, and as a function of tissue differentiation. Conversely, the pathogens that colonize the mucosal surfaces are also rich in carbohydrates, and lectin-like activities are often characteristic of pilus-containing gram-negative and eukaryotic microorganisms. Mannose is frequently present in N-linked saccharides of glycoproteins and is recognized by a large number of bacteria known to bind mannose. However, pathogens may also express glycolipid-specific adhesins, indicating that a combination of recognition specificities confers bacterial tropism. Cell surface glycoconjugates thus function as nonselective coreceptors, and pathogen internalization is only achieved together with the primary bacterial receptor.

Lectins, Proteoglycans, and Mucins

Two major families of animal lectins, the calcium-dependent C-type lectins and the S-type lectins, have been classified based on their carbohydrate-binding structures and functional properties. The C-type lectins are either soluble extracellular proteins or integral membrane proteins and include receptors involved in the uptake of plasma glycoproteins in the liver, such as the asialoglycoprotein receptor, adhesion receptors such as selectins involved in leukocyte homing, and collectins, such as the serum mannan-binding protein involved in complement activation and phagocytosis. High-affinity binding is conferred by multivalent interactions between multimeric C-type lectins and complementary complex saccharides on the bacterial surface. The mannan-binding proteins present on granulocytes and macrophages and in serum are the best characterized, with the latter being able to activate complement in an antibody-independent manner after encountering cognate carbohydrates on microbial surfaces. Complex mannose structures present on bacteria, yeasts, and host-derived oligosaccharides of viruses play a role in pathogenesis by binding to mannan-binding proteins. Evidence for the biological relevance and host defense property of these interactions is based on the fact that individuals with a genetic deficiency in the mannan-binding protein have increased susceptibility to certain infectious diseases. Additional members of the C-type lectins (also designated the collectin family) include conglutinin and the lung-associated proteins surfactant proteins A and D. Attachment of *Pneumocystis carinii* or *S. aureus* to specialized epithelium or alveolar macrophages, respectively, is mediated by these collectins through interactions between their C-type lectin domains and bacterial carbohydrates. Because the collagen-type helices of these collectins are recognized by the collectin receptor (also designated C1q-receptor), which is present on the lung epithelium and on many other (phagocyte) cell types, they serve a bridging function in bacterium-host cell adherence.

The S-type lectins include several soluble proteins that are characterized by their affinity for lactose and other β-galactosides. They are composed of either two identical or two different carbohydrate-binding subunits, which serve as bridges between bacterial and host cell carbohydrates at various mucosal epithelial surfaces in a calcium-independent manner. Due to the widespread appearance of β-galactosides as structural determinants of bacterial carbohydrates, S-type lectins found in the intestines, lungs, and kidneys, as well as on macrophages, are able to recognize various pathogens.

By analogy to ECM proteoglycans, membrane-associated cell surface proteoglycans of the integral membrane type (syndecan family), as well as the glycosylphosphatidylinositol-anchored type (glypican family), directly interact with matrix proteins, proteinase inhibitors, and their target enzymes, growth factors, or lipoproteins. Consequently, cellular proteoglycans are thought to contribute to cell adhesion, cell growth, and the control of proteolytic and lipolytic pathways, particularly in the vasculature, by influencing the local concentration, stability, conformation, activity, and clearance of various protein ligands.

In particular, cell-associated heparan sulfate proteoglycans are used by several microbes, such as *Plasmodium*, *Leishmania*, and *Trypanosoma* species, to enter host tissue. Bacterial binding is significantly reduced by secreted heparinases that may either dissociate cell-adherent microorganisms or weaken cell-ECM interactions and decrease the stability of the ECM. All these processes may increase bacterial penetration into host tissues. Moreover, soluble heparan sulfate (or heparin) effectively interferes with *Trypanosoma cruzii* invasion mediated by the adhesin penetrin, indicating that successful bacterial invasion depends on the interplay between soluble and immobilized host factors. *Chlamydia trachomatis* expresses a heparan sulfate-like glycan that links the bacterium to host cell heparin-binding proteins, thereby using a trimolecular complex for adherence and invasion. Finally, the epithelial-cell mucosal barriers in the body provide different high-molecular-weight mucin glycoproteins (containing 50 to 80% carbohydrate) which exhibit considerable genetic polymorphism among individuals. In addition to their protective properties, particular mucins, such as episialin (MUC1), serve cell adhesive functions in normal and tumor tissues or mediate selective binding of, e.g., *Haemophilus influenzae* with airway epithelium.

Glycolipids

Several hundred eukaryotic cell surface glycolipid structures are known, and they belong to different core saccharide series such as the lacto-, the ganglio-, the globo-, or the muco-series. These core structures often contain terminal antigenic determinants, including the blood group ABO(H) and Lewis antigens or sialic acid and sulfate groups. Despite the high diversity of saccharide structures linked to the membrane-anchored lipid tail, each glycolipid has only one saccharide, as opposed to glycoproteins and proteoglycans, which often contain several different oligosaccharides linked to the same polypeptide. Because blood group antigens are also expressed on cells comprising the mucosal surfaces of the intestine and other organs, increased bacterial adherence and also risk of infectious diseases may be associated with the appearance of these glycolipid components, as exemplified by the binding of *Helicobacter pylori* to the mucous cell Lewis[b] blood group antigen. In addition, most common urinary tract infections with *E. coli* involve galactose-specific adherence. The phagocytosis (termed lectinophagocytosis) of bacteria by animal cells such as peritoneal macrophages and polymorphonuclear leukocytes is mediated by bacterial surface lectins behaving as glycolipid receptors. The specificity of these interactions is undefined.

In particular, lactosyl-ceramide is present in the colon epithelium but not in the small intestinal epithelium and serves to establish the adhesion of gram-negative and gram-positive bacteria to their specific host cell binding factors. In contrast to the terminally placed recognition sequence for bacteria, glycolipid "isoreceptors" contain an internal recognition structure that may affect the overall binding of microorganisms with regard to their affinity and specificity. Conversely, some gram-negative mucosal pathogens express glycolipids that are identical in their chemical and antigenic properties to carbohydrates of glycosphingolipids present in host cells or tissues. The availability of capsular polysaccharides of the lactoneo-series type with similar antigenicity to host glycolipids suggests that these structures are involved in the mechanism of survival or invasion of bacteria in human tissues.

Cell Surface Adhesion Receptors

Physiological processes in the host that are related to cell activation, proliferation, differentiation, or cellular motility require cell-cell or cell-ECM contacts and are differentially mediated by adhesion receptors, including selectins, cadherins, immunoglobulin family members, as well as ubiquitously expressed heterodimeric integrins. Integrins provide a physical linkage between the extracellular environment and the intracellular cytoskeleton; they are intimately involved in signal transduction pathways and thereby directly control the principal cellular processes and cell survival. Integrins appear to be a common target for pathogens and for subsequent manipulation of the host cell machinery to gain microbial entry. The general strategy of microbes to engage these essential host adhesion molecules during the course of pathogenesis may result in the loss of important cellular functions.

Integrins of the β_1 subclass are present on epithelial, mesenchymal cells and on circulating blood cells, β_2-integrins are expressed exclusively on leukocytes and are up-regulated during inflammatory processes and infectious diseases, and β_3-integrins along the vasculature and on platelets are the major target receptors for interaction with pathogenic microorganisms (Table 3.3). Three different strategies may be used by bacteria to infiltrate the host system and thereby promote intracellular survival. (i) Microorganisms express an RGD-containing or RGD-like peptide that serves as recognition site for integrins (ligand mimicry), such as the *E. coli* outer mem-

Table 3.3 Microbial interactions with host adhesion receptors

Microorganism and ligand	Integrins ($\alpha\beta$) and other receptors
Yersinia invasin	$\alpha_3\beta_1$, $\alpha_4\beta_1$, $\alpha_5\beta_1$, $\alpha_6\beta_1$, $\alpha_v\beta_1$
Histoplasma capsulatum	$\alpha_L\beta_2$, $\alpha_M\beta_2$, $\alpha_x\beta_2$
Leishmania LPG, LPS	$\alpha_M\beta_2$, $\alpha_x\beta_2$
Borrelia burgdorferi	$\alpha_{IIb}\beta_3$
Neisseria meningitidis Opc protein	NCAM
Neisseria gonorrhoeae	CD66, proteoglycans
Plasmodium falciparum	ICAM-1, VCAM-1, E-selectin, CD36
Bordetella pertussis	P-selectin, E-selectin, $\alpha_M\beta_2$
Shigella flexneri	$\alpha_5\beta_1$
Listeria monocytogenes	E-cadherin
Adenovirus penton base	$\alpha_v\beta_3$, $\alpha_v\beta_5$
HIV-1[a] Tat protein	$\alpha_5\beta_1$, $\alpha_v\beta_3$

[a]HIV-1, human immunodeficiency virus type 1.

brane protein intimin, which interacts with β_1-integrins, or *B. burgdorferi* proteins, which bind to activated integrins on platelets. Some microorganisms, such as *Mycobacterium avium* and *Candida* strains, express proteins whose structure resembles that of integrin (receptor mimicry). Alternatively, complex mimicry of host cellular recognition molecules is found in bacterial surface structures (such as the filamentous hemagglutinin of *Bordetella pertussis*) that may interact with glycoconjugates, heparin, and β_2-integrin or its ligands. (ii) The microorganism binds to a non-RGD recognition site on the integrin (ancillary recognition). (iii) The bacterium binds to a natural integrin ligand, leading to bacterial entry (masking), as exemplified by *Neisseria meningitidis*. Bacterial mimicry of the natural ligands for leukocyte integrins also takes into account the fact that endotoxin, coagulation factor X, and complement C3bi serve as RGD-independent ligands for β_2-integrins. Once the bacterium is attached to the host cell, the subsequent fate of the pathogen is determined by the activation of the receptor, the ligand used by the bacterium, the signal transduction pathway from the integrin to the cytoskeleton, or other host cell responses such as degranulation. As a prerequisite for entry of *Yersinia* species into host cells, high-affinity binding of the bacterial adhesin invasin to various β_1-integrins allows effective competition with low-affinity binding of host ligands, such as fibronectin, and subsequent invasion can interfere with signal transduction via accessory proteins.

The β_2-integrins on circulating blood cells are crucial for the arrest of leukocytes at the endothelium and their subsequent recruitment into tissues toward an inflammatory stimulus. In the rare condition leukocyte adhesion deficiency, patients suffer from severe bacterial infections, affecting the oral, respiratory, and urogenital mucosa as well as the intestine and skin. Microorganisms that must survive within tissue macrophages are often internalized via β_2-integrins. Pathogens do so by either synthesizing a ligand(s) that is recognized by β_2-integrins or binding to integrin ligands such as C3bi (e.g., *Legionella* and *Leishmania* species), which subsequently induce phagocytic uptake without activating the macrophage antimicrobial cytotoxic machinery. Several lines of evidence have documented that integrin-mediated internalization of microorganisms involves the host cell cytoskeleton, differing from surface adhesion of microbes on phagocytic or nonphagocytic cells.

Additional cell surface adhesion molecules that are recognized by microbes include counterreceptors of integrins, such as ICAM-1, VCAM-1, or E-selectin; CD36; and the hyaluronan receptor CD44 (Table 3.3). The pathogens may either directly bind to these cytoskeleton-associated host receptors or interact with host-derived secreted polysaccharides or adhesion proteins, which subsequently interact with the above-mentioned receptors and thereby indirectly mediate the binding of the microorganism or support internalization. CD44 mediates bacterial phagocytosis in human neutrophils by a cytoskeleton-dependent mechanism. This process was augmented by the CD44 ligand hyaluronan. Calcium-dependent adhesion molecules, such as cadherins involved in homotypic intercellular junction interactions, are utilized by *Listeria* species for invasion. E-cadherin on epithelial cells was shown to be targeted by surface-located internalin of *L. monocytogenes* as a prerequisite for cellular entry.

Conclusion

The emergence of pathogens resistant to conventional antimicrobial substances and the possible harmful side effects of cross-reactivities between, e.g., streptococcal M proteins and host factors or the large number of distinct serotypes require novel strategies for the prevention and treatment of infectious diseases. Based on the identification of adhesion and invasion mechanisms of pathogens and the relevant components involved in these processes at different host niches, novel therapeutic interventions can be designed. These are summarized as "anti-adhesion therapies" for microbial diseases and include substances such as soluble bacterial adhesins or host receptor analogues, antibodies against bacterial adhesins, and low molecular weight adhesion/invasion antagonists (Figure 3.2). The ECM-binding bacterial proteins or adhesins can be used as candidates in an antiadhesin vaccine. These strategies are, however, complicated by the fact that not all adhesin genes are present in every strain and that environmental factors can affect the expression of these genes. Because each microorganism can

Figure 3.2 Schematic representation of bacterial interactions with host cells and ECM mediated by adhesion molecules. COL, collagens; FN, fibronectin; LN, laminin; VN, vitronectin; SFBI, fibronectin-binding streptococcal adhesin.

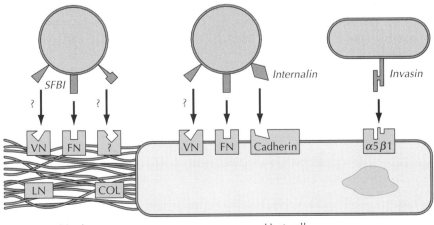

Novel antimicrobial strategies to circumvent the emergence of antibiotic resistance

The emergence of pathogens resistant to conventional antimicrobial therapy, as well as harmful side reactions due to cross-reactivities between bacterial surface proteins and host factors, requires novel strategies for the prevention and treatment of infectious

diseases. Based on the identification of adhesion and invasion mechanisms of pathogens and the relevant components involved in these processes at different host niches, novel "antiadhesion therapies" can be designed that include substances such as soluble bacterial adhesins or host receptor analogues, antibodies against bacterial adhesins, and low-molecular-weight antagonists. Since each microorganism can use multiple mechanisms of adhe-

sion to initiate infection, effective antiadhesion drugs should contain cocktails of different inhibitors. These strategies are, however, complicated by the fact that not all bacterial adhesin genes are present in every strain and that environmental factors can affect the expression of these genes. Thus, the knowledge of strategies used by microorganisms for their tissue and cellular entry may be converted into beneficial therapeutic regimens.

use multiple mechanisms of adhesion to initiate infection, effective antiadhesion drugs may contain cocktails of different inhibitors (Box 3.2).

In several vascular diseases, vessel injury promotes the development of wounds and the predisposition toward bacterial infections, as observed in diabetes-associated progressive nonenzymatic glycation of proteins and lipids. Certain invasive bacteria may utilize these modified tissues for binding, as was shown for *Pseudomonas* soil strains. Moreover, the presence of bacteria in degenerated vessels suggests an association with atherosclerosis, and indeed, dental infections pose a serious risk for acute myocardial infarction. Although causal relationships are not sufficiently clarified in these cases, antibacterial interventions may well also be of benefit for the associated vascular complications.

Selected Readings

Cheung, A. L., M. Krishnan, E. A. Jaffe, and V. A. Fischetti. 1991. Fibrinogen acts as a bridging molecule in the adherence of *Staphylococcus aureus* to cultured human endothelial cells. *J. Clin. Investig.* **87:**2236–2245.

Chhatwal, G. S., K. T. Preissner, G. Müller-Berghaus, and H. Blobel. 1987. Specific binding of the human S protein (vitronectin) to streptococci, *Staphylococcus aureus,* and *Escherichia coli. Infect. Immun.* **55:**1878–1883.

Coleman, J. L., J. A. Gebbia, J. Piesman, J. L. Degen, T. H. Bugge, and J. L. Benach. 1997. Plasminogen is required for efficient dissemination of *B. burgdorferi* in ticks and for enhancement of spirochetemia in mice. *Cell* **89:**1111–1119.

Hynes, R. O. 1992. Integrins: versatility, modulation, and signaling in cell adhesion. *Cell* **69:**11–25.

Karlsson, K. A. 1995. Microbial recognition of target-cell glycoconjugates. *Curr. Opin. Struct. Biol.* **5:**622–635.

Kuusela, P., T. Vartio, M. Vuento, and E. B. Myhre. 1984. Binding sites for streptococci and staphylococci in fibronectin. *Infect. Immun.* **45:**433–436.

Mengaud, J., H. Ohayon, P. Gounon, R.-M. Mege, and P. Cossart. 1996. E-cadherin is the receptor for internalin, a surface protein required for entry of *L. monocytogenes* into epithelial cells. *Cell* **84:**923–932.

Ofek, I., I. Kahane, and N. Sharon. 1996. Toward anti-adhesion therapy for microbial diseases. *Trends Microbiol.* **4:**297–299.

Patti, J. M., and M. Höök. 1994. Microbial adhesins recognizing extracellular matrix macromolecules. *Curr. Opin. Cell Biol.* **6:**752–758.

Relman, D., E. Tuomanen, S. Falkow, D. T. Golenbock, K. Saukkonen, and S. D. Wright. 1990. Recognition of a bacterial adhesin by an integrin: macrophage CR3 (αMβ2, CD11b/CD18) binds filamentous hemagglutinin of *Bordetella pertussis. Cell* **61:**1375–1382.

Roberts, D. D., J. A. Sherwood, S. L. Spitalnik, L. J. Panton, R. J. Howard, V. M. Dixit, W. A. Frazier, L. H. Miller, and V. Ginsburg. 1985. Thrombospondin binds falciparum malaria parasitized erythrocytes and may mediate cytoadherence. *Nature* **318:**64–66.

Virji, M. 1996. Microbial utilization of human signalling molecules. *Microbiology* **142:**3319–3336.

4

Bacterial Adherence to Cell Surfaces and Extracellular Matrix

B. Brett Finlay and Michael Caparon

An essential step in successful colonization and production of disease by microbial pathogens is their ability to adhere to host cell surfaces and the underlying extracellular matrix. Although different pathogens can adhere to nearly any site in the body, each pathogen usually colonizes one or a few particular sites in the host. The ability to adhere to specific host molecules enables a pathogen to target itself to a particular tissue within the body, thereby giving the pathogen tissue tropism. In addition to localizing pathogens to particular sites, adherence can be the first step in penetrating further into the body by initiating invasion through activation of host cell signaling pathways (see chapter 5). If the pathogen also produces toxins that cause tissue damage (see chapter 11), destruction of host cells will expose the extracellular matrix, to which many pathogens can bind. Finally, most pathogens possess many molecules (adhesins) that mediate adherence to host cells and the extracellular matrix. Often these adhesins are synergistic in their function, thereby enhancing adherence. Alternatively, they can be differentially regulated such that specific adhesins are expressed under different environmental conditions, enabling a pathogen to express different adhesins as it progresses through different sites within its host. Not all adhesins are essential virulence factors, and the specific role of a particular adhesin in disease has been surprisingly difficult to define due to the presence of multiple adherence factors in most pathogens. Many adhesins are found in both nonpathogenic and virulent strains. However, in most cases adhesins do play some role in disease, often by increasing the virulence of a pathogen. If more than one adhesin is deleted from a pathogen, the effect on virulence is usually more pronounced than when any single adhesin is deleted.

The choice of host cell substrate that a pathogen can adhere to is large. The mammalian cell surface contains many proteins, glycoproteins, glycolipids, and other carbohydrates that could potentially serve as a receptor for an adhesin. Additionally, the extracellular matrix provides a rich source of glycoproteins for adhesins to bind to, and implanted devices are major factors in bacterial adherence. Although many bacterial adhesins have been identified, only a few receptors have been identified. These receptors encompass a wide array of cell surface and matrix molecules, although they generally include a carbohydrate moiety exposed on the host molecule.

Bacterial adhesins are usually proteins, either located at the tip of a scaffold-like structure on the bacterial surface (called a pilus or fimbria) or anchored in the bacterial membrane but surface exposed (nonfimbrial or afimbrial adhesins). It is becoming apparent that the molecular machinery needed to build pili or to transport an afimbrial adhesin to the bacterial surface is often conserved. Despite this conservation, the receptor specificity of the adhesin is dictated by a portion of the adhesin exposed at the pilus tip or on the bacterial surface. These concepts are illustrated below by examining a few well-characterized examples of adhesins, both fimbrial and nonfimbrial. Alternatively, many pathogens are capable of binding soluble host molecules such as matrix proteins and complement. These molecules can then serve as a bridge between the bacterium and host cell surface when these molecules bind to their natural receptor on host cell surfaces (Figure 4.1).

Fimbriae of Gram-Negative Bacteria

P Pili: Model Fimbriae

One of the best-studied examples of pili and its associated assembly machinery is that of P pili (pyelonephritis-associated pili), which are encoded by *pap* genes. *Escherichia coli* strains that express P pili are associated with pyelonephritis, which arises from urinary tract colonization and subsequent infection of the kidneys. It is thought that P pili are essential adhesins in this disease process.

The *pap* operon is a useful paradigm since it contains many conserved features that are found among various pilus operons (Figure 4.2). Two molecules guide newly synthesized pilus components to the bacterial surface. PapD, a conserved chaperone molecule, has an immunoglobulin-like fold which is necessary to transport several pilus subunits from the cytoplasmic membrane to the outer membrane. PapC accepts molecules from PapD and has been proposed to serve as an outer membrane usher. The major subunit of the pilus rod is PapA, which is anchored in the outer membrane by PapH. At the distal end of the pilus rod is the tip fibrillum, composed of PapE and the actual molecule that mediates adherence (i.e., tip adhesin), PapG. Two other proteins, PapF and PapK, are involved in tip fibrillum synthesis.

P pili bind to the α-D-galactopyranosyl-(1-4)-β-D-galactopyranoside moiety present in a globoseries of glycolipids which are found on host cells lining the upper urinary tract. However, there are three adhesin variants of PapG, PapG-I, PapG-II, and PapG-III, which recognize three different but related Galα-(1-4)-Gal receptors. It is thought that different hosts and tissue may contain differences in distribution of these receptors, and differential expression of the PapG adhesins at the pilus tip could enhance tissue and host specificity for adherence.

Figure 4.1 Bacterial adherence to host cell surfaces. Bacteria can adhere to host cell surfaces either directly by using a surface-anchored adhesin or indirectly by binding a soluble host component that then serves as a bridge between the bacterial adhesin and the natural host receptor for that molecule.

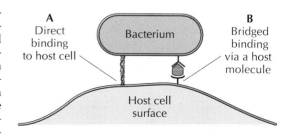

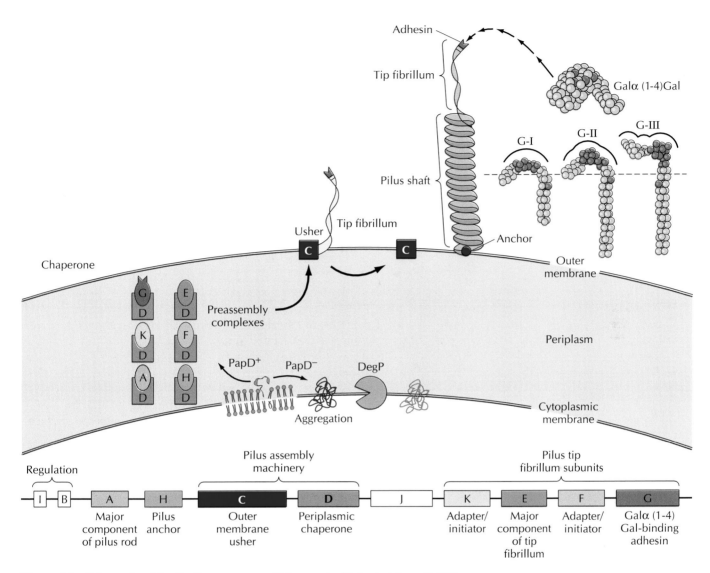

Figure 4.2   Schematic of the P pilus operon and biogenesis. (Adapted from *Cell* **73:** 887–901, 1993.)

Although the host receptor varies for different bacterial pili, the general concepts provided by the P pilus operon are conserved in many other pilus systems and the components are often interchangeable. For example, the PapD chaperone also can modulate the assembly of type 1 pili, which mediate binding to mannose-containing molecules on the host cell surface. There is a large (>13-member) family of such periplasmic PapD-like chaperones that are necessary for the assembly of several pili, including K88, K99, and *Haemophilus influenzae* pili. Additionally, homologous chaperones are needed for several afimbrial adhesins, including filamentous hemagglutinin (FHA) and the pH 6 antigen from *Yersinia pestis*. The gene organization among such pilus operons also is usually conserved. Thus, type I and P pili have very similar operons and functionally analogous sequences that can be aligned. However, they bind to quite different carbohydrates on the cell surface.

Type IV Pili

Although *pap*-like sequences are common throughout adhesins of gram-negative bacteria, there are other families of pili that use alternative biogenesis and assembly machinery to form a pilus. One such group is the type IV pili found in diverse gram-negative organisms (Figure 4.3). This family includes pili from *Pseudomonas aeruginosa*, *Neisseria* species, *Moraxella* species, enteropathogenic *E. coli*, and *Vibrio cholerae*. Type IV pilus subunits contain specific features, including a conserved, unusual amino-terminal sequence that lacks a classic leader sequence and instead usually utilizes a specific leader peptidase that removes a short, basic peptide sequence. Several possess methylated amino termini on their pilin molecules and usually contain pairs of cysteines that are involved in intrachain, disulfide-bond formation near their carboxy termini. The pilus assembly genes are also members of the type II secretion system, a general secretion system used to transport molecules across the gram-negative bacterial envelope, which is needed to assemble this complex organelle on the bacterial surface. It has been proposed that the pilin molecules located at the tip have different sequences exposed from those that are packed into repeating structures

Figure 4.3 Schematic of adhesins of various gram-negative bacteria.

Afimbrial adhesion:

Bald bacterial surface, afimbrial adhesins embedded in surface.

Pap fimbriae:

Very hairy surface, thin filaments protruding from surface.

Type IV bundle-forming pilus:

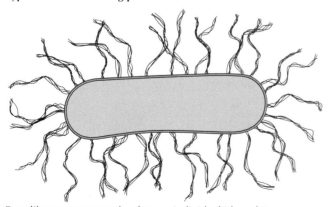

Ropelike structures made of many individual "threads" intertwined; then ropes are tangled.

Curli:

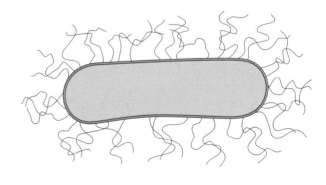

Coiled surface structure intertwined. Not ropelike like type IV. Curved/curled

within a pilus and that these exposed regions may function as the adhesins. However, analogous to the P pilus tip adhesin, a separate tip protein (PilC for *Neisseria gonorrhoeae*) may function as a tip adhesin for these pili. A membrane glycoprotein (CD46) has been proposed as the receptor for *N. gonorrhoeae* pili, but the pilus component that binds this molecule has not been established. Alterations in the pilus subunit can also affect adherence levels. For example, although *P. aeruginosa* strains usually express only one pilus subunit, this subunit can vary significantly between strains, which affects their adhesive capacities. At least for one strain of *P. aeruginosa*, the pilus mediates binding to the disaccharide β-GalNAc(1-4)βGal in asialo-GM_1 gangliosides on host cell surfaces. Additionally, the pilin subunit of *N. gonorrhoeae* that is expressed is always changing due to genetic switching of the expressed pilin gene, which leads to antigenic variation. Different pilin sequences may lead to tissue tropism for adherence and different invasion capacities into epithelial cells. The recent crystallographic structure of the *N. gonorrhoeae* pilin may begin to provide additional clues about how this molecule functions in adherence.

Curli

Diarrheagenic and other *E. coli* strains produce thin aggregative fimbriae termed curli, a descriptive name based on their curved appearance (Figure 4.3). In addition, curli are homologous to thin aggregative fimbriae (SEF17) produced by *Salmonella enteritidis,* and they appear to have a common ancestry. These fimbrial structures are assembled by a mechanism that is surprisingly different from that for other fimbrial assembly systems and that does not employ the complex assembly machinery typical of most fimbriae. Instead, they use a self-assembly mechanism which involves the production of a bacterial cell surface-bound nucleator followed by polymerization of curlin (the structural repeating subunit of curli which is secreted out of the bacteria) condensing on the nucleator. Curlin can be supplied by the nucleator-presenting cell or by adjacent bacteria, driven by a diffusion gradient. Once assembled, curli are quite stable structures and are difficult to dissociate. Curli have been reported to bind to several host proteins including fibronectin, laminin, plasminogen, and tissue-type plasminogen activator.

Pathogens and Their Nonfimbrial Adhesins

Escherichia coli Afimbrial Adhesins

E. coli contains many adhesins and can cause several diseases including meningitis, diarrhea, and urinary tract infections. There is a family of adhesins that are associated with *E. coli* and that cause the last two diseases. Members of this family include the fimbrial adhesins F1845 and Dr and two nonfimbrial adhesins, Afa-I and Afa-III (Figure 4.3). All four of these adhesins use the Dr^a blood group antigen present on decay-accelerating factor (DAF) on erythrocytes and other cell types as their receptor, although they appear to recognize different epitopes of the Dr antigen. The Dr adhesin, but not the other three, also binds type IV collagen.

Much like other fimbrial adhesins, expression and production of these adhesins require five or six gene products. These include a periplasmic chaperone, an outer membrane anchor protein, one or two transcriptional regulators, and the adhesin. At least for Afa-III, two adhesins (AfaD and AfaE) appear to be encoded within the operon. One perplexing question has been why some adhesins in this family form fimbriae (F1845 and Dr) yet others form afimbrial adhesins on the bacterial surface (Afa-I and Afa-III). Recent work indicates that the sequence of the adhesin molecule dictates whether it will be assembled into a fimbria and that genetically switching the genes encoding these adhesins switches the adhesin type. It is possible that the *E. coli* afimbrial adhesins have evolved from the related fimbrial adhesins but have been altered such that the properties needed to polymerize a pilus are missing yet the adhesin domain remains anchored on the bacterial surface. It is now possible to test the role in bacterial disease of assembly of an adhesin into a fimbrial structure versus the presence of an afimbrial adhesin.

Helicobacter pylori

Helicobacter pylori is a human-specific pathogen that adheres to epithelial cells from the gastrointestinal tract but not to cells from the nervous system or urogenital tract. Additionally, it shows preferential adherence to the gastric mucosa rather than the colon, which correlates with the site of *H. pylori* infections (the stomach). This organism also preferentially adheres to surface mucous cells of the gastric mucosa and does not bind to host cells that are found deeper within the mucosal layer. Recent work has shown that *H. pylori* binds to the Lewis[b] blood group antigen that is expressed on cells in the stomach epithelium, presumably explaining the tissue specificity of this adherence. Interestingly, *H. pylori* preferentially infects people that are Lewis[b] positive, but other individuals can also be infected. *H. pylori* also has several other adhesins which mediate adherence to other host receptors, although the details of these adhesins and their receptors have not been fully characterized. The bacterial molecule responsible for mediating Lewis[b] binding is an outer membrane protein that belongs to a large family of related *H. pylori* outer membrane proteins recently identified by using the entire genome sequence. Other members of this family also appear to function as adhesins and porins. Thus, this family of adhesins requires no specialized assembly machinery, and the adhesins are anchored directly in the outer membrane with the adhesin domain exposed to the extracellular surface. Interaction of *H. pylori* with gastric mucous cells results in intimate attachment to the host cell surface, and it has been reported that such attachment results in localized actin polymerization and host cell signaling beneath the adherent organisms, much like that seen with enteropathogenic *E. coli* (see chapter 5).

Bordetella pertussis

Bordetella pertussis, the causative agent of whooping cough, is a respiratory mucosal pathogen that possesses several potential adherence factors that exemplify the complexity of bacterial adherence to host cell surfaces. Potential *B. pertussis* adherence factors include at least four fimbrial genes and several nonfimbrial adhesins, including FHA, pertactin, pertussis toxin, and BrkA. FHA is a large (220-kDa) secreted molecule with several domains that are homologous to other bacterial adherence molecules or to eukaryotic sequences that mediate cell-cell adhesion. FHA is homologous to two high-molecular-weight, nonpilus adhesins from *Haemophilus influenzae*, another pathogen that adheres to respiratory surfaces. FHA also contains an RGD tripeptide sequence, which is a characteristic eukaryotic recognition motif that binds to host cell surface integrins. This RGD sequence appears to be involved in FHA binding to the leukocyte integrin CR3, which mediates bacterial uptake into macrophages without triggering an oxidative burst. It has been proposed that by possessing this RGD sequence, FHA mimics host molecules (at a molecular level). Additionally, this RGD sequence induces enhanced *B. pertussis* binding to monocytes by activating a host signal transduction complex that normally up-regulates the CR3 binding activity. (Most bacterial molecules that mediate adherence to integrins, such as *Yersinia* invasin, do not possess RGD sequences, although their receptor affinity is often greater than that of the native host ligand.) Both FHA and one of the binding subunits of pertussis toxin (S2) have homologous sequences that mediate binding to lactosylceramides, which suggests that these two proteins might use similar motifs to bind to host molecules. Pertussis toxin also mimics eukaryotic adhesive molecules. For example, the S2 and S3 subunits have several features in common with eukaryotic selectins. By mimicking host molecules, these bacterial adhesins can effect responses in the host that enhance "desired" (from the bacterial viewpoint) interactions with host cells. These examples highlight the ability of bacteria to mimic host molecules and to use this mimicry to enhance their pathogenesis. At least two other *B. pertussis* molecules are involved in adherence to host cells. Both pertactin and BrkA, which are 29% identical, contain RGD motifs that may be involved in adherence. BrkA is also involved in serum resistance.

Neisseria Species

N. gonorrhoeae and *N. meningitidis* are two mucosal pathogens that have developed sophisticated and overlapping mechanisms to adhere to host cell surfaces. As discussed above, these bacteria produce a type IV pilus which mediates adherence to cell surfaces. In addition to pili, they produce a family of outer membrane proteins called opacity proteins (Opa), which also mediate adherence and invasion. *Neisseria* species can express up to 12 different Opa proteins, with each being expressed independently of the others. This variable expression is thought to contribute to the antigenic variation and possibly altered adherence characteristics of this genus. It has recently been shown that Opa proteins bind to members of the carcinoembryonic (CEA or CD66) family. Members of this eukaryotic transmembrane glycoprotein family show high levels of similarity and belong to the immunoglobulin superfamily. Although most Opa proteins bind to the CEA proteins, their affinity for individual members of this family vary, and these differences may contribute to differential adherence and invasion of these pathogens in different tissues. Lipooligosaccharide (LOS) also contributes to *Neisseria* adherence and invasion into host cells, possibly by interacting with the asialoglycoprotein receptor. LOS can be sialyted, which inhibits bacterial invasion. Thus, this pathogen expresses several different adhesins which are involved in binding to host cell surfaces.

Yersinia Species

Pathogenic *Yersinia* species contain several adhesins which mediate adherence to the host cell surface or matrix. Perhaps the best-studied adhesin, which can also mediate invasion under certain conditions, is invasin. This outer membrane protein adheres tightly to members of the β_1-integrin family, including the fibronectin receptor (see chapter 5 for details). In addition to invasin, *Y. enterocolitica* expresses another adhesin which mediates low levels of invasion called Ail (for "attachment invasion locus"). The host receptor for this outer membrane protein has not been identified. Finally, YadA of *Y. enterocolitica* mediates binding to cellular fibronectin but not to plasma fibronectin, which may provide a possible mechanism for the organism to adhere to tissue rather than to bind to circulating molecules. YadA also mediates adherence to various collagens and laminins. The loss of YadA decreases virulence in mice by 100-fold, which suggests that adherence to these molecules may potentiate disease (see below). Yersiniae also express a flexible fimbria whose assembly genes are homologous to those encoding Pap pili and other related fimbriae. In *Y. pestis,* this organelle is produced at pH 6 and is thus called the pH 6 antigen (a homologue [Myf] is also found in *Y. enterocolitica*). Studies with the pH 6 antigen indicate that it is needed for full virulence of *Y. pestis* in mice, although its exact role in disease and its receptor have not been defined.

Mycobacterium Species

Mycobacterium species also bind fibronectin by using three related bacterial molecules (called the BCG85 complex). At least two other fibronectin-binding molecules also have been described for *Mycobacterium* species. One of these has been described for *M. avium* and *M. intracellulare* as a 120-kDa protein recognized by antibodies against β_1-integrins, which indicates that these bacteria may express integrin-like molecules that may mediate matrix adherence. A fibronectin attachment protein from *M. avium* has been characterized and shown to be highly conserved in *M. leprae, M. tuberculosis,* and other mycobacteria.

M. leprae has a specific neural tissue tropism to Schwann cells of the peripheral nerves. Recently it has been established that this tropism is due to bacterial binding of a host matrix molecule (laminin) which serves as a bridge between the bacterium and a host receptor, β_4-integrin. Although the bacterial component was not identified, it was shown that *M. leprae* binds specifically to the G domain of the soluble laminin α_2 chain. Laminin in turn binds to its cell surface receptors, including $\alpha_6\beta_4$-integrin. Not only does this provide a good model for determining tissue tropism mediated by bacterial adherence, but also it provides an excellent example of a pathogen binding a soluble matrix protein which serves as a bridge to its host cell receptor, thereby mediating bacterial adherence.

Staphylococcus and *Streptococcus* Species

As described for *M. leprae,* a common theme in the adherence of many gram-positive cocci, including many species of staphylococci and streptococci, involves bacterial recognition of host proteins that, in turn, associate with the surface of target host cell or tissue. The host proteins that are most often bound by these bacteria include components of the extracellular matrix. Binding to fibronectin, collagen, fibrinogen, vitronectin, laminin, bone sialoprotein, elastin, and thrombospondin has been reported for various staphylococcal and streptococcal species (Table 4.1). Numerous adhesins that recognize a specific extracellular-matrix component and several that can recognize multiple components have been characterized at the molecular level. In some cases, the bacterium does not discriminate between recognition of soluble, immobilized, and tissue-specific forms of the extracellular-matrix component and will bind to all forms equally. However, in other cases, a specific form of the matrix component is exclusively bound. This latter instance almost always involves preferential recognition of tissue or immobilized forms of the component over soluble forms. As described above for YadA of *Yersinia enterocolitica,* this level of discrimination may be useful in situations where the adhesin encounters both soluble

Table 4.1 Selected examples of the interactions between some gram-positive bacteria and various components of the extracellular matrix

ECM[a] molecule	Microorganism	Adhesin
Fibronectin	*Staphylococcus aureus*	FnbA, FnbB
	Coagulase-negative staphylococci	
	Group A streptococci	PrtF1/Sfb1, PrtF2, GAPDH,[b] ZOP, LTA, SOF/SfbII, FNB54, 28-kDa antigen, M3 protein
	Group B streptococci[c]	
	Group C streptococci[c]	FnBA, FnBB
	Group G streptococci[c]	FnB, GfbA
	Streptococcus pneumoniae[c]	
	Streptococcus gordonii[c]	CshA
	Enterococcus faecalis	
Collagen	*Staphylococcus aureus*	Cna
	Group A streptococci	57-kDa protein
	Streptococcus mutans	
Fibrinogen	*Staphylococcus aureus*	ClfA, Fib
	Group A streptococci	M proteins, FNB54
	Streptococcus parasanguis[c]	FimA
Vitronectin	*Staphylococcus aureus*	60-kDa protein
	Group A streptococci	
Laminin	*Staphylococcus aureus*	
	Group A streptococci	
	Streptococcus gordonii	145-kDa protein
Bone sialoprotein	*Staphylococcus aureus*	
Elastin	*Staphylococcus aureus*	EbpS
Thrombospondin	*Staphylococcus aureus*	
Multiple-binding activity	*Staphylococcus aureus*	Map
	Group A streptococci	GAPDH
	Group C streptococci	FIA

[a]ECM, extracellular matrix.
[b]GAPDH, glyceraldehyde-3-phosphate dehydrogenase.
[c]Binding is preferentially to an immobilized form of the extracellular matrix component.

and immobilized forms of the extracellular-matrix component. Preferential recognition of the immobilized form would prevent competition for binding with the soluble form to allow the microorganism to most efficiently adhere to the target substrate.

Binding to extracellular-matrix components may contribute to the tropism of streptococcal and staphylococcal infections, particularly when the bacteria recognize extracellular-matrix components that are distributed only among defined compartments within the host (e.g., laminin in basement membrane, or bone sialoprotein in bone). In other cases, the extracellular-matrix components that are recognized are broadly distributed. An example of this latter class is binding to fibronectin. This extracellular-matrix glycoprotein is present in most tissues and fluids of the host, is found in many secretions, and is a prominent component of wounds. Fibronectin is found as a soluble form, as an immobilized form bound to receptors on various cell types, or as a form incorporated into many different types of extracellular matrices. Given this broad distribution, it is not surprising that many different species of streptococci, staphylococci, and enterococci have evolved the ability to interact with fibronectin.

Perhaps the best-characterized fibronectin-binding proteins of gram-positive cocci are members of a large family that includes protein F1, Sfb, protein F2, and serum opacity factor produced by the group A streptococcus; FnbA and FnbB of *Staphylococcus aureus;* FnBA and FnBB of *Streptococcus dysgalactiae* (a group C streptococcus); FnB of *Streptococcus equisimilis* (a group G streptococcus); and GfbA of human-associated group G streptococcal isolates. In general, these proteins are of the class that does not discriminate between binding to soluble and immobilized forms of fibronectin, their binding is highly specific for fibronectin, and they bind to fibronectin in an essentially irreversible manner. Furthermore, they bind fibronectin with very high affinity, and apparent K_d values in the 1.0-nM range are not uncommon. When the corresponding genes are cloned and expressed in *Escherichia coli,* the proteins retain the ability to bind fibronectin (although they are not expressed on the *E. coli* cell surface), and when purified from *E. coli,* the proteins can competitively inhibit the binding of fibronectin not only to the homologous strain but often also to other gram-positive cocci which express a member of this family. Finally, the binding of fibronectin by the members of this adhesin family involves protein-protein interaction rather than recognition of a carbohydrate moiety of the fibronectin glycoprotein.

Comparison of the DNA sequences of the genes which encode these adhesins has revealed that the deduced structures of the corresponding proteins have a similar domain architecture (the structure of protein F1 is shown in Figure 4.4). Sequences at their carboxy termini are characteristic of proteins from gram-positive cocci that are sorted and displayed on the cell surface by a pathway first described for the M protein of the group A streptococcus and protein A of *S. aureus.* These include a proline- and lysine-rich domain (W in Figure 4.4) that is thought to be important for interaction with cell walls of gram-positive bacteria, a domain rich in hydrophobic amino acids (M in Figure 4.4) that may be involved in interaction with the membrane, and a short, positively charged carboxy terminus. A characteristic highly conserved pentapeptide LPXTG motif is located between the cell wall interaction domain and the membrane interaction domain. Studies with protein A have shown that this sequence is recognized by the sorting pathway following secretion of the nascent polypeptide across the membrane, with the result that the protein is cleaved following the threonine residue of the LPXTG motif. The threonine then becomes covalently coupled by a pentaglycine bridge to a lysine residue of the cell wall peptidoglycan. The end product of this sorting pathway is a cell surface protein that is covalently tethered to the cell wall by its carboxy terminus and has its amino terminus exposed to the environment.

The amino termini of the proteins in the protein F family contain a signal sequence characteristic of gram-positive bacteria. This is followed by a large domain that contains unique sequences that are not shared between the different members of this family and that can constitute up to 80% of the total sequence of the protein (U in Figure 4.4). This region can actually be quite divergent among different alleles of the same family member. For the most part, its function is unknown. A notable exception is found in serum opacity factor, where this region contains the catalytic domains that are involved in cleaving apolipoprotein AI in serum, which then initiates a cascade of events that renders the serum opalescent. The unique amino-terminal domain is often followed by another domain of unknown function

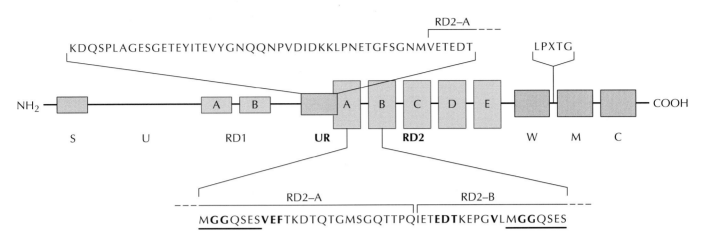

Figure 4.4 Domain architecture of protein F. Protein F is a member of a large family of related surface proteins found among the gram-positive cocci that bind to fibronectin. The signature feature of this family is a repetitive domain consisting of 32 to 44 amino acids that is repeated two to six times in tandem. This domain consists of 37 amino acids in protein F and is known as RD2. Sequences highly conserved in the repetitive region from different relatives of protein F are highlighted in bold type. This repetitive domain binds to the N-terminal 29-kDa fibrin-binding domain of fibronectin, and the minimal functional binding domain of RD2 includes 44-amino-acid residues derived from the carboxy-terminal segment of one repeat and the N-terminal segment of the adjacent repeat, as shown underneath RD2. This binding unit begins and terminates with the amino acid motif MGGQSES (underlined). Other common features include a signal sequence (S), a long N-terminal nonrepetitive region of sequence that is unique to different family members (U), and a second repetitive region of unknown function (called RD1 in protein F). The proteins also have features common to many surface proteins of gram-positive cocci, including a proline- and lysine-rich domain that may interact with the cell wall (W), a hydrophobic domain that may interact with the membrane (M), a short tail of charged amino acids (C) and an LPXTG motif (LPATG in protein F) that serves as a processing site, in which the protein becomes cleaved after the threonine and is cross-linked to a lysine residue of the peptidoglycan cell wall. Some members of this family contain an additional domain for binding to fibronectin, and this domain can be located in various regions N-terminal to the repetitive domain. In protein F, this domain is known as UR and is composed of 43 amino acids immediately N-terminal to RD2, as shown above UR. The ability of UR to bind to fibronectin also has an absolute requirement for the first 6 amino acids from the first repeat unit of RD2. UR binds to a region of fibronectin that includes the N-terminal 29-kDa domain and the adjacent 40-kDa collagen-binding domain. The various domains are not shown to scale.

that frequently consists of a sequence motif that is repeated several times in tandem (RD1 in Figure 4.4).

The signature feature of the protein F family is a repetitive domain located just amino-terminal of the cell wall attachment domains (RD2 in Figure 4.4). This signature domain consists of a 32- to 44-amino-acid motif that can be repeated up to six times in tandem. There is considerable variation in the number of repeats between different members of the family, and the number can even vary extensively between alleles of the same gene. This level of variation is not uncommon in repetitive regions of other surface proteins from gram-positive cocci. For example, the M proteins of the group A streptococci exhibit a similar variation. A large body of evidence, much of which initially came from the analysis of FnbA of *S. aureus,* has strongly implicated this repetitive domain in binding to fibronectin. However, further study has suggested that the mechanisms by which this domain interacts with fibronectin may not be identical for all family members. While a single 37-amino-acid repeat unit of *S. pyogenes* Sfb was an effective inhibitor of fibronectin binding to the native protein, a single repeat unit of the very closely related protein F1 was not. Further analysis of protein F1 revealed that the minimal functional binding unit consists of a 44-amino-acid region derived from the carboxy-terminal end of one repeat and the amino-terminal region of the adjacent repeat (Figure 4.4). This binding element begins and ends with a direct repeat of the sequence motif MGGQSES.

Additional studies of protein F1 have revealed that the binding of fibronectin to this protein was much more complicated than originally thought and may involve an additional domain of the adhesin. This was based on the observation that while the full-length protein could completely inhibit the binding of fibronectin to streptococcal cells, the repeat region was capable of only partial inhibition, even when tested at very high concentrations. A second domain was subsequently found which could also partially inhibit binding, and when a mixture of the two domains was tested, inhibition similar to that obtained with the full-length protein F1 was obtained. This second domain, called UR (Figure 4.4), contains the 43 amino acids located immediately N-terminal to the repetitive domain. In addition, 6 amino acids derived from the first repeat of the repetitive domain are absolutely required for the binding activity of UR. Other than these six residues, there is no obvious homology between UR and the repetitive domain. Protein F2 was also found to use a two-domain mechanism to bind fibronectin. However, while its second binding domain is also located amino terminal to the repetitive domain, the two domains are separated by 100 amino acids. Also, there is no obvious homology between the second domains of protein F2 and UR.

Investigation of the domains of fibronectin that are recognized by these proteins has strongly implicated the very N terminus of the protein, which is contained on a 29-kDa protease cleavage fragment that also includes binding sites for fibrin and heparin. The isolated 29-kDa domain will bind to all staphylococci and streptococci that express adhesins of the protein F family, and there is considerable evidence that this domain of fibronectin recognizes the repetitive domain of the microbial adhesins. In contrast, this 29-kDa region of fibronectin does not bind to the UR domain of protein F1. Instead, UR interacts with a region of fibronectin that must include both the 29-kDa domain and the adjacent 40-kDa collagen-binding domain on the same polypeptide fragment. Thus, not only does protein F1 contain two domains that can bind independently to fibronectin, but also these two domains recognize distinct sites in the fibronectin molecule.

The principal function of fibronectin and many other extracellular matrix proteins is to serve as a substrate for the adherence of eukaryotic cells. It is interesting that the gram-positive cocci have exploited the host's own cell adhesion molecules to mediate their adherence to the host (Table 4.1). This strategy has an additional advantage that makes use of the fact that most extracellular-matrix components themselves have a broad range of ligands with which they can interact. By binding an extracellular matrix protein, which can in turn bind a large number of different cells, structures, and tissues, the bacterium can gain an extensive adhesive potential. For example, by binding to fibrinogen, *S. aureus* gains the capacity to bind to endothelial cells, and by binding to soluble fibronectin, *S. pyogenes* can bind much more efficiently to a collagen-containing structure like the dermis.

Role of Adherence in Disease

Adhesins play an important role in disease and represent the interface between the pathogen and the host cell. In many cases, the precise role that individual adhesins play in the pathogenesis of specific diseases has been established. For example, inactivation of the gene which encodes Cna, a collagen-binding adhesin of *S. aureus*, results in a mutant with a considerably diminished capacity to cause septic arthritis in an animal model. Similarly, construction of a mutant of *E. coli* which can express a P pilus that is intact except for the loss of the PapG tip adhesin produces a bacterium which can infect the bladder but has lost the ability to infect the kidneys. Also, studies which have successfully used adhesins like PapG as vaccines have shown that novel antimicrobial therapies can be based on understanding the precise role that adhesins play in the pathogenic process.

However, in many instances, the precise role that adhesins play in disease has been less clear. One reason may be that due to the importance of adherence, many pathogens express multiple adhesins, each of which may contribute unique or overlapping specificities of host cell recognition. This level of complexity is illustrated for infection by the invasive pathogen *Salmonella typhimurium* (Box 4.1). *Y. enterocolitica* and *Y. pseudotuberculosis* provide similar examples of the complexity of bacterial adherence and its involvement in disease. In *Y. enterocolitica*, invasin is needed for efficient penetration of the intestinal epithelium. However, YadA is also needed for persistence in Peyer's patches and for the bacterium to cause disease in mice infected either orally or intraperitoneally. The role of Ail in intestinal colonization is not as pronounced. Invasin mutants of *Y. pseudotuberculosis* are unable to efficiently penetrate murine Peyer's patches and instead colonize on the luminal surface of the intestinal epithelium. Mutants with mutations in YadA or the pH 6 antigen also show decreased binding to the luminal surface, indicating that these two molecules also participate in intestinal colonization.

The complexity of the adherence process, as illustrated by the examples of *Yersinia* and *Salmonella*, has frustrated most attempts to base novel antimicrobial therapies on intervening in this critical step of host cell-pathogen interaction. In addition to the complex issue of multiple adhesins, individual adhesins themselves often have evolved strategies to evade host immune responses. Examples of these are antigenic variation of the major

---

BOX 4.1

The complexity of *Salmonella* adherence

While adherence plays a key role in establishing infection by *Salmonella typhimurium*, this pathogen can contain up to five distinct fimbrial operons (*fim, lpf, agf, sef,* and *pef*), each of which may participate in adherence. In addition, the *S. typhimurium* invasion loci (*inv*) can mediate adherence, which is a prerequisite to invasion of nonphagocytic cells. The *lpf*-encoded fimbriae provide tissue specificity by mediating adhesion to Peyer's patches in mice, the site of bacterial entry in the intestine. Although strains containing *lpf* mutations are attenuated, a much stronger attenuation and lack of colonization occurs if strains containing mutations in both *lpfC* and *invA* are used in the oral infections. However, if this strain is delivered by the intraperitoneal route, it is fully virulent, indicating that these adhesins and invasins are involved in intestinal colonization and not the later systemic events. Attenuation and lack of colonization are also seen following oral challenge with strains containing a *pefC* mutation. Thus, it seems that several adhesins are involved in intestinal colonization, rather than being mediated only by a single bacterial adhesin, and that they work in concert to mediate adherence to host tissues.

immunodominant domains of the adhesins, as occurs in the type IV pili of the neisseriae, and the localization of a single copy of the adhesin subunit as a minor component at the very tip of a large heteropolymeric fiber, as occurs with P pili and type 1 pili. However, unraveling the molecular details of adhesin structure, organization, and assembly is leading to increased knowledge about the process of adherence. Perhaps the best example of this has come from studies on type 1 pili. Knowledge of the structure of the adhesive organelle, identification of the adhesive subunit of the organelle (see above), and advances in expression and purification of the unstable adhesive subunit have led to the ability to test the efficacy of a vaccine composed solely of the adhesive subunit. Initial results from trials in animal models of infection of the bladder have demonstrated that this approach can be successful. Thus, continued study of other adhesins and this critical component of pathogen-host cell interaction holds similar promise for augmenting our arsenal of antimicrobial agents.

Selected Readings

Baumler, A. J., R. M. Tsolis, P. J. Valentine, T. A. Ficht, and F. Heffron. 1997. Synergistic effect of mutations in *invA* and *lpfC* on the ability of *Salmonella typhimurium* to cause murine typhoid. *Infect. Immun.* **65:**2254–2259.

Boren, T., P. Falk, K. A. Roth, G. Larson, and S. Normark. 1993. Attachment of *Helicobacter pylori* to human gastric epithelium mediated by blood group antigens. *Science* **262:**1892–1895.

Finlay, B. B., and S. Falkow. 1997. Common themes in microbial pathogenicity. II. *Microbiol. Mol. Biol. Rev.* **61:**136–169.

Hultgren, S. J., S. Abraham, M. Caparon, P. Falk, J. St. Geme, and S. Normark. 1993. Pilus and nonpilus bacterial adhesins: assembly and function in cell recognition. *Cell* **73:**887–901.

Langermann, S., S. Palaszynski, M. Barnhart, G. Auguste, J. S. Pinkner, J. H. Burlein, P. Barren, S. Koenig, S. Leath, C. H. Jones, and S. J. Hultgren. 1997. Prevention of mucosal *Escherichia coli* infection by FimH-adhesin-based systemic vaccination. *Science* **276:**607–611.

Ozeri, V., A. Tovi, I. Burstein, S. Natanson-Yaron, M. G. Caparon, K. M. Yamada, S. K. Akiyama, I. Vlodavsky, and E. Hanski. 1996. A two-domain mechanism for group A streptococcal adherence through protein F to the extracellular matrix. *EMBO J.* **15:**989–998.

Patti, J. M., B. L. Allen, M. J. McGavin, and M. Höök. 1994. MSCRAMM-mediated adherence of microorganisms to host tissues. *Annu. Rev. Microbiol.* **48:**585–617.

Pepe, J. C., M. R. Wachtel, E. Wagar, and V. L. Miller. 1995. Pathogenesis of defined invasion mutants of *Yersinia enterocolitica* in a BALB/c mouse model of infection. *Infect. Immun.* **63:**4837–4848.

5

Molecular Basis for Cell Adhesion and Adhesion-Mediated Signaling

Benjamin Geiger, Avri Ben-Ze'ev, Eli Zamir, and Alexander D. Bershadsky

This chapter addresses the complex molecular cross talk between cell adhesion and the transduction of transmembrane signals in cells and shows that adhesion sites such as focal contacts and cell-cell adherens junctions contain multimolecular protein complexes, which participate both in the assembly of the adhesion sites and in their interaction with the cytoskeleton, and in the transduction of long range signals. The network of molecular interactions of the different adhesions, the involvement of these interactions in constructing the molecular linkages with the cytoskeleton, and their role in adhesion-mediated signaling are also discussed.

Structural and Functional Diversity of Cell Adhesions

Adhesive interactions of cells exert major short- and long-term effects on cell shape and fate and thus play a central role in the assembly of individual cells into multicellular organisms. Cell adhesion in metazoan organisms is a very complex and molecularly diversified process involving a multitude of molecular systems. There are many distinct types of extracellular surfaces with which cells interact in vivo. These include networks of extracellular matrix (ECM) molecules as well as the membranes of adjacent cells. Within each group of adhesions, there is further molecular complexity and diversity. These consist of numerous distinct ECM molecules including their isoforms, which interact with cells via families of specific adhesion receptors. Cell-cell adhesion is also mediated by a multitude of molecules, some of which are Ca^{2+} dependent, some Ca^{2+} independent, some involved in homophilic interactions, and some involved in heterophilic interactions. Moreover, the long-term effects of adhesive interactions on cell activity and fate may differ greatly from one adhesion site to the other. For example, some adhesions promote cell growth while others may suppress growth, adhesion to different matrices can have distinct effects on cell differentiation and cell survival, and apoptosis can also be differentially regulated by various forms of cell adhesion. The objective of this chapter is to discuss the structure-signaling relationships in two molecularly related adhesions, focal contacts and cell-cell adherens junctions. Recent studies have shed much light on the molecular structure of these adhesive sites, providing an insight not only into their mechanical properties but also into their interactions with the signal transduction machinery.

Focal contacts and adherens junctions obviously differ in the nature of the adhesion surface and adhesion receptors involved in their formation. Focal contacts and related matrix adhesions are widely present in tissues, involving different ECM molecules and receptors of the heterodimeric integrin family (Figure 5.1). In adherens junctions, on the other hand, cell-cell adhesion is mediated by members of the homophilic, Ca^{2+}-dependent cadherin family (Figure 5.2). The integrin- and cadherin-dependent adhesions are both associated with the actin cytoskeleton via a submembrane network or "plaque" comprised of the "anchor proteins." Examples of these relationships are provided in Figure 5.3, demonstrating that in cultured cells actin-containing stress fibers terminate at vinculin- and paxillin-containing focal contacts. Double immunofluorescent labeling for vinculin and $\beta_3\alpha_v$-integrin indicated that the two proteins were colocalized in the matrix adhesion sites, yet careful examination of such images reveals vinculin-rich structures which are devoid of integrin, corresponding to cell-cell adhesions. In cell-cell adherens junctions, the membrane receptors are various members of the cadherin family that are associated with the sub-

▶ *For Figure 5.3, see color insert.*

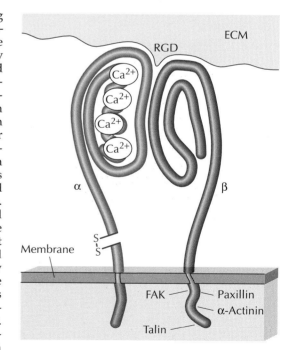

Figure 5.1 Scheme depicting the structure of a typical integrin molecule. Integrins are composed of two noncovalently associated subunits designated α and β. Sixteen different α variants and eight different β variants have been identified in mammals to date. The integrin heterodimer has a globular head containing a ligand-binding pocket formed by both subunits. Two extended stalks correspond to the C-terminal parts of the α and β subunits. Some α subunits are cleaved posttranslationally to give heavy (extracellular) and light (transmembrane) chains linked by S–S bonding. The majority of integrin ligands contain the Arg-Gly-Asp (RGD) motif as the minimal sequence necessary for the integrin binding. Divalent cations, especially calcium, play an important role in the regulation of ligand binding, and the extracellular domain of the α subunit contains multiple repeats of EF-hand Ca^{2+}-binding sites. The adhesion plaque proteins talin, α-actinin, paxillin, and FAK interact with the short cytoplasmic domain of the β subunit.

A

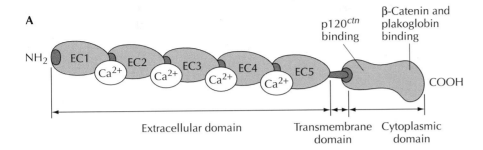

B

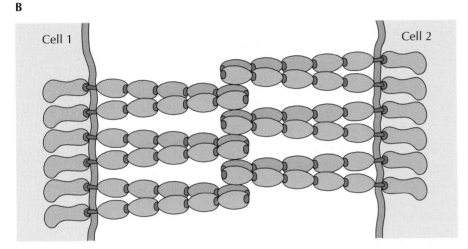

Figure 5.2 Scheme depicting classical cadherins involved in the formation of cell-cell adherens junctions. **(A)** A classical cadherin molecule contains five homologous extracellular domains, denoted EC1 through EC5. Ca^{2+} ions bind to the regions between the EC domains and probably contribute to the rigid elongated shape of the molecule. The cytoplasmic domain of the classical cadherin is highly conserved and contains binding sites for the armadillo family proteins β-catenin and plako-globin, while another family member, $p120^{ctn}$ protein, binds to the juxtamembrane region of the cadherin tail. **(B)** Cadherin molecules exist as parallel dimers. It is thought that cadherin dimers on one cell make homophilic contacts with the dimers on a neighboring cell, forming a zipper-like structure. The EC1 domain participates both in dimer formation and in homophilic adhesive interactions. Adhesive specificity depends on this domain of cadherin.

membrane plaque via β-catenin and the homologous protein plakoglobin (Figure 5.4). As pointed out above, some of the plaque components are shared by focal contacts and adherens junctions, while others are unique to each type of adhesion site (Figure 5.3).

    An interesting aspect of the structure and function of the submembrane plaque, which physically links the adhesion receptors of the integrin or cadherin families to the actin cytoskeleton, is its association with various signaling molecules, either constitutively or transiently. While it is still common to distinguish between "structural" and "signaling" molecules at adhesion sites, it becomes increasingly difficult to justify such a distinction. For example, several molecules (e.g., focal adhesion kinase [FAK] and β-catenin) may be directly involved both in the mechanical interactions present at the adhesion sites and in the generation and transduction of long-range adhesion-mediated signals.

▶ *For Figure 5.4, see color insert.*

Transmembrane Interaction of the Extracellular Matrix with the Actin Cytoskeleton

Electron and immunofluorescence microscopy examination (Figure 5.3) reveals an abundance of actin microfilaments at the cytoplasmic side of focal contacts. Elucidation of the molecular structure of these sites is based on two major types of data. The first includes immunocytochemical (usually immunofluorescence) localization of the various proteins, and the second is based on direct biochemical analysis of the molecular interactions between the various plaque components and between them and the cytoskeleton. Such approaches provided the basis for "interaction maps" similar to the one shown in Figure 5.5, in which a network of interactions, indirectly linking the actin cytoskeleton to the ECM or to cadherin of an adjacent cell (see below), is presented. In this scheme, we may distinguish several molecular domains of the adhesion sites.

The external surface to which the membrane is attached at focal contacts and related matrix adhesions may contain several types of ECM molecules such as fibronectin, vitronectin, and collagens. This diversity is quite intriguing, since it is still not known whether matrix adhesions formed with molecularly different surfaces direct the assembly of molecularly and functionally different adhesion sites. Moreover, despite the extensive work invested in the characterization of focal contacts over the last several decades and numerous publications describing their molecular properties, a satisfactory definition of these adhesion sites (either structural or molecular) is not available yet.

The transmembrane domain of matrix adhesions consists of adhesion receptors, mainly different members of the integrin superfamily (Figure 5.1). As may be expected from the fact that these receptors can interact with different matrix molecules, this domain is also quite heterogeneous with respect to the integrin composition. Such differences may exist between different cells, each expressing a different set of integrins, or may even be detected between individual adhesion sites within single cells.

Another domain of focal contacts is the submembrane plaque, which harbors a multitude of proteins, some of which directly bind to the cytoplasmic faces of different integrins (i.e., talin, α-actinin, focal adhesion kinase [FAK], and probably tensin). Other proteins of the plaque, such as vinculin or zyxin, can bind to these components either directly or indirectly. Interestingly, several of the "anchor proteins" which reside in the submembrane plaque (named after their capacity to anchor actin filaments in the membrane) have multiple binding sites and may interact alternatively or simultaneously with several proteins. For example, one of the major components of the plaque, vinculin, was shown to interact with talin, paxillin, tensin, α-actinin, vasodilator-stimulated phosphoprotein (VASP), and actin. In addition, it can interact with other vinculin molecules or form intramolecular interactions between its "head" and "tail" domains (Figure 5.6). Zyxin, another plaque component, interacts with α-actinin and VASP. VASP, in turn, interacts with profilin, vinculin, zyxin, and actin (Figure 5.5). Various actin-binding proteins, like profilin, fimbrin, and members of the ezrin-radixin-moesin (ERM) family, mainly radixin, are enriched in adhesion plaques. The presence of such diverse binding proteins suggests that the submembrane plaque is indeed a tightly packed three-dimensional meshwork, which may be highly diversified in its molecular composition.

As pointed out above, some of the plaque proteins are signaling molecules, including serine/threonine- and tyrosine-specific protein kinases

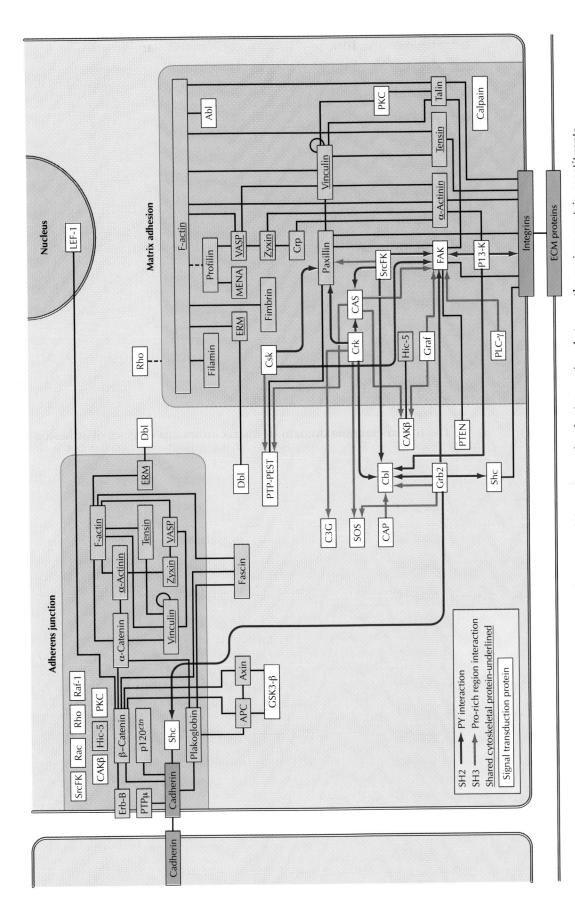

Figure 5.5 "Interaction map" showing some of the major molecular interactions between the various protein constituents of cell-matrix adhesions, like focal contacts, and cell-cell adherens-type junctions. Lines connecting protein boxes indicate biochemical interactions between the two proteins; SH2 interactions are marked by a red arrow pointing to the phosphotyrosine (PY)-containing target; SH3-mediated interactions are marked with a dotted gray line and a box pointing to the proline-rich target. The black print on a red-shaded background indicates unique components of either cell-matrix or cell-cell adhesions, while the underlined black print on a red-shaded background indicates proteins associated with both sites. The black print on a white background marks putative signal transduction proteins, including enzymes (mainly kinases), their substrates, and adapter proteins. Proteins reported to be physically associated with adhesion sites appear within the adhesion compartment. Effects on adhesion sites without a proven interaction are indicated by broken lines.

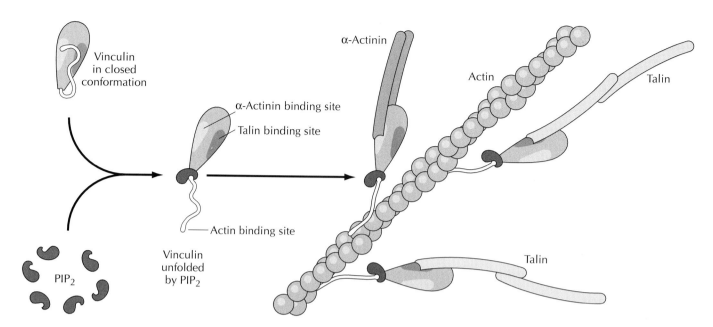

Figure 5.6 Conformational changes in vinculin which can regulate the formation of adhesion plaques. The vinculin molecule consists of a globular head domain, with the binding sites for talin and α-actinin, and a C-terminal tail domain, with the binding sites for actin and several other proteins. In closed conformation (left), the vinculin tail is attached to the head in such a way that the binding sites to talin, α-actinin, and actin are masked. The signaling molecule PIP$_2$ binds vinculin at its "neck" region and induces a transition from the closed conformation to an extended one, rendering the binding sites accessible (middle). Vinculin in the opened conformation can efficiently link actin to talin or to α-actinin (right), promoting the assembly of adhesion complexes.

and their substrates and adapter proteins. The involvement of these proteins in adhesion-dependent signaling is discussed below, but it is noteworthy that such molecules are involved not only in signaling events but also in the physical cross-linking of different components of the adhesion sites. For example, FAK is a multidomain protein that can interact with several focal contact proteins, including paxillin, c-Src, p130^{cas}, and Grb2 (Figure 5.5). While some of these interactions are apparently constitutive, others (for example, those mediated by Src homology 2 [SH2] domains on one protein and phosphotyrosine groups on a target protein) are regulated by specific phosphorylation and dephosphorylation events. It is interesting that despite the broad interest in tyrosine phosphorylation and the assembly of adhesions, it is still largely unclear whether such phosphorylation events occur locally at focal contacts and are involved in the recruitment of SH2 domain-containing proteins to the adhesion sites or whether such phosphorylation occurs, at least in part, outside the focal contact area and leads to the targeting of the phosphorylated molecules to the SH2 domain-containing plaque.

Another potential mechanism for regulating the assembly of the submembrane plaque involves conformational changes in different plaque proteins. One example of such a mechanism is displayed by vinculin, which apparently is folded "head-to-tail," forming a "closed" conformation in which its binding sites to several proteins (talin, α-actinin, actin, etc.) are hidden, while in the extended conformation these sites are exposed and available for binding (Figure 5.6). It was proposed that transition from the folded to the extended state can be induced by phosphatidylinositol-4,5-bisphosphate (PIP_2). This possibility is attractive since it may constitute an important mechanism for the cross talk between the signaling processes which stimulate PIP_2 formation and the assembly of cell adhesions. This property of conformational changes that lead to alterations in binding and functional properties is not unique to vinculin; other cytoskeletal and signaling molecules, such as ezrin and a Src, may also exhibit similar properties.

Finally, the assembly of cell adhesions may be modulated by regulated alterations in the cytoplasmic levels of various junction-associated proteins. Such modulation may be achieved either by controlled expression or by degradation. Thus, it was shown that changes in cell adhesion and cytoskeletal organization can have a profound effect on the synthesis of such proteins as vinculin, α-actinin, and actin. The mechanisms underlying this process are regulated at both the transcriptional and posttranscriptional levels. Another process which can have a marked effect on the levels of junctional proteins is proteolytic degradation. One of the recent and detailed documentations of such regulation is the degradation of β-catenin by the APC-ubiquitin proteasomal system. It has been shown that free β-catenin (i.e., β-catenin that is not associated with cadherin in the junction or with transcription factors in the nucleus) can bind to the tumor suppressor protein adenomatous polyposis coli (APC) and to members of the axin family, and is phosphorylated by glycogen synthase kinase 3β. This is followed by ubiquitination and degradation by the proteasome system. This process, which is controlled by the Wg/Wnt signaling pathway, plays a central role in regulating the signaling activity of β-catenin (Figure 5.7) (see below). Other examples of regulation of junctional molecules by controlled proteolysis have also been described. The level of FAK in some systems is regulated by degradation. This degradation is carried out by a calcium-activated neutral protease, calpain. Calpain was shown to localize in adhesion plaques and can specifically cleave components including talin and integrin, as well as FAK.

Signaling Components in Matrix Adhesions

The notion that adhesion to the ECM can trigger signaling events has long been recognized, since it was demonstrated that the spreading of cells on appropriate matrices is essential for cell growth, differentiation, macromolecular metabolism, and survival. However, an insight into the molecular basis for such adhesion-mediated signaling was obtained only recently, with the discovery that focal adhesions contain a multitude of signaling molecules including kinases, kinase substrates, and adapter proteins. Moreover, cell adhesion to the ECM was shown to trigger a cascade of signaling events that together convert changes in integrins and integrin-associated complexes into specific changes in gene expression and in the stability of target proteins.

The major signaling molecules detected in focal contacts are depicted in Figure 5.5. A major signal transduction component is the tyrosine kinase FAK, whose activation is an early event in adhesion-mediated signaling. FAK was reported to bind to the cytoplasmic domain of β-integrin and to other proteins localized at focal adhesions, including paxillin, talin, and p130^{cas}. Other protein-tyrosine kinases that are similar in their molecular

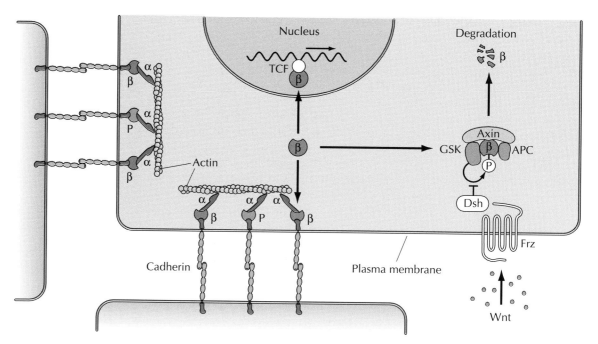

Figure 5.7 Alternative fates of β-catenin and its control by the Wg/Wnt signaling pathway. In adherens junctions, β-catenin is directly associated with the cytoplasmic domain of cadherin molecules, mediating their interaction with the actin cytoskeleton. In the cytoplasm, β-catenin can interact with the APC protein and axin and can be phosphorylated by GSK-3β. This phosphorylation marks β-catenin for degradation by the ubiquitin-proteasome system. In the nucleus, β-catenin associates with transcription factors of the TCF/LEF family, transactivating the expression of specific target genes (mostly still unknown). The degradation by the APC pathway can be regulated by wnt signaling.

structure to FAK were recently discovered; however, their role in cell adhesion is still unknown. The main function of FAK is suggested to be the transduction into the cell of signals generated at focal adhesions downstream. FAK binds Grb2 and most probably participates in the integrin-dependent activation of the Ras/mitogen-activated protein kinase (MAPK) signaling pathway. Interestingly, a recent study suggests that an activated (membrane-tethered) form of FAK can rescue adhesion-deprived epithelial cells from apoptosis, and when overexpressed, can even transform MDCK cells.

Protein kinases of the Src family constitute another group of kinases associated with focal adhesions. pp60[v-src] was the first protein tyrosine kinase localized in the residual focal adhesions of Rous sarcoma virus-transformed cells. The association of v-src with focal adhesions can induce hyperphosphorylation of resident proteins and, consequently, disrupt focal contact organization. The normal, proto-oncogenic counterpart of v-src, pp60[c-src], was later identified in association with adhesion plaques of normal fibroblasts. The association of c-Src with focal adhesions is apparently not constitutive, since its phosphorylation on tyrosine 527 by Csk, another focal adhesion-associated tyrosine kinase, causes c-Src dissociation from the plaque. It was also reported that c-Src is activated in an integrin-dependent manner after cell adhesion and that fibroblasts derived from c-Src[−/−] mice exhibit a reduced rate of spreading on fibronectin. In addition to c-Src, other members of the Src family such as Fyn can be associated with focal adhesions. Kinases of the FAK and Src families can interact with each other. FAK can undergo autophosphorylation at Tyr 397, enabling its binding to the SH2 domains of $pp60^{src}$ and $pp59^{fyn}$. In addition, Src can phosphorylate FAK, and this phosphorylation increases the kinase activity of FAK and enhances Grb2 binding and subsequent MAPK activation.

Another signaling/oncogenic protein displaying a complex interaction with focal adhesions is the Abl tyrosine kinase. The activated viral form of Abl ($pp160^{gag/v-abl}$) is localized at focal adhesions, similarly to v-*src*. Another activated form of Abl, the Bcr-Abl oncoprotein ($pp210^{bcr/abl}$), induces the formation of multimeric complexes containing paxillin, $p130^{cas}$, and talin in transformed cells. Recent studies show that cell adhesion to fibronectin triggers the transient recruitment of c-Abl from the nucleus to focal adhesions and activation of its tyrosine kinase activity.

Tyrosine phosphatases can also be associated with focal adhesions. One member of this family associated with focal adhesions is LAR, a transmembrane tyrosine phosphatase. In addition, the closely related receptor tyrosine phosphatases of the HPTP family were also suggested to be involved in cell-ECM interactions. These kinases and phosphatases regulate the phosphorylation levels of tyrosine on a number of focal contact components. A variety of tyrosine-phosphorylated proteins in the cell are localized at focal adhesions and cell-cell contact sites, as demonstrated by the staining of cells with fluorescent antiphosphotyrosine antibody, either without treatment or following a short inhibition of tyrosine dephosphorylation by vanadate. Upon adhesion of cells to the ECM, there is a group of proteins whose level of phosphorylation on tyrosine increases conspicuously. These include FAK, Src, $p130^{cas}$, paxillin (a target for FAK), and tyrosine kinases of the Src family. Paxillin was also shown to undergo strong serine phosphorylation following cell adhesion to the substrate.

In addition to tyrosine kinases and phosphatases, protein kinase C (PKC) has been localized in focal adhesions in association with talin and vinculin. In addition, by using the yeast two-hybrid system, a new serine/threonine kinase, ILK, that binds to the cytoplasmic domain of β-integrin and can induce anchorage-independent growth was discovered. Focal contacts were reported to contain phosphatidylinositol 3-kinase (PI3-K). The regulatory p85 subunit of this kinase was suggested to interact with phosphotyrosine residues on certain focal adhesion proteins containing SH2 domains. The serine/threonine kinase Akt (also known as PKB), a close relative of PKC, participates in the PI3-K-mediated signaling. PI3-K is required for integrin-stimulated Akt and Raf-1/MAPK pathway activation. Phospholipase C-γ was also localized in focal adhesions, and membranes enriched in focal adhesions were shown to contain its substrate, PIP_2.

The interactions of these signaling molecules with each other, or with other components of the submembrane plaque, are mediated by specialized motifs present on these molecules. These include SH2, which mediates binding to tyrosine-phosphorylated sequences on partner proteins. It was proposed that phosphorylation events may not only trigger long-range signaling but also play a central role in regulating the assembly and reorganization of cell-matrix adhesions (see below). Interactions of SH3 domains with polyproline motifs may also play a role in the formation of multimolecular junctional complexes (Figure 5.5). Profilin binds to a polyproline stretch of VASP, and VASP binds to the polyproline stretch of vinculin. Interactions of zyxin with cysteine-rich protein (CRP) and the localization of paxillin at focal adhesions depend on protein-protein interactions mediated by LIM zinc finger motifs.

Molecular Interactions in Adherens Junctions

The adhesion receptors in adherens junctions are cadherins. These transmembrane molecules belong to a multigene superfamily that consists of "classical cadherins" having a highly conserved cytoplasmic domain, desmosomal cadherins, T-cadherins lacking a cytoplasmic tail, and a novel family of cadherins expressed in neurons at the synaptic complex.

The immediate partners of cadherins in the submembrane junctional plaque are catenins, in particular β- and γ-catenin (plakoglobin), which bind to the same domain in the cadherin cytoplasmic tail, and p120^{ctn} catenin, which associates with the juxtamembrane region of cadherin (Figure 5.2). Additional components of the junctional plaque include α-catenin, vinculin, and α-actinin. Similar to what was described above for focal contacts, many of the junctional plaque components are capable of forming multiple interactions with other molecular residents of the junctions (Figure 5.5); for example, α-catenin, a vinculin homologue, can interact with β-catenin, plakoglobin, vinculin, α-actinin, and actin. Some of the molecular interactions present in these adhesion sites appear to be constitutive and depend mainly on the availability of the partner proteins. Other interactions, mainly those depending on phosphorylation events, i.e., interaction of cadherin with the adapter protein Shc, can be transient and subject to modulation by external stimulation or internal changes in cytoskeletal interactions.

Signaling from Cell-Cell Adherens Junctions

Similar to focal contacts, adherens junctions contain a battery of plaque proteins, some of which are essentially the same ones found in focal contacts (i.e., vinculin, α-actinin, tensin, and ERM family proteins) while others (α- and β-catenin, plakoglobin, and p120^{ctn}) are apparently specific for cell-cell adhesions. It is noteworthy that adherens junctions also contain signaling molecules, such as transmembrane receptor tyrosine kinases, including ErbB-1 (epidermal growth factor receptor) and ErbB-2 and receptor tyrosine phosphatases, as well as Src-family tyrosine kinases.

Another mode of signaling from adherens junctions involves the junctional molecules β-catenin and plakoglobin, which belong to a highly conserved family of proteins showing a central domain consisting of multiple "arm repeats" (named after the *Drosophila* homologue of β-catenin, armadillo). In adherens junctions, β-catenin and plakoglobin play an essential role in interconnecting the cadherin cytoplasmic domains, via α-catenin, to the actin cytoskeleton. In addition, β-catenin and its homologue in *Drosophila* armadillo play a major role in signal transduction by the Wg/Wnt pathway to regulate morphogenesis. When expressed in excess, or upon release from the membrane, β-catenin and plakoglobin can translocate into the nucleus, where, together with specific transcription factors of the LEF/TCF family, they activate the transcription of specific target genes (Figure 5.7). Uncontrolled regulation of gene expression by β-catenin is suggested to be a major cause of its oncogenic action in various tumors. The regulation of catenin-mediated signaling may occur at several different levels. Best studied is the regulation of β-catenin levels by its controlled degradation by the ubiquitin-proteasome system via the Wg/Wnt pathway.

Interestingly, cadherins may play a dual role in controlling the fate of β-catenin. On the one hand, they may sequester β-catenin by binding it to the plasma membrane (away from the nucleus), thus potentially suppressing the β-catenin-driven transactivation. Conversely, cadherin binding may protect β-catenin from degradation, leading to an increase in its levels. It remains to be shown how β-catenin-driven transactivation is regulated physiologically (for example, by a controlled "slow release" from the junctional membrane of transactivation-competent β-catenin).

Cross Talk between Adhesion-Dependent Signaling and the Rho Family GTPases

One of the most potent regulators of focal contact assembly is the small G-protein Rho. Rho, while not found associated constitutively with focal adhesions, is necessary for the formation and maintenance of focal adhesions in cells attached to the ECM and for adhesion-mediated signaling. In particular, integrin-dependent tyrosine phosphorylation of FAK and paxillin and activation of the MAPK Erk2 depend on Rho activity. Other members of the Rho family are also known to affect the membrane-associated actin filament system. For example, Rac is involved in the formation of the lamellar, actin-rich cell extensions known as ruffles and lamellipodia, while Cdc42 is required for the formation of finger-like filopodia or microspikes. The activation of the Rho family of G-proteins requires extracellular ligands, including serum factors and hormones such as lysophosphatidic acid, which is a major Rho activator in serum; platelet-derived growth factor and epidermal growth factor, which activate Rac; and bradykinin, which activates Cdc42.

More recently, it was shown that the activities of both Rac and Rho are also required for the assembly and maintenance of cell-cell adherens junctions and that, unlike in focal adhesions, both Rac and Rho are localized at cell-cell adhesion sites. It was also demonstrated that the selective inhibition of the endogenous Rho or Rac eliminates the assembly of cadherin complexes with cell-cell junctions whereas desmosomes are not perturbed by such conditions. Rac activity was shown to be sufficient for the recruitment of actin to the clustered cadherin receptors, while both Rac and Rho are apparently necessary for the formation of fully developed junctions.

Since Rho family proteins interact with a variety of target molecules, several suggestions were put forward to explain the possible role of downstream components in the Rho-dependent signaling that modulates the assembly of focal adhesions and the associated actin filament bundles. It was suggested that Rho is involved in the regulation of PIP_2 production by the activation of phosphatidylinositol-4-phosphate-5-kinase. Increased PIP_2 production could be involved in focal contact formation, since this signaling phospholipid is enriched in focal adhesions, and α-actinin and vinculin are PIP_2-binding proteins. It was further suggested that PIP_2 induces the conformational change in vinculin that exposes cryptic sites and thus promotes the interaction with actin, talin, and α-actinin (Figure 5.6). In addition, polyphosphoinositides, including PIP_2, can uncap actin filaments and promote actin polymerization in cells. Another Rho target that may promote actin polymerization is p140^{mDia}, a mammalian homolog of *Drosophila* diaphanous, a ligand for profilin. Profilin can supply actin monomers to the growing ends of actin filaments.

Another component of the Rho-mediated signaling that involves the actin cytoskeleton is the Rho-activated serine/threonine kinase (Rho kinase). It was shown that microinjection of Rho kinase into cells can mimic the effect of Rho activation on the formation of actin bundles and focal adhesions. Rho kinase phosphorylates (and activates) the regulatory light chain of myosin II and, together with Rho, inactivates myosin light-chain phosphatase by its phosphorylation. This, in turn, increases the level of phosphorylation of the myosin II regulatory light chain and stimulates cell contractility. In several cellular systems including smooth muscle, neural cells, fibroblasts, and macrophages, activation of Rho induces cell contraction. The way in which activation of cell contractility is related to adhesion-mediated signaling is discussed in the following section.

Rho-induced formation of focal adhesions is regulated by ERM, a special family of actin-binding proteins. Members of this family—ezrin, radixin, and moesin—are found in adherens junctions and focal adhesions (Figure 5.5), as well as in other structures formed by membrane-cytoskeleton associations, such as microvilli, lamellipodia, and the cleavage furrow. "Closed" and "open" conformations of these proteins have been described; in the open conformation, these proteins can bind to both actin filaments and the cytoplasmic domain of the transmembrane protein CD44, forming a direct link between the actin cytoskeleton and the membrane. The association of ezrin with other transmembrane proteins, for example, the cystic fibrosis transmembrane conductance regulator, is mediated by another protein, EBP50. Phosphorylation of ERM family proteins by Rho kinase promotes their transition to the "open" conformation and thereby regulates signal transduction involving these proteins.

Tension-Dependent Activation of Adhesion-Mediated Signaling

The above discussion suggests that adhesion-mediated signaling is driven primarily by the coassembly of various structural and signaling components at cell junctions. According to this view, the assembly of the junctional plaque brings the signaling molecules into sufficient proximity to allow their interaction and the induction of signaling events. Recent studies, however, suggest that adhesion to the ECM, per se, is insufficient to induce adhesion-mediated signaling. Thus, Rho activity, in addition to integrin-mediated interaction, is necessary for focal-adhesion formation and adhesion-dependent signaling. Rho activation, via Rho kinase, stimulates actomyosin contractility. It has been shown that an increase in cell contractility, induced by Rho activation, is critical for adhesion-dependent signaling and that inhibition of myosin II activity blocks the formation of focal adhesions and stress fibers that are normally induced after Rho activation in ECM-attached cells.

Another line of investigation indicated that disruption of the microtubule network can trigger integrin signaling by inducing cellular contractility. Microtubule disruption increases myosin II-driven contractility in various cell types, such as cultured fibroblasts and endothelial cells or *Xenopus* oocytes. Disruption of microtubules by specific drugs is sufficient to induce a rapid assembly of focal adhesions and stimulation of adhesion-dependent signaling. An increase in cell contractility is an indispensable step in the activation of adhesion-dependent signaling in this case also.

A plausible suggestion is that the maintenance of focal adhesions depends on the tension developed by the attached actomyosin system or that focal adhesions are "tensegrity" structures, preserving their integrity only under tension. It was further suggested that nascent actin-associated contacts may operate as tension-sensing "devices" converting mechanical signals into protein modification, such as tyrosine phosphorylation, and hence activating intracellular signaling pathways.

The importance of tension for triggering adhesion-dependent signal transduction is supported by recent findings where external forces were applied to cell-ECM adhesion sites by the stretching of an elastic substrate or the trapping of cell surface-attached beads covered with ligands to the adhesion receptors. This adhesion-associated tension was shown to affect the strength of adhesion, the organization of the cytoskeleton, the phosphorylation of focal adhesion proteins, and the progress of downstream signaling events. In addition, the plating of cells on flexible substrates suppressed tension development and inhibited the formation and tyrosine phosphorylation of focal adhesions. Collectively, these studies imply that a tension response at focal adhesions may play an important regulatory role in eukaryotic cells.

Nonconventional Interactions Mediated by Specific Adhesion Receptors

The adhesion machinery and associated signaling systems are predominantly responsible for cell adhesion to other cells or to ECM surfaces. In recent years, however, it was shown that integrins and cadherins may also mediate attachment to other ("nonconventional") surfaces. In particular, bacteria exploit cellular mechanisms of adhesion and signaling for entry into mammalian cells. Since this subject is discussed in detail in other chapters of this volume, it is only briefly addressed here. *Listeria monocytogenes*, for example, uses a surface protein named internalin to adhere to and then enter cells. The receptor for this bacterial protein was shown to be E-cadherin. The process of internalization, or induced phagocytosis (for normally nonphagocytic cells), depends on the cytoplasmic part of the E-cadherin molecule and can be blocked by inhibitors of tyrosine phosphorylation and by actin-depolymerizing drugs. Both the PI3-K and MAPK pathways are involved in this bacterial entry. Moreover, contact with bacteria was shown to stimulate the MAPK pathway and PI3-K activity. Stimulation of PI3-K was shown to depend on the interaction of another bacterial protein, internalin B, with an as yet unidentified cellular receptor. Various microbial pathogens bind to host cell integrin receptors. For example, *Yersinia* adheres to cells by using the adhesion receptor $\alpha_5\beta_1$-integrin, which interacts with the bacterial protein invasin. Internalization of *Yersinia* is largely similar to that of *Listeria* and is also sensitive to tyrosine kinase inhibitors and cytochalasin D. Attachment and endocytosis of BCG (bacillus Calmette-Guérin) by human bladder tumor cells is mediated by $\alpha_5\beta_1$-integrin and is dependent on fibronectin. Binding of vitronectin to gonococci (*Neisseria gonorrhoeae*) was recently shown to trigger bacterial internalization into epithelial cells. Blocking integrin function by antibodies specific to $\alpha_v\beta_5$- or $\alpha_v\beta_3$-integrin resulted in the abrogation of vitronectin-triggered bacterial internalization.

Shigella and *Salmonella* use a different method of entry into cells. These bacteria trigger the formation of numerous ruffles on the cell surface, engulfing the bacterium in a macropinocytic process. It was shown that IpaA, a *Shigella* protein secreted upon cell contact, rapidly associates with vinculin during bacterial invasion. An IpaA mutant defective for cell entry differs from wild-type *Shigella* in its ability to recruit vinculin. IpaA-vinculin interaction was suggested to initiate the formation of focal adhesion-like structures required for efficient invasion. The process of *Shigella* invasion is Rho dependent. The three Rho isoforms are recruited into bacterial entry sites with differential localizations relative to the membrane. A Rho-specific inhibitor abolished *Shigella*-induced membrane folding and impaired the entry of bacteria into cells. After entering epithelial cells, bacteria escape into the cytoplasm, move intracellularly, and pass from cell to cell. By using a sarcoma cell line that does not produce cell adhesion molecules and cell lines transfected with E- or N-cadherin, it was demonstrated that expression of a cadherin is required for the cell-to-cell spreading of *Shigella*.

Finally, enteropathogenic *Escherichia coli* induces epithelial-cell actin rearrangements, resulting in pedestal formation beneath adherent bacteria. This requires the secretion of specific proteins needed for signal transduction and adherence. One of these proteins, Hp90 (also known as Tir), is inserted into the host cell membrane. Hp90 is a receptor to which the bacterium adheres to trigger additional host signaling events and the nucleation of actin assembly. Hp90 interacts with intimin, an outer membrane protein of EPEC that induces tyrosine phosphorylation of Hp90. This is apparently essential for tight attachment of bacteria and pedestal formation. In this case, therefore, the bacterial pathogen inserts its own receptor into the mammalian cell surface.

Selected Readings

Ben-Ze'ev, A., and B. Geiger. 1998. Differential molecular interactions of β-catenin and plakoglobin in adhesion, signaling and cancer. *Curr. Opin. Cell Biol.* **10:**629–639.

Bershadsky, A., A. Chausovsky, E. Becker, A. Lyubimova, and B. Geiger. 1996. Involvement of microtubules in the control of adhesion-dependent signal transduction. *Curr. Biol.* **6:**1279–1289.

Chothia, C., and E. Y. Jones. 1997. The molecular structure of cell adhesion molecules. *Annu. Rev. Biochem.* **66:**823–862.

Daniel, J. M., and A. B. Reynolds. 1995. The tyrosine kinase substrate p120cas binds directly to E-cadherin but not to the adenomatous polyposis coli protein or alpha-catenin function. *Bioessays* **19:**883–891.

Gumbiner, B. M. 1996. Cell adhesion: the molecular basis of tissue architecture and morphogenesis. *Cell* **84:**345–357.

Humphries, M. J., and P. Newham. 1998. The structure of cell-adhesion molecules. *Trends Cell Biol.* **8:**78–83.

Matsui, T., M. Maeda, Y. Doi, S. Yonemura, M. Amano, K. Kaibuchi, and S. Tsukita. 1998. Rho-kinase phosphorylates COOH-terminal threonines of ezrin/radixin/moesin (ERM) proteins and regulates their head-to-tail association. *J. Cell Biol.* **140:**647–657.

Parsons, J. T., and S. J. Parsons. 1997. Src family protein tyrosine kinases: cooperating with growth factor and adhesion signaling pathways. *Curr. Opin. Cell Biol.* **9:**187–192.

Pelham, R. J., Jr., and Y. Wang. 1997. Cell locomotion and focal adhesions are regulated by substrate flexibility. *Proc. Natl. Acad. Sci. USA* **94:**13661–13665.

Takeichi, M. 1995. Morphogenetic roles of classic cadherins. *Curr. Opin. Cell Biol.* **7:**619–627.

Tapon, N., and A. Hall. 1997. Rho, Rac and Cdc42 GTPases regulate the organization of the actin cytoskeleton. *Curr. Opin. Cell Biol.* **9**:86–92.

Yamada, K. M., and B. Geiger. 1997. Molecular interactions in cell adhesion complexes. *Curr. Opin. Cell Biol.* **9**:76–85.

Yap, A. S., W. M. Brieher, and B. M. Gumbiner. 1997. Molecular and functional analysis of cadherin-based adherens junctions. *Annu. Rev. Cell Dev. Biol.* **13**:119–146.

6

Cell Adhesion Molecules and Bacterial Pathogens

GUY TRAN VAN NHIEU AND PHILIPPE J. SANSONETTI

The development of an infectious disease depends on the capacity of the causative microorganism to multiply within and to colonize specific host tissues. Pathogenic microorganisms do so by virtue of specific virulence factors interacting with host cell components. The nature and the specificity of these interactions determine the physiopathology of the disease. If the pathogen grows extracellularly, for example on the apical surface of an epithelium, a prerequisite for the infection will be the ability of the pathogen to adhere to epithelial cells. This property allows the microbe to colonize host tissues by resisting mechanical clearing mechanisms or by conferring a selective advantage to the pathogen over the endogenous host flora. Adhesion could be considered a superficial interaction allowing the pathogen to use host tissues as a nutrient-providing matrix. This is probably a reductionist view, however, since some adhesion mechanisms result from a sophisticated dialogue between the pathogen and the host cell. Also, even for pathogens growing extracellularly, the establishment of disease is dependent on the ability of the pathogen to control the host defense response. More elaborate schemes of pathogen-host cell interactions are provided by intracellular microorganisms. Invasive and intracellular pathogens show a profound adaptation to the host and subvert host cell functions starting from the initial steps of invasion. Entry into normally nonphagocytic cells can be used to multiply intracellularly or just as a means of crossing an epithelial layer and penetrating deeper within host tissues, where further interactions are established. Important partners of these interactions include cell adhesion molecules as primary targets of bacterial ligands.

This chapter will not give an extensive review of interactions between pathogens and cell adhesion molecules described in the literature. The reason for this is that even though considerable progress has been made in recent years, many of these studies are still in their infancy. Also, the identification of the respective partners of an interaction between a pathogen and host cells is only the first step in dissecting the molecular mechanisms involved. In most instances, especially when an interaction is not limited to simple microbial adhesion to the host cell surface, little is known about the roles performed by each of the molecular partners. Rather, this chapter focuses on a few examples of dialogues between bacteria and cell adhesion molecules during infection and in interactions involved in processes such as adhesion or invasion or in controlling the process of host inflammation.

Gaining Access to the Targets

Some pathogens can rapidly gain access to their host niche (Figure 6.1). This is the case for a variety of microbes which are injected into the bloodstream by a biting insect vector. For most bacterial pathogens, however, the initial stage of the infectious process starts with interaction with a host epithelium. Pathogens that multiply extracellularly on the surface of an epithelium have to establish the interactions required for binding to and colonizing the epithelium (Figure 6.1). It may therefore not be by chance that such interactions very often consist of association between bacterial lectins and carbohydrate moieties of host cell surface glycoproteins. Such carbohydrate moieties are readily accessible to pathogens on the apical surface of the epithelium; they may be largely represented on glycoproteins which are expressed on the cell apical surface, or they may constitute part of the mucin layer coating the epithelium.

Intracellular pathogens, on the other hand, are confronted with the problem of invading or crossing the epithelium, which also means finding their specific receptors to induce the uptake process. A normal epithelium consists of one or several layers of cells that are tightly sealed by different junctional structures. These cell junctions are involved in maintaining the polarization between the apical and the basolateral surfaces of the cell,

Figure 6.1 Interactions of invading pathogens with an intestinal epithelium. The open arrows show potential interactions between the pathogen and cell adhesion molecules on the surface of various cell types during the invasion process. Bacteria that adhere to the apical surface of the epithelium often do so by establishing interactions with carbohydrate moieties or by a bridging mechanism (see the text). Invasive pathogens may enter via specialized cells, such as M cells present in the intestinal epithelium, and have elicited various ways to survive ingestion by professional phagocytes. Invasive pathogens that bind to cell receptors that are basolaterally distributed (open boxes) may then enter cells via the basolateral side or, after receptor redistribution during trauma or polymorphonuclear leukocyte (PMN) transmigration, via the apical side.

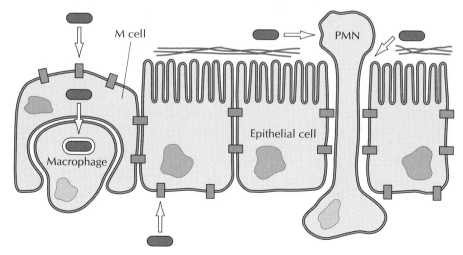

which are distinct in terms of their integral membrane component contents. Bacterial engagement of the proper receptor may therefore be a limiting factor for receptors such as cadherins or integrins, which are preferentially expressed basolaterally in a polarized epithelium. In these latter cases, internalization of the invasive bacterium by polarized epithelial cells would require receptor recruitment to the apical cell surface. Such receptor redistribution from the basolateral surface to surfaces exposed apically has been reported for integrins following a mechanical trauma or during leukocyte transmigration across the epithelium (Figure 6.1). It is therefore conceivable that some factors, such as local inflammation, might paradoxically favor infection.

For a healthy epithelium, however, these events probably do not account for efficient infection, and invasive pathogens that use cell receptors expressed on the basolateral side must use other routes of entry for successful infection. Such routes may consist of specific structures of the epithelium. A mucosal epithelium not only acts as a barrier against microorganisms; it also needs to allow exchange with the lumen. These exchanges occur at the level of specialized structures that may also represent an Achilles heel in terms of defense against pathogens. For the intestinal epithelium, for instance, sampling of luminal contents takes place at the level of M cells (Figure 6.1). M cells are specialized absorbtive cells with a poorly organized brush border and overlying solitary lymph nodes as well as lymphoid follicles or Peyer's patches: they play a role in the presentation of antigens sampled from the lumen to the immune system. Although markers for M cells have not been characterized in detail, it is likely that potential receptors for pathogens are accessible on the apical surface of M cells. This would explain the preferential invasion of enteropathogens such as *Yersinia, Listeria, Shigella,* and some *Salmonella* species via M cells rather than via neighboring enterocytes. In order not to be killed following invasion of M cells, intracellular pathogens also need to devise schemes to survive the encounter with host defense cells, in particular macrophages. For *Shigella,* this encounter results in proinflammatory cytokine production, which attracts monocytes and polymorphonuclear cells to the site of infection. This influx of inflammatory cells results in destabilization of the epithelium and favors further invasion of the enterocytes from the lumen of the colon.

Adhesion Molecules as Pathogen Receptors: Defining a Pathogen Strategy

Adhesion molecules are cell surface receptors that establish cell-cell interactions or interactions between cells and the extracellular matrix (ECM) (Box 6.1). They are involved in a wide variety of processes such as embryogenesis, cell growth, and differentiation. Adhesion molecules are classically divided into four main groups: the integrins, the cadherins, the immunoglobulin (Ig) superfamily, and the selectins (Figure 6.2). Other surface receptors such as the proteoglycans can also act as adhesion molecules, but their functions have been less well studied.

Probably the most extensively studied adhesion molecules are integrins, which are ubiquitous receptors involved in processes such as cell adhesion to the extracellular matrix and cell-cell interaction; in vivo studies have implicated them in various functions such as cell migration, wound healing, and embryogenesis. Integrins consist of heterodimers

Cell adhesion molecules and bacterial pathogens

Adhesion molecules participate in various fundamental processes such as cell adhesion, cell migration, and cell differentiation. They consist of various families of receptors that allow the cell to sense and to respond to its environment, and engagement of these receptors can result in signaling that leads to short-term cell responses such as cytoskeletal reorganization or to long-term responses such as de novo gene transcription and cytokine induction. Cell adhesion molecules are used by some pathogenic microorganisms to attach to host cells or to invade normally nonphagocytic cells, and these early interactions determine the physiopathology associated with these pathogens. During these very initial phases of host-pathogen interaction, diversion of host cell processes can occur, for example, by induction of a phagocytic process by an invasive organism, but it may also translate into signaling via cell adhesion molecules to neighboring tissue in attempts by the pathogen to control the inflammatory response. In many instances, the cell responses involved do not appear to have physiological equivalents, and the study of these interactions is likely to shed new light on the function of these receptors.

bearing an α chain and a β chain, several α chains can associate with a given β chain, and a specific combination between an α subunit and a β subunit confers ligand specificity. For instance, $\alpha_5\beta_1$-integrin is a fibronectin receptor whereas $\alpha_6\beta_1$-integrin preferentially binds laminin. This specificity, however, is relative, since a given integrin heterodimer can often bind several ligands and can show overlapping function with other integrins. The original classification of integrins based on the nature of the β chain probably needs to be revised because of the association of several β chains with the same α chain. This structural classification, however, appears to reflect functional requirements; β_1-integrins are usually receptors for the ECM, whereas β_2-integrins are expressed on the surface of leukocytes and participate in cell-cell interactions during inflammatory processes.

The cytoplasmic domain of the integrin β_1 chain associates with the actin cytoskeleton via actin-binding protein such as talin or α-actinin, thereby allowing anchorage of the cell to the ECM at the level of focal adhesions. The formation of these β_1-integrin clusters localizing with F-actin rich structures is subject to complex regulation involving the recruit-

Figure 6.2 Cell adhesion molecules. β_1-Integrins bind the ECM via their extracellular domain and associate with the actin cytoskeleton via the cytoplasmic tail of the β_1 subunit. Cadherins establish homotypic interactions at cell-cell junctions and also associate with the cytoskeleton via their cytoplasmic tail. Members of the Ig superfamily bind to a counterreceptor such as the LFA-1 ($\alpha_L\beta_2$)-integrin or VCAM-1 on the surface of neutrophils and monocytes. Selectins bind to carbohydrate residues on the surface of leukocytes.

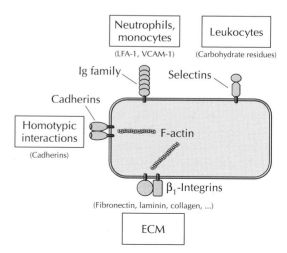

ment of actin-binding proteins and tyrosine phosphorylation of focal adhesion substrates. Src kinases and the small G-protein RhoA have also been implicated as upstream regulators of focal adhesions, although the molecular mechanisms by which the regulation occurs remain to be characterized. Integrin function can be regulated by cytosolic signals (inside-out signaling) which modulate the affinity of integrins for ligand. The interaction between fibrinogen and the platelet integrin GpIIb/IIIa (αIIbβ3) is an example of such a regulation; stimulation of platelets by thrombin or collagen during blood clotting results in a cascade of cytosolic signals leading to activation of the GpIIb/IIIa integrin. This activation leads to a change of conformation of the receptor with a much higher affinity for fibrinogen. Also, various signals that determine processes such as apoptosis or cell proliferation can be transmitted after ligand binding to integrins (outside-in signaling).

The cadherins are involved in cell-cell interactions at the level of intercellular junctions, where they establish calcium-dependent homotypic interactions with cadherins of the adjacent cell. The classical (P, F, and N) cadherins consist of homodimers present at the adherens junctions of polarized cells. These molecules can interact with the cytoskeleton via binding of their cytoplasmic tail to catenins. Formation of cadherin junctions appears to be negatively regulated by tyrosine phosphokinases, whereas activation of the small G proteins Rho and Rac is required for adherens junction formation. Formation of intercellular junctions appears to influence focal adhesions and cell adhesion to the ECM, perhaps by regulating Src kinase activity and by intersecting pathways involved in growth-factor receptors signaling.

The Ig superfamily of adhesion molecules is mostly involved in cell-cell interactions and consists of receptors with extracellular domains having homology to Igs. These molecules are involved in homotypic interactions such as NCAM-NCAM binding in nerve cells, but they also establish heterotypic interactions, in particular during inflammatory processes. For example, the interaction of intercellular cell adhesion molecule 1 (ICAM-1) with the LFA-1 integrin ($\alpha_L\beta_2$) on the surface of lymphocytes is critical for the antigen-dependent activation of T cells, whereas ICAM-1/Mac1($\alpha_M\beta_2$) and VCAM-1/$\alpha_4\beta_1$ interactions on the surface of endothelial cells and leukocytes participate in the attachment and extravasation of polymorphonuclear cells through blood vessels during inflammation. Besides their role in inflammation, members of the Ig superfamily can also regulate growth factor responses. For example, association of ICAM-1 with the fibroblast growth factor receptor modulates the tyrosine kinase activity of this receptor.

Finally, the selectins are a family of lectin receptors that interact with carbohydrate residues on the surface of endothelial cells and leukocytes. As opposed to the families of adhesion molecules described above and as inferred from knockout studies with mice, the function of selectins appears to be limited to the vascular system. Therefore, the main role of selectins appears to be the mediation of the attachment and the rolling of leukocytes on endothelial tissues during inflammatory processes, although they may also participate in lymphocyte homing in Peyer's patches.

As shown in Table 6.1, interactions between pathogens and the different families of adhesion molecules have been reported. Although this list is far from being exhaustive, we have tentatively regrouped these examples into three main categories of interactions, promoting (i) bacterial adhesion and/

Table 6.1   Cell adhesion molecules as receptors for bacterial pathogens

Pathogen	Ligand	Receptor
Adhesion/low levels of invasion		
Direct interaction		
Gram-negative bacteria	Pili, fimbriae	Carbohydrate residues (ECM, integrins, Ig superfamily)
Bordetella pertussis	FHA	CR3
	FimD	RGD binding integrins
	Pertussis toxin	P- and E-selectins
Neisseria meningitidis	Capsule/polysialic acid	NCAM
Neisseria gonorrhoeae	Opa	Proteoglycans, CD66
Bridging mechanism		
Staphylococcus	LTA, FnBP	Fibronectin
Streptococcus	LTA, M protein, FnBP	Fibronectin
Yersinia	YadA	Fibronectin, collagen
Neisseria	Opc	Vitronectin, fibronectin
Mycobacterium		Fibronectin, laminin
Invasion		
Listeria	Internalin	E-cadherin
Shigella	IpaB, IpaC	$\alpha_5\beta_1$
Salmonella	?	?
Yersinia	Invasin	$\alpha_3, \alpha_4, \alpha_5, \alpha_6, \alpha_v\beta_1$
Uptake by phagocytic cells		
Legionella	Momp	C3b
Mycobacterium	?	C3b
Neisseria gonorrhoeae	Opa	CD66

or low levels of internalization in host cells, (ii) efficient bacterial internalization by host cells, and (iii) bacterial interaction with adhesion molecules of phagocytic cells. The distinction between the interactions leading to low or efficient levels of internalization by normally nonphagocytic cells may seem arbitrary, but as discussed below, it is based on the assumption that interactions leading to adhesion often consist of lectin-carbohydrate associations or indirect associations of the bacteria to β_1-integrins via components of the ECM. On the other hand, integrin or cadherin receptors can also allow the uptake of bacterial pathogens after direct interaction with a bacterial surface ligand. This latter situation may reflect a requirement for deeper subversion of the adhesion molecules function, since none of these receptors have been implicated in phagocytic processes. It also suggests a role for the cell cytoskeleton during bacterial uptake, since the cytoplasmic moiety of integrins and cadherins binds to actin filaments; this role, however, is not defined and may be indirect. Also, some bacterial mechanisms of entry into normally nonphagocytic cells, such as those used by *Chlamydia* or *Campylobacter* species, are insensitive to cytochalasin, an inhibitor of F-actin, and do not require the actin cytoskeleton. The molecular basis of pathogen-host cell interaction leading to entry has not been characterized in these latter cases.

Pathogen Adhesion as a Means of Host Tissue Tropism

Direct Interaction

A widespread feature of pathogens is their ability to directly bind cell surface receptors via adhesive structures (Table 6.1). These structures can consist of pili (fimbriae), which are filamentous multimeric structures on the surface of gram-negative bacteria or surface proteins in the case of grampositive bacteria. The vast majority of adhesins expressed at the tips of fimbriae act as lectins and recognize sugar moieties of cell adhesion glycoproteins. This is true for type 1 fimbriae, expressed by various pathogenic strains of *Escherichia coli*, which preferentially recognize mannose residues on the surface of the ECM as well as on adhesion molecules. Since these residues are found on numerous glycoproteins, these types of interactions are not usually considered to be specific for a particular receptor. Interestingly, they appear to determine the tropism of a microorganism for a specific host tissue, due to the preferential expression of certain carbohydrate moieties in some tissues and the differential specificity of the bacterial adhesins.

Carbohydrate-lectin interactions can also mediate microbial attachment via cell surface proteoglycans. Proteoglycans, which are ubiquitous on the surface of many cell types, are membrane glycoproteins with two major types of glycosaminoglycan chains, heparan sulfate or chondroitin sulfate. Although their precise physiological role remains largely undefined, they have been implicated in cell-cell interactions as well as in adhesion to the extracellular matrix. Sugar inhibition studies show that various bacterial species adhere to cells via proteoglycans. These species include the spirochete *Borrelia burgdorferi*, the agent of Lyme disease, and *Helicobacter pylori*, the agent of peptic ulcers. The Opa proteins on the surface of *Neisseria gonorrhoeae* have also been reported to bind heparan sulfate proteoglycans, but it is not clear if these interactions are by themselves responsible for subsequent internalization of the bacterium. In fact, it is possible that interactions of proteoglycans with bacterial ligands participate in the initial stages of adhesion and that other types of interactions are involved either in stabilization of the bacterial adhesion processes or in invasion. For example, one of the *Neisseria* Opa proteins also binds to CD66 carcinoembryonic antigens, which are members of the Ig superfamily, and this interaction mediates bacterial internalization into host cells. This may represent a novel example of bacterial internalization via a noncadherin or integrin adhesion molecule.

Adherence via ECM Proteins: Bridging Mechanism

Several pathogens have the ability to bind to proteins of the ECM. Often, bacteria have devised several means to achieve this. This is true for grampositive bacteria, such as *Staphylococcus* or *Streptococcus* species, which can bind to fibronectin or collagen via specific fibronectin-binding proteins expressed at the bacterial surface as well as via their lipotechoic acid. It is also true for some gram-negative bacteria such as *Pseudomonas* or *Yersinia* species. The YadA protein of *Yersinia* binds fibronectin and collagen. This interaction appears to be important for bacterial attachment to insoluble forms of ECM proteins found on the surface of an epithelial cell. Beside allowing the attachment of the pathogen to the matrix, binding to matrix proteins can also promote cell adhesion by a bridging mechanism in which the pathogen interacts with the ECM protein and in turn associates with receptors for ECM protein such as integrins. For *Mycobacterium leprae,* the agent of leprosy, it has been suggested that such a mechanism is involved in neural tropism. Although the bacterial ligand has yet to be identified, it appears that *M. leprae* specifically associates with merosin, a laminin variant found around Schwann cell axons. This interaction then may allow bacterial association with a laminin receptor present on the surface of Schwann cells. Both *Yersinia* and *Mycobacterium* have the ability to invade normally non-phagocytic cells, but it is not clear if this type of bridging mechanism can result in internalization. It has been reported that fibronectin-coated particles can be ingested by cultured epithelial cells, presumably via integrin receptors, but in this case uptake does not appear to be very efficient. It is believed that this type of bridging association with cells generally results in extracellular localization and that bacterial invasion usually results either from mechanisms which superimpose on such adherence mechanisms or from independent specific mechanisms.

EPEC Adhesion: an Alternate Adhesion Mechanism

Some bacterial pathogens may not need cell surface receptors to adhere to host cells. This is the case for enteropathogenic *Escherichia coli* (EPEC) strains, which have developed an original way to adhere intimately to host cells. EPEC strains, commonly responsible for diarrhea in children, have the ability to adhere to intestinal epithelial cells. Adhesion of EPEC to cultured cells is characterized by effacing lesions, corresponding to the disruption of microvilli and followed by the formation of a pedestal-like structure at the site of bacterial interaction with the host cell. Initial adherence of EPEC to cells is mediated by BFP, a type IV pilus. This first step is followed by an intimate adherence of the microorganism to the cell surface, which depends on the activity of a specialized bacterial secretory apparatus induced upon cell contact (type III secretion).

As discussed below, this type of secretory apparatus is widely represented in various pathogenic species and appears to allow the translocation of bacterial effectors into the cell cytosol upon cell contact, although these effectors may perform various functions once in the cell cytosol. For EPEC strains, bacterial effectors that are translocated via this secretory apparatus contribute to the formation of a self-designed bacterial receptor. Two of these effectors, EspA and EspB, are required for activation of cell signaling pathways and tyrosyl phosphorylation of a third (90-kDa) bacterial protein called Tir. The Tir protein inserts into the cell membrane and acts as a receptor for intimin, a bacterial adhesin responsible for intimate adhesion. The role of the tyrosyl phosphorylation is still unclear, but it may allow the recruitment of cell components involved in the reorganization of the cytoskeleton and actin polymerization in the pedestal structure.

Intracellular Pathogens and Adhesion Molecules: Surviving the Macrophage or Breaching an Epithelium

The first line of defense that a pathogen has to face during host tissue invasion usually consists of polymorphonuclear neutrophils or resident macrophages. Once phagocytosed by these cells, invading microorganisms are killed by fusion of the bacterial phagosome with lysosomes containing hydrolytic enzymes and toxic products and also by the generation of toxic oxygen radicals during the respiratory burst. Intracellular microbial pathogens therefore need to elicit ways to survive the encounter with these cells, and some do this so successfully that they even grow within macrophages. Examples include *Mycobacterium* and *Legionella* species, which inhibit phagolysosome fusion and multiply intracellularly within specific "replicative phagosomes."

It remains unclear to what extent the pathway of uptake is important for the fate of the bacterial vacuole, but it is clearly established that it can modulate the generation of the oxidative burst. An invasion mechanism via specific adhesion molecules on the surface of professional phagocytes may represent a way for the pathogen to avoid being killed by oxygen radicals. For example, *N. gonorrhoeae* internalization by PMNs does not appear to be accompanied by the strong oxidative response that is usually associated with phagocytosis of opsonized particles. This particular feature may result from the specific Opa-mediated internalization via CD66 antigens. Similarly, many other intracellular pathogens such as *Legionella* and *Mycobacterium* species have the ability to bind the complement component C3 and are internalized by monocytes via the complement receptor CR3 ($\alpha_M\beta_2$). This may also be a means of avoiding the generation of a strong oxidative response, which appears to require both CR3 activation and Fc receptor ligation during opsonized phagocytosis.

Other pathogens have the ability to enter normally nonphagocytic cells. This property allows them to multiply within a local niche or to breach an epithelium to gain access to deeper tissues. The invasin-mediated uptake of *Yersinia* is probably among the best-characterized bacterial systems for invasion of epithelial cells. Enteropathogenic *Yersinia* species such as *Y. enterocolitica* and *Y. pseudotuberculosis* are responsible for enteric diseases after ingestion of contaminated food by the host. *Yersinia* can cross the epithelium at the level of M cells of the terminal ileum to reach the lamina propria, where it can multiply. *Yersinia* multiplies mostly extracellularly during these later stages of infection, but the invasin protein is believed to be critical for efficient bacterial internalization by M cells and for crossing of the epithelium during the initial steps of the infectious process.

Invasin is a bacterial surface protein that binds at least five members of the β_1-integrin family. The protein by itself is sufficient to promote cell internalization, since inert particles coated with invasin are readily internalized by cultured epithelial cells. These receptors include the fibronectin receptor $\alpha_5\beta_1$, but as opposed to other bacteria which bind the same receptor after associating with fibronectin (*S. aureus, Streptococcus* [see "Adherence via ECM proteins: bridging mechanism," above]), binding of invasin to β_1-integrin is followed by the internalization of the invasin-coated particle. Competition experiments suggested that the difference between internalization or uptake after binding to β_1-integrins is not due to binding of invasin or fibronectin to different domains on the receptor. Interestingly, mutational analysis of the invasin protein indicates that an aspartate residue at position 911 of the protein is critical for binding to β_1-integrins, perhaps by playing a similar role to the aspartate residue in the RGD sequence found in numerous ECM ligands and involved in direct interaction with β_1-integrins. An important factor in determining the fate of the bound particle after integrin binding is the affinity of the ligand-receptor interaction. Indeed, invasin was shown to bind to $\alpha_5\beta_1$-integrin, with a dissociation constant 2 orders of magnitude lower than that of fibronectin for the same receptor. It is believed that this strong affinity allows recruitment of β_1-integrin on the bacterial surface, which results in uptake of the bacterium in a zipper-like process. The cellular processes involved in bacterial uptake via β_1-integrins are not well understood. Although these processes involve host cell tyrosine kinase activity, as well as actin polymerization, it appears that signals transmitted via β_1-integrins during this process differ from classical signals leading to focal adhesion assembly, and bacterial internalization may have signals in common with receptor-mediated endocytosis involving clathrin.

Listeria monocytogenes, an enteroinvasive gram-positive bacterium responsible for meningitis and fetal systemic infection, provides another example of a single bacterial surface-ligand promoting entry into nonphagocytic cells. Like many invasive pathogens, *L. monocytogenes* has elicited several pathways to promote its uptake by epithelial cells involving a family of related bacterial proteins, the internalins. One of these products, the InlA protein (internalin), mediates internalization of the bacterium by binding to E-cadherin. As observed for *Yersinia* invasin binding to integrins, *L. monocytogenes* internalization occurs in a "zippering" process, with the formation of a tight phagosome surrounding the internalized bacterium (Figure 6.3A). The mechanism of internalin-mediated entry also involves tyrosine phosphorylation and actin polymerization. The cascade of events leading to internalization is not well understood but involves the activation of the phosphatidylinositol 3-kinase (PI3-K) p85/p110.

For both invasin and internalin, the entry process appears to be driven by incremental interactions between bacterial surface ligands and cell adhesion molecules. These interactions result in receptor recruitment and clustering at the bacterium-cell interface and transduction of signals controlling the cytoskeletal reorganization required for the completion of the uptake process. As discussed below, there may be other instances where adhesion molecules are less directly involved in invasive pathogen internalization by host cells. Rather, these molecules may allow the formation of cellular structures required for bacterial entry.

Signaling via Adhesion Molecules and Pathogens

Focal Adhesions and Pathogen Entry

Some pathogens have elicited ways to bypass classical activation via surface receptors and can potentially interfere directly with the cell machinery regulating the function of adhesion molecules. For example, the formation of focal adhesions after integrin binding to the matrix involves a reorganization of the cell cytoskeleton with the recruitment of several components at the level of these structures. Focal adhesion components include actin-binding proteins as well as substrates for the Src tyrosine kinase. It has been proposed that a hierarchy of events occurs during the formation of focal adhesions; binding of β_1-integrins to the ECM induces receptor clustering. The focal adhesion tyrosine kinase and tensin rapidly associate with the cytoplasmic domain of β_1-integrins and are specifically tyrosyl phosphorylated. This initial response is then followed by a massive ligand occupancy-dependent recruitment of actin-binding proteins such as vinculin, talin, and α-actinin, followed by a third wave of recruited components such as other Src substrates (e.g., cortactin and p120) and Src kinases requiring actin polymerization and tyrosine kinase activity. The process of integrin clustering appears to be dependent on activation of the small GTPase Rho. Consistent with this, introduction into the cell of a constitutively activated form of Rho results in increased focal adhesions and stress fiber formation; the molecular intermediates involved in Rho activation of

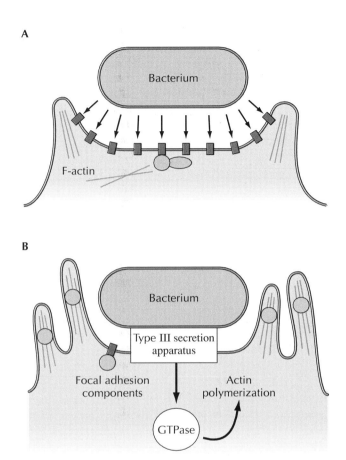

Figure 6.3 Cellular internalization of invasive bacteria by a "zippering" (A) or "macropinocytic-like" (B) process. **(A)** Internalization of *L. monocytogenes* or *Y. pseudotuberculosis* results in bacterium surrounded by a tight phagosome. For *Y. pseudotuberculosis,* high-affinity binding of invasin to integrins is critical to drive the phagocytic process. **(B)** For *Salmonella* or *Shigella* entry, it is believed that bacterial effectors translocate via a type III secretion apparatus in the cell cytosol. These bacterial effectors may activate small G proteins regulating the actin cytoskeleton and inducing membrane extensions.

focal adhesion formation are not well characterized. Rho could stimulate focal-complex formation by inducing cytoskeletal structures necessary for the translocation of Src kinase to the membrane or by directly activating lipid kinases or serine/threonine kinases modulating the recruitment of focal adhesion components.

Interestingly, the enteroinvasive pathogens *Salmonella* and *Shigella* induce an important reorganization of the actin cytoskeleton at the site of bacterium-cell contact during invasion of epithelial cells. The formation of this invasion structure by *Shigella* implies the formation of a focal adhesion-like structure with recruitment of several focal adhesion components as well as Src and activation of Rho. The precise role of the bacterial effectors participating in such a structure during the entry process is still unclear; it has been reported that the *Shigella* IpaB and IpaC proteins can associate with the $\alpha_5\beta_1$-integrins, but this association does not promote adhesion of the bacterium on the cell surface. This may be because the majority of the *Shigella* Ipa proteins are not found on the bacterial surface but are secreted into the extracellular medium. In fact, there is evidence that bacterial effectors may act from within the cell, thus bypassing classical activation schemes via cell surface receptors. Several pathogens, including *Shigella*, express a cell contact-dependent secretion apparatus, which in some instances allows them to translocate bacterial effectors into the cell cytosol. For instance, the *Yersinia* Yop proteins are translocated in the macrophage cytosol from the extracellularly bound bacterium, preventing phagocytosis by the macrophage. Similarly, upon cell contact, *Salmonella* is also able to translocate the SipB and SipC proteins within epithelial cells. In this latter case, however, bacterial effectors stimulate bacterial uptake presumably after activation of the small GTPase Cdc42. For *Shigella*, it appears that Rho activation is required for bacterial uptake. These findings are surprising since the cellular extensions at the levels of the invasion structure induced by *Shigella* are more reminiscent of rufflings induced by growth factors (Figure 6.3B). This paradoxical situation may be explained by different levels of modulation of the cell response by various *Shigella* effectors. Some of these effectors, i.e., IpaB and IpaC (homologs of SipB and SipC), are responsible for the initial events leading to actin polymerization, whereas another *Shigella* invasin, the IpaA protein, interacts with vinculin and would be responsible for the formation of a focal adhesion-like structure during the entry process.

Up-Regulation of Adhesion Molecules

Up-Regulation as a Means of Enhancement of Bacterial Attachment or Uptake

Bordetella pertussis, the causative agent of whooping cough, binds to ciliated respiratory cells and leukocytes. This property is a prerequisite for *B. pertussis* infection, because it allows for colonization and destruction of the lung epithelium by the microorganism. Different bacterial factors are involved in *B. pertussis* adhesion to cells. These factors include the FimD fimbrial protein, which binds β_1-integrins; pertussis toxin, which binds carbohydrate residues on cell surface glycoproteins; and filamentous hemagglutinin (FHA), which binds the CR3 integrin. These different factors appear to act in a cooperative fashion to promote cell attachment. For instance, binding of the pertussis toxin to leukocytes appears to result in up-regulation of CR3, reinforcing bacterial attachment via FHA-CR3 interactions (Figure 6.4A). The situation is, in fact, more complex, since FHA binding to CR3 by itself also amplifies CR3 up-regulation. Indeed, FHA bears the RGD tripeptide sequence that is found in numerous ECM proteins and that is involved in directly interacting integrins. It had been suggested that by analogy to ECM proteins, FHA binds CR3 via its RGD sequence. In fact, it appears that FHA binding to CR3 occurs via an FHA domain which does not contain the RGD sequence and that the RGD sequence is actually involved in directly interacting with the integrin-associated LRI-IAP complex, which has been implicated in activation of neutrophil adhesive functions. This direct interaction would result in cross-linking of the LRI-IAP complex and up-regulate CR3 binding to FHA. A similar mechanism of up-regulation of CR3 following cross-linking of the $\alpha_5\beta_1$-integrin by the *B. pertussis* FimD protein may also contribute to bacterial adhesion to monocytes. In the case of *B. pertussis*, such a receptor up-regulation contributes to strong attachment to the cell surface. In the case of *Mycobacterium avium*, however, cross-linking of the $\alpha_v\beta_3$-integrin via bacterial ligands induces up-

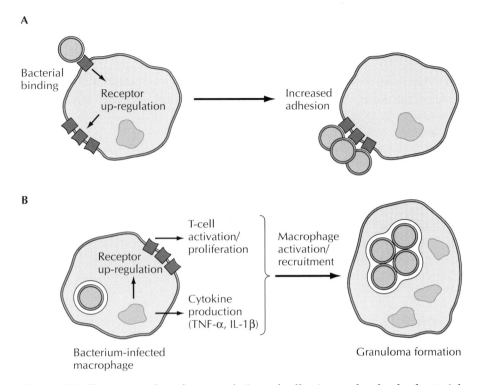

Figure 6.4   Two examples of up-regulation of adhesion molecules by bacterial pathogens. **(A)** Up-regulation of the CR3 molecule as a means of enhancing attachment of *B. pertussis* to leukocytes. Initial interaction of the bacterium (solid circle) with the cell leads to up-regulation of CR3 (hatched box), which strengthens bacterial adhesion via FHA-CR3 interactions. **(B)** Up-regulation of ICAM-1 (hatched box) on the surface of a granuloma during *M. tuberculosis* infection participates in activation of T cells and recruitment of macrophages that fuse to form mature granulomas.

regulation of CR3 and results in enhanced uptake via the CR3 pathway as well as enhanced bacterial attachment.

Up-Regulation as Part of the Physiopathological Schemes

Besides short-range dialogue with host cells, pathogens can also manipulate cell adhesion molecule expression in the course of the infection and influence cell-cell interactions. Perhaps the most important examples of such manipulations are seen during the host inflammatory response to the invading organism. Up-regulation of adhesion molecules such as selectins or members of the Ig family often occurs in endothelial tissues as a result of local inflammation induced at the site of infection. Modulation of expression of adhesion molecules at the cell surface can be induced by various bacterial components such as surface proteins, bacterial lipopolysaccharide, or secreted toxins and also via the release of cytokines by inflammatory cells. In some instances, the host inflammatory response and the changes in cell adhesive properties are used by the pathogen for the recruitment of specific host cells participating in the establishment of the disease. For example, *M. tuberculosis* is able to survive and replicate within macrophages. The infection process of *M. tuberculosis* correlates with the formation of granulomas, which consist of large multinucleated cells that probably correspond to the fusion of macrophages as well as T lymphocytes. The formation of these granulomas can be viewed as a host defense response to clear out invading microorganisms, but in cases where bacterial clearing is not successful, granulomas can become a niche where *M. tuberculosis* persists or multiplies. Production of several cytokines, including interleukin-1 and tumor necrosis factor alpha by inflammatory cells, and T-cell activation in response to antigen presented by accessory cells are essential for granuloma formation. Also, ICAM-1 is up-regulated in granulomas either indirectly in response to cytokine production or directly by stimulation by bacterial components (Figure 6.4B). Since the association of ICAM-1 with LFA-1 (the $\alpha_L\beta_2$-integrin) on the surface of lymphocytes determines T-cell activation, the up-regulation of ICAM-1 may contribute to recruitment of T cells, leading to the sustained inflammatory reaction induced by *Mycobacterium* infection and granuloma formation.

Similarly, up-regulation of ICAM-1 expression on endothelial cells by *Chlamydia pneumoniae* is thought to favor interactions between leukocyte and blood vessel endothelial cells which could account for local vascular inflammatory alterations linked to *C. pneumoniae* infection.

Conclusion

This chapter discusses a few examples of bacterial manipulation of cell adhesion molecules, or adhesion molecule functions, that illustrate the vast diversity of strategies used by pathogens during the infectious process. Bacterial adhesion to cell adhesion molecules is often the result of interactions between lectins and sugar moieties—a bridging mechanism by which the pathogen binds a protein from the ECM. Bacterial invasion, on the other hand, appears to result from high-affinity interaction with specific receptors linking the cytoskeleton or from a combination of an adhesion mechanism with specific invasion determinants. The function of bacterial interactions with adhesion molecules is not limited to adherence and invasion but can also play a role in the ability of the pathogen to control and subvert the host defense system. These examples reflect only a minute portion of what has yet to be learned from the perfect understanding that microorganisms have of the host physiology. As always, general rules are to be treated with caution, since the characterization of a particular interaction between an adhesion molecule and a pathogen also needs to be integrated into the general scheme of microbial infection of a particular host. Determining the precise relevance of such interactions—interactions that are most often characterized in in vitro systems—may prove challenging. Such issues depend mostly on the availability of animal models that faithfully reproduce the disease. A promising area of research, along with the molecular characterization of pathogen-host interactions, is the use of transgenic animals expressing heterologous adhesion molecules that are specific receptors for a given pathogen in animals that are innately insensitive to the pathogen. These types of studies should allow a better understanding of the role of a specific interaction in the development of the infection.

Selected Readings

Finlay, B. B., and P. Cossart. 1997. Exploitation of mammalian host cell functions by bacterial pathogens. *Science* **276:**718–725.

Isberg, R. R. 1994. Discrimination between intracellular uptake and surface adhesion of bacterial pathogens. *Science* **252:**934–938.

Ishibashi, Y., S. Claus, and D. A. Relman. 1994. *Bordetella pertussis* filamentous hemagglutinin interacts with a leukocyte signal transduction complex and stimulates bacterial adherence to CR3 (CD11b/CD18). *J. Exp. Med.* **180:**1225–1233.

Rambukkana, A., J. L. Salzer, P. D. Yurchenco, and E. I. Tuomanen. 1997. Neural targeting of *Mycobacterium leprae* mediated by the G domain of the laminin-α2 chain. *Cell* **88:**811–821.

Virji, M. 1996. Microbial utilization of human signalling molecules. *Microbiology* **142:**3319–3336.

7

Membrane Trafficking in the Endocytic Pathway of Eukaryotic Cells

Raluca Gagescu and Jean Gruenberg

Membrane Trafficking in Eukaryotic Cells

Eukaryotic cells need to be in constant communication with their environment to perform most of their functions, such as the transmission of neuronal, metabolic, and proliferative signals, the uptake of nutrients, and the ability to protect the organism from microbial invasion, to name only a few. During a process called endocytosis, cell surface receptors and their ligands, as well as particles or solutes present in the extracellular space, are taken up by vesicles which form at the plasma membrane and are delivered to early endosomes. Macromolecules destined for degradation are then targeted to late endosomes and finally to lysosomes, where they are digested by lysosomal hydrolases. The resulting metabolites are subsequently released into the cytoplasm, where they can be recycled by incorporation into newly synthesized macromolecules.

Cellular homeostasis is maintained by balancing degradation and cell division with biosynthesis (Figure 7.1). Proteins destined for secretion into the extracellular space as well as membrane proteins of the vacuolar apparatus are synthesized de novo on ribosomes and cotranslationally translocated into the lumen of the endoplasmic reticulum (ER). Once correctly folded by luminal chaperones, they are transported through the different stacks of the Golgi apparatus, where they acquire their mature sugar composition. Finally, upon arrival in the *trans*-Golgi network (TGN), they are sorted, packaged into specific vesicles, and forwarded to their final destination, endosomes or the plasma membrane. Fusion of TGN-derived vesicles with the plasma membrane occurs during a process called exocytosis. It results in either the insertion of transmembrane proteins into the plasma membrane or the secretion of soluble proteins into the extracellular space. Exocytosis can be either constitutive or regulated. All eukaryotic cells carry out constitutive exocytosis, but only a small subset of cells, which specialize in the secretion of hormones, neurotransmitters, or digestive enzymes, display a regulated secretory pathway.

A consequence of this subcellular organization is that the lumen of each organelle of the vacuolar apparatus is topologically equivalent to the extracellular space. Thus, endocytosed macromolecules can be selectively transferred from one organelle to the next without ever being in contact

113

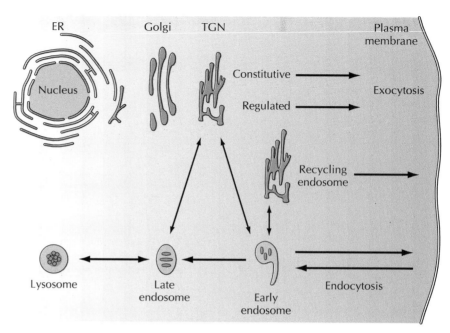

Figure 7.1 Intracellular compartments involved in endocytic and biosynthetic membrane trafficking. Newly synthesized molecules are transported from the ER through the Golgi apparatus to the plasma membrane. In the endocytic pathway, molecules are internalized at the plasma membrane and transported first to early endosomes, then to late endosomes, and finally to lysosomes. The endocytic and biosynthetic pathways are interconnected. Transport intermediates and most recycling routes are not depicted.

with the cytoplasm. Forward transport is balanced by recycling or retrieval pathways back to the donor organelle to replenish donor membranes with lipids and proteins, including components regulating transport itself. Box 7.1 describes two mechanisms by which intracellular transport may occur.

An unwanted communication with the extracellular space comes from the capacity that some parasites have evolved to mediate their own uptake by endocytosis. Some bacteria, such as *Listeria* and *Shigella*, induce their own internalization by the host cell and then escape from the endosomal system into the cytoplasm, where they multiply. Others, such as *Mycobacterium,* modify the dangerous environment of the endosome to make themselves a congenial home after being internalized.

Organelles and Membrane Dynamics

Endocytic and biosynthetic organelles exhibit a wide variety of shapes and structures, which can be easily visualized by classical electron microscopy (Figure 7.2). They range from the network organization of the ER, extending throughout the cytoplasm, to the pancake-like structure of the Golgi complex, and from the clusters of thin, long tubules (often several micrometers long) of recycling endosomes to late endosomes containing onion-like sheets of internal membranes, tubules, or vesicles. At present, little is known about the molecular mechanisms controlling organelle shape and biogenesis or the functional significance of such diversity. A snapshot image

Endocytic membrane transport: maturation or vesicular transport?

The last decade has witnessed an intense debate concerning two models for membrane trafficking in endocytosis. The discussion also applies to traffic through the Golgi apparatus, which is not addressed in this chapter. According to the vesicular transport model, internalized cargo molecules are packaged into transport vesicles that bud from preexisting early endosomes and fuse with preexisting late endosomes, thus transferring the contents from one organelle to the other. The maturation model, on the other hand, assumes that the cargo always remains in the same compartment, which is formed de novo by coales-

cence of vesicles originating from the plasma membrane. The resulting early endosome then undergoes functional changes, as molecules that define its properties are progressively removed by recycling to the plasma membrane and molecules conferring late endosome function are added from the TGN.

The high selectivity of membrane transport implies that specific receptors are present on endosomes to enable the docking of specific transport vesicles. It is clear that each model makes different predictions about how this selectivity is achieved. According to the maturation model, the specific receptors must first be inactivated and then either removed by recycling or degraded in lysosomes during the

maturation process. In contrast, the stable-compartment model predicts that these receptors may remain as resident components of the early endosomal membrane.

It has not been possible to distinguish whether transport occurs as proposed by one or the other model by using the in vivo and in vitro transport assays developed so far. Thus, although endocytic organelles can be readily distinguished morphologically, it has not been possible to extrapolate to the mechanisms that give rise to the observed structures. This may be because the two models, although conceptually opposed, can both explain the distribution and cycling of the known key regulators of membrane traffic (rab proteins, SNAREs, etc.).

of the cell reveals that organelles of the biosynthetic pathway, such as the Golgi apparatus or the ER, appear as single copies whereas both early and late endosomes are present in multiple copies. On the other hand, the observation of living cells, together with in vitro transport studies, has shown that both endocytic and biosynthetic organelles exhibit a high degree of plasticity at interface or during mitosis, which is at least in part due to homotypic fusion and fission events. The equilibrium between fission and fusion reactions most probably determines the steady-state number of apparent copies for each organelle. Thus, the collection of individual elements of early and late endosomes may, in fact, each form single functional compartments.

Beyond clear shape differences, the precise boundaries between different compartments are often blurred at the molecular level (Box 7.1). Thus, the nature of the intermediate compartment, which is situated between the ER and the Golgi complex, or the boundary between sorting early endosomes and recycling early endosomes remains illdefined and controversial. This situation is largely due to the high dynamics of membrane flow, which result in the localization of certain molecules to more than one organelle at steady state, making it difficult to use the distribution of these proteins to define precise boundaries between compartments.

Ways of Entry into the Cell

Endocytic Pathways

The term "endocytosis" refers to the cellular uptake of solutes, lipids, and membrane proteins, including receptor-ligand complexes, whereas the

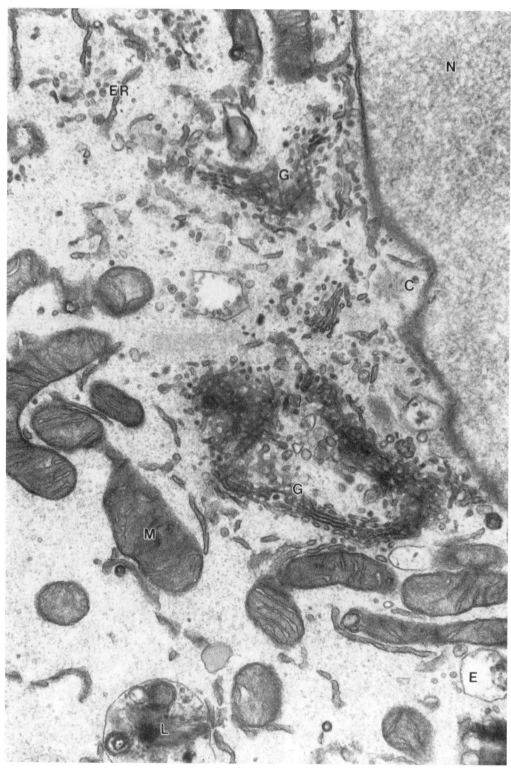

Figure 7.2 Intracellular compartments of the higher eukaryotic cell. This electron micrograph shows a perinuclear region of a Vero cell. C, centriole; N, nucleus; G, Golgi; E, possible early endosome; L, late endosome or lysosome; M, mitochondrion; ER, endoplasmic reticulum. Courtesy of Rob Parton.

term "pinocytosis" was originally used to refer to the uptake of solutes exclusively. This distinction is no longer useful, since it is clear that solutes can be internalized by all types of invaginations forming at the plasma membrane. In contrast, membrane proteins, as well as some specialized lipids, are internalized in a highly selective manner. The vast majority of cell surface receptors are internalized via a clathrin-dependent pathway. Clathrin and adapter proteins assemble from a cytoplasmic pool to form specialized regions on the plasma membrane called clathrin-coated pits (Figure 7.3). Some receptors are concentrated 10- to 20-fold in these regions by a direct interaction of their cytoplasmic tails with the adapter proteins. Indeed, these receptors contain a specific recognition marker or signal in their amino acid sequence which interacts specifically with adapter proteins. The pits then invaginate and pinch off to form 100- to 150-nm-diameter clathrin-coated vesicles. Both clathrin-coated pits and clathrin-coated vesicles can be easily visualized by electron microscopy, due to the regular lattice formed by clathrin.

The most frequent endocytosis signal is made of a single tyrosine residue placed in the context of a tight turn contained in the cytosolic tail of

Figure 7.3 Two types of invaginations at the cytosolic side of the plasma membrane visualized by freeze-etch electron microscopy. **(A to C)** Different states of clathrin assembly, which presumably reflect different stages of vesicle formation from a flat lattice (A) to a deeply invaginated pit (C). **(D to G)** Different types of caveolae with the typical spiraling appearance, perhaps corresponding to different stages of invagination. Courtesy of John Heuser.

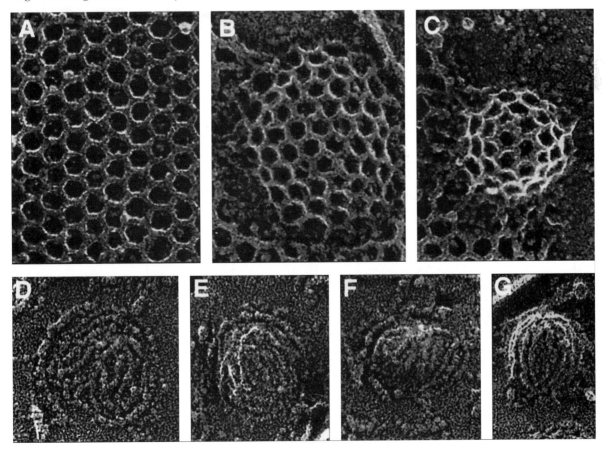

cell surface receptors. The required consensus sequence is rather flexible, as long as it results in a tight turn preceding the tyrosine residue. The position of the tyrosine also does not appear to be important. This endocytosis signal has been shown to interact with the medium chain of the AP2 adapter complex that acts to recruit clathrin at the plasma membrane. Another sorting signal for endocytosis consists of two juxtaposed leucines. No functional difference has been demonstrated between the two types of signals, but dileucine motifs may interact with different regions of the AP2 adapter complex. Both tyrosine and dileucine motifs are also active in the biosynthetic pathway.

Morphometric analyses, combined with biochemical studies, have shown that the clathrin-mediated pathway can account for most of the fluid-phase endocytosis, at least in BHK cells, suggesting that this pathway accounts for the bulk of cell surface turnover. However, the use of agents which perturb the clathrin-dependent pathway indicates that one or more clathrin-independent, alternative pathways of solute internalization also exist. Evidence for the existence of a nonclathrin pathway was obtained after inhibition of the clathrin pathway by expression of a dominant-negative mutant of dynamin, a protein involved in clathrin-coated vesicle formation. In these experiments, fluid-phase uptake is inhibited immediately upon expression of the mutated dynamin gene but recovers later, indicating that an adaptable, alternative mechanism of uptake exists. It is not clear to what extent this alternative pathway functions constitutively or as a rescue mechanism induced by the disruption of the main pathway. Until now, this alternative pathway has remained mysterious, in part because of the absence of specific molecular markers and in part because of the preponderant role of the clathrin-mediated pathway. It is, however, clear that molecules internalized by clathrin-dependent or -independent mechanisms are all targeted to endosomes.

Although the mechanism by which fluid-phase uptake continues after inhibition of the clathrin pathway remains to be elucidated, two types of vesicles that could account for the alternative pathway have been observed. Macropinosomes are large, noncoated vesicles of different sizes (0.5 to 2 μm in diameter) that preferentially form at the leading edge of the cell in response to growth factors (see also "Phagocytosis," below). Their formation is due to the extension of membrane protrusions accompanying a rearrangement of the subcortical actin cytoskeleton. As these protrusions form, they enclose a given volume of extracellular fluid, which is consequently endocytosed. A second type of noncoated vesicles has been observed. These vesicles have diameters of approximately 100 nm and have been proposed to carry fluid-phase markers and toxins into the cell.

Caveolae, which represent a fourth type of invagination at the plasma membrane, were discovered long before clathrin-coated pits. As shown in Figure 7.3, caveolae have no clearly visible cytoplasmic coat and exhibit a distinctive spiraling appearance. They are highly enriched in the cellular lipid cholesterol and in the integral membrane protein caveolin-1 (VIP21-caveolin). Involvement of caveolae in transcytosis in endothelial cells is well documented. However, for other cell types, it is still controversial whether caveolae are static invaginations at the plasma membrane or whether they can pinch off and act as vesicular endocytic carriers. In addition to playing a potential role in endocytosis, caveolae are most probably involved in cell signaling, since they contain a variety of signaling molecules, such as

G-protein-coupled receptors and tyrosine kinases, which are down-regulated by direct interaction with caveolin-1.

Phagocytic Pathways

Phagocytosis in mammals is carried out mainly by polymorphonuclear granulocytes, monocytes, and macrophages and involves the internalization of large particles. These professional phagocytes specialize in the defense against pathogens and the clearance of old cells or cell debris. As opposed to endocytosis, phagocytosis occurs via clathrin-independent mechanisms and involves major actin polymerization and rearrangement. Various particles larger than 0.5 μm in diameter, ranging from inert pollutants to invasive bacteria, can be ingested by phagocytosis. Particles can be recognized directly, or, alternatively, recognition can be mediated by the interaction of opsonins coating the surface of the particle with receptors present on the phagocyte. In serum, the most common opsonins are antibodies and complement, and professional phagocytes express on their surface Fc receptors and complement receptors accordingly.

According to the zipper model, opsonized particles specifically interact with surface receptors present on the phagocytic cell. The particle is progressively surrounded by the plasma membrane of the phagocyte, as more and more of its opsonins, distributed all around its surface, bind to their receptors. To complete phagocytosis of the particle, all of its surface needs to interact with the surface of phagocytic cells. Phagocytosis of opsonized particles results in the formation of tight vacuoles, where the membrane originating from the phagocyte is closely apposed to the particle. Alternatively, phagocytosis can occur by a different mechanism, which involves the formation of large macropinosomes. Following local stimulation by growth factors, the cell sends out unguided pseudopodia, or ruffles, that fold back against the cell surface and accidentally engulf extracellular fluid as well as the particle. Consequently, this mechanism results in the formation of a phagosome where the membrane resulting from the phagocyte will be rather loose around the particle.

The zipper mechanism is by far the most frequent, but macropinosome formation has also been documented in the context of phagocytosis, in particular for *Salmonella typhimurium* invasion. Both mechanisms are actin dependent, since pseudopod formation requires actin polymerization. The signal transduction cascades causing actin polymerization as a result of an opsonin-receptor interaction, or in response to a growth factor, are still being elucidated, although it is already clear that tyrosine phosphorylation plays a key role in Fc receptor-mediated phagocytosis. The two signaling cascades probably merge at an as yet unknown point, since they both result in actin polymerization and the formation of pseudopodia.

Endosomes at the Crossroads of Membrane Traffic

Biosynthetic Pathway

The endocytic pathway is in constant interaction with other compartments of the vacuolar apparatus via vesicular traffic, in particular with the biosynthetic pathway. Newly synthesized lysosomal hydrolases arrive in the TGN together with many other proteins destined for the plasma membrane. To allow their sorting from other secreted proteins in the TGN, lysosomal enzymes carry a unique marker in the form of the sugar mannose-6-phosphate added onto their oligosaccharide side chains. In the TGN, this sugar is recognized by the mannose-6-phosphate receptors, which can interact with adapters of the clathrin coat present on the TGN. Functionally, TGN- and plasma membrane-derived clathrin-coated pits and vesicles are mirror images of each other. They contain distinct but similar adapter complexes, AP2 on the cell surface and AP1 on the TGN. The interaction with the AP1 adapter complex ensures specific sorting of the lysosomal enzymes into clathrin-coated vesicles, which then fuse with endosomes. It is not completely clear whether this clathrin-dependent pathway leads to the early or the late endosome. Once in endosomes, the receptors release their cargo of lysosomal hydrolases at the low endosomal pH. Then receptor molecules are recycled from late endosomes back to the TGN for reutilization whereas lysosomal enzymes are packaged into lysosomes. Recycling routes from endosomes to TGN are likely to play a center-stage role in the control of vacuolar membrane dynamics, since membrane components of the biosynthetic transport machinery which are required during trafficking from the TGN to the plasma membrane or to endosomes must be returned to the TGN for reutilization. Recent studies have revealed the existence of another adapter complex, AP3, which may or may not be associated with clathrin. Although the function of this complex is not known, several lines of evidence suggest that it is also involved in protein sorting and vesicular traffic between the biosynthetic and the endocytic pathway.

Autophagy

Endosomes are also connected to the autophagic pathway, which is responsible for the turnover of some intracellular organelles, particularly mitochondria. It is believed that immature autophagic vacuoles are generated when membranes derived from the ER enclose an organelle and cytosol within a sealed vacuole. These vacuoles have a double membrane and are devoid of lysosomal enzymes and are therefore nondegradative. After acquisition of lysosomal hydrolases from the endocytic pathway, autophagosomes lose the internal membrane and become degradative. The level at which the meeting of the two pathways occurs is still under debate; however, the acquisition of lysosomal enzymes is believed to occur by direct fusion of either late endosomes or lysosomes with immature autophagocytic vacuoles. Recent studies with yeast have identified some key components involved in the regulation of autophagy, and their mammalian counterparts are starting to be found.

Phagocytosis

Finally, and most importantly, complex types of interactions appear to occur between endosomes and phagosomes. It was long thought that phagolysosomes form through direct fusion of a lysosome with a phagosome. This view has changed over the last few years, and it is now believed that phagosomes undergo progressive changes through multiple exchanges with endocytic organelles. In this process, a large segment of the plasma membrane, which becomes the phagosomal membrane upon internalization, may gradually acquire endosomal and then lysosomal characteristics. Indeed, with increasing time after phagocytosis, phagosomes progressively lose plasma membrane markers and acquire first endosomal and later lysosomal markers, indicating that membrane is continuously added and removed. In vivo and in vitro observations suggest that phagosomes interact at more than one level with elements of the endocytic pathway and that the pH of phagosomes decreases in parallel with the acquisition of lysosomal hydrolases. Thus, the resulting phagolysosomes exhibit all the lysosomal characteristics required for degradation of the ingested particles or microorganisms. One of the challenges in the field will be to understand the precise molecular mechanisms which regulate interactions between endosomes and lysosomes and thus control the biogenesis of phagolysosomes.

Endocytic Pathway

Sorting in Early Endosomes

Cell surface proteins and lipids, as well as solutes, typically internalized by clathrin-coated pits, are first delivered to early endosomes (Figure 7.4). These organelles consist of tubular and cisternal elements associated with vesicular regions 0.3 to 0.5 µm in diameter, are located in the cell periphery, and exhibit a slightly acidic pH of 6.0 to 6.2 due to the presence of the vacuolar proton pump. A surface equivalent to the entire plasma membrane is internalized every hour, and the vast majority (more than 90%) of internalized lipids and proteins are recycled back to the cell surface with rapid kinetics ($t_{1/2} \approx 5$ min). Only a minor part of the endocytosed molecules is forwarded to degradation. This tremendous flow of recycling membrane has been estimated to be ca. 10-fold higher than biosynthetic or degradative membrane traffic at steady state.

Figure 7.4   Electron micrograph of an early endosome labeled with LDL-gold. LDL-gold was internalized into endosomes in vivo before the cells were fractionated. The endosomal fraction was mounted onto mica plates and processed for freeze-etch electron microscopy. The image shows the typical organization of an early endosome consisting of tubular and vesicular elements connected to a cisternal region. The large vesicular element containing LDL-gold (top right) may correspond to a forming endosomal carrier vesicle. Courtesy of John Heuser.

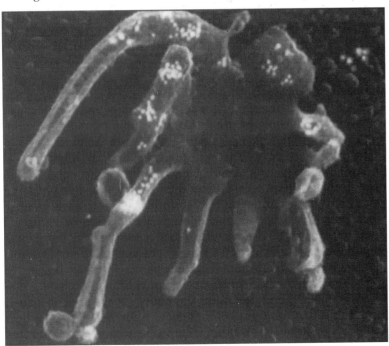

Early endosomes are important sorting stations along the endocytic pathway, since incoming receptors have to be sorted to different destinations. The low-density lipoprotein (LDL) receptor, the transferrin receptor, and the epidermal growth factor (EGF) receptor are all internalized at the plasma membrane by clathrin-coated pits, but their intracellular fates are different. Both the LDL receptor and the transferrin receptor are internalized constitutively, whether they bind ligand or not. At the acidic pH of the early endosome, LDL dissociates from its receptor and transferrin loses its two ferric ions but remains associated with its receptor. Whereas LDL and the ferric ions are transported to late endosomes and lysosomes to be digested by lysosomal acid hydrolases and used for cellular metabolism, respectively, the receptors are recycled back to the plasma membrane to undergo further rounds of endocytosis. Typically, a cell surface receptor like the transferrin receptor or the LDL receptor can cycle more than 100 times between the plasma membrane and early endosomes before being degraded. In contrast, internalization of the EGF receptor does not serve nutritional purposes. This receptor is internalized mainly upon EGF binding, which initiates a signaling cascade. When the ligand is abundant in the extracellular space, both receptor and ligand are transported to late endosomes and lysosomes for degradation, in order to down-regulate cell surface expression.

Recycling Pathway

Early and late endosomes exhibit major differences in their topological organization, ultrastructure, acidification properties, function, and protein composition. These sections address the general organization and characteristics of endosomes along the recycling and degradation pathways.

Once internalized, recycling receptors, such as the transferrin receptor, rapidly leave the sorting early endosome. In many cell types, they are found, together with other recycling molecules, within narrow tubules (diameter, ca. 50 nm), which are believed to be similar to the tubular regions of early endosomes. They often form intricate networks, which localize to the pericentriolar region, where they are sometimes associated with microtubules. The purpose of this association is not clear, since microtubules appear not to be essential for transferrin receptor recycling. Strikingly, these tubules are devoid of molecules destined to be degraded, e.g., LDL, and exhibit a pH of ca. 6.4, slightly higher than that of sorting early endosomes. Although these tubular elements were first observed long ago, their properties and function have just begun to be elucidated. It is not clear, in particular, whether passage through the recycling endosome is an obligatory route for all recycling molecules or whether rapid recycling to the cell surface can also occur directly from the sorting elements of the early endosome. It also remains to be established whether membrane flow along the recycling route occurs strictly in a nonselective manner, by bulk flow, or whether sorting signals are necessary to identify proteins which need to be recycled.

Although the boundaries between sorting and recycling endosomes are still ill defined, progress has been made in understanding the molecular mechanisms regulating protein trafficking along the recycling pathway. Whereas the small GTPase rab5 and its effectors rabaptin5 and EEA1 are involved in homotypic fusion of early endosomes, rab4, together with the SNARE cellubrevin, is thought to regulate transferrin recycling (see below).

Thus, functional differences exist between early sorting and recycling endosomes.

Degradation Pathway

In contrast to the recycling pathway, a small fraction only (<10%) of all internalized components are transported from early to late endosomes and lysosomes. However, this route is very efficiently followed by down-regulated receptors, demonstrating that transport to late endosomes is highly selective and presumably signal mediated. Although a consensus sequence responsible for lysosomal targeting has not yet been identified, candidate signals have been reported. These include different regions identified in the cytoplasmic domains of P-selectin, the mannose-6-phosphate receptor, and the EGF receptor.

Proteins destined to be degraded are first collected within the vesicular portion of early endosomes, which then detach and become free vesicles. Typically, these vesicular elements, whether attached to early endosomes or free, are relatively large (0.3 to 0.5 μm in diameter) and contain internal tubules and vesicles; they are therefore called multivesicular bodies (MVBs). Since these vesicles mediate transport to late endosomes, they have also been termed endosomal carrier vesicles (ECVs). In this chapter they are referred to as ECVs/MVBs. Accumulation of internal membranes, as seen in Figure 7.5, is a hallmark of all endosomes along the degradation pathway, including late endosomes, which often exhibit a multilamellar appearance. Whether internal and limiting membranes are physically connected, at least to some extent, or whether they interact dynamically via internal fission and fusion events is still unclear. Internal membranes selectively incorporate some proteins, including down-regulated EGF receptor and the mannose-6-phosphate receptor, as well as the lipid lysobisphosphatidic acid (LBPA), but the functional significance of this spatial segregation is still unknown.

Concomitantly with ECV/MVB formation, the luminal pH of endosomes rapidly drops to ca. 5.0, which corresponds to the optimal pH for the activity of many lysosomal enzymes. ECV/MVB formation appears to depend on endosome acidification and on an endosomal subcomplex of the COPI coat proteins. Once formed, ECVs/MVBs move from the cell periphery toward late endosomes, which are often clustered in the perinuclear region. In vitro and in vivo studies indicate that this movement occurs on microtubule tracks and requires the activity of the minus-end-directed motor protein cytoplasmic dynein. Then ECVs/MVBs dock onto and fuse with late endosomes in a process which requires N-ethylmaleimide-sensitive factor (NSF), an ATPase involved in most but perhaps not all steps of intracellular membrane traffic. Transport from early to late endosomes may also depend on the small GTPase rab7.

Late endosomes often appear as large multilamellar elements, which sometimes seem to be interconnected via tubular or cisternal regions. They are depleted of recycling receptors and in some cell types contain large amounts of mannose-6-phosphate receptors. Lysosomal glycoproteins are major constituents of their membranes, which also contain the small GTPases rab7 and rab9. Late endosomes play an essential role as the last sorting station before lysosomes. The mannose-6-phosphate receptor, for instance, is recycled from late endosomes back to the TGN in a process which depends on rab9 and on its putative effector p40. In addition, late

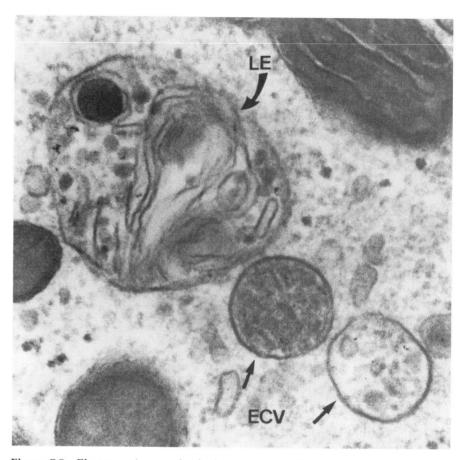

Figure 7.5 Electron micrograph of a late endosome (LE) and two endosomal carrier vesicles (ECV). Internal membranes, characteristic of organelles in the degradative pathway, are visible in both types of structures. Courtesy of Rob Parton.

endosomes correspond to the major convergence point between the endocytic, phagocytic, and autophagic pathways.

Lysosomes exhibit a distinct vesicular and electron-dense appearance and are devoid of the mannose-6-phosphate receptor, rab7, and rab9. Lysosomal glycoproteins accumulate along the degradation pathway to reach their maximal concentration and activity in late endosomes and lysosomes. This organelle is thus perfectly equipped to complete the degradation process that was initiated at earlier stages of the pathway. It is still unclear whether entry into lysosomes is an irreversible process or whether molecules can be returned to late endosomes.

Sorting in Polarized Cells

Polarized epithelial cells exhibit two plasma membrane domains with different protein and lipid compositions and separated from each other by tight junctions. In an intact organism, the apical plasma membrane directly faces internal cavities whereas the basolateral plasma membrane can indirectly communicate with the blood circulation. Polarity is established during biosynthesis by sorting newly synthesized molecules in the TGN to the apical or basolateral plasma membrane domain. A second sorting platform is the early endosome, which can either recycle molecules to the same plasma membrane domain from which they were first internalized or transcytose them to the opposite pole of the cell. Thus, establishment and maintenance of polarity result from the interplay of two sorting stations in the cell, whose relative importance strongly depends on the cell type.

Upon internalization from either plasma membrane domain, macromolecules first arrive in distinct sets of apical and basolateral early endosomes. The apical and basolateral degradative endocytic pathways both resemble the endocytic pathway in nonpolarized cells and converge at the level of the late endosome. Over the last few years, it has become apparent that receptors recycling from the basolateral plasma membrane domain can be found, together with transcytosed receptors, in a common, apically located endosome with the characteristic tubular morphology of the recycling endosome observed in nonpolarized cells. It is therefore attractive to speculate that transport, whether recycling or transcytosis, to either plasma membrane domain may occur from these apical tubules.

Neurons represent another cell type that exhibits a high degree of polarization. They possess a long axon, through which they send out signals, and shorter dendrites, which in turn receive signals from surrounding cells. Axonal and somatodendritic plasma membrane domains of neurons appear to correspond functionally to the apical and basolateral plasma membrane domains of epithelial cells, respectively. Accordingly, the overall organization of the endocytic pathway is similar in both cell types.

Protein signals responsible for the apical targeting of the polymeric immunoglobulin A receptor have been identified. Conversely, basolateral sorting signals that are very similar to the classical tyrosine endocytosis signals have been described. In fact, most basolateral sorting signals act for both biosynthetic and endocytic basolateral sorting, as well as stimulating endocytosis. The functional significance of this similarity is still unknown. Here again, dileucine motifs have similar functions for some receptors.

Specialized Routes of Endocytic Membrane Transport

In certain cell types, endocytic membrane traffic carries out highly specialized functions. One of the best-studied examples is the recycling of synaptic vesicle components in neurons. Newly synthesized neurotransmitters are included into synaptic vesicles and exocytosed into the synaptic cleft. From there, they are rapidly endocytosed by the presynaptic nerve terminal and reach synaptic vesicles again, allowing a new exocytosis cycle to start. This process is extremely rapid compared to constitutive endocytosis in nonneuronal cells. The precise relationship of the synaptic vesicle recycling route to the housekeeping recycling pathway remains to be clarified.

Another important example is antigen presentation by cells of the immune system. To trigger an immune response, exogenous protein antigens must be converted into short peptides and loaded onto major histocompatibility complex (MHC) class II molecules in order to be presented by antigen-presenting cells to helper T cells. Newly synthesized MHC class II molecules are transported through the biosynthetic pathway, and, in parallel, exogenous antigens are endocytosed and processed into peptide fragments. It is not clear whether the loading of the peptide occurs in standard late endosomes or in a specialized endosomal compartment.

Molecular Mechanisms

This section briefly reviews some proteins and molecular mechanisms, which are believed to regulate endocytic membrane traffic.

Coat Proteins

Most of the budding events characterized so far involve coat proteins, which may be responsible for cargo selection and/or mechanical deformation of the membrane. Clathrin coats act in both the biosynthetic and endocytic pathways and represent the best-characterized proteinaceous coat. Clathrin triskelions consist of three 180-kDa clathrin heavy chains and three 30-kDa clathrin light chains. Many triskelions come together to form hexagonal flat lattices or hexagonal or pentagonal cages. Clathrin cages can form spontaneously in vitro in the absence of membranes and adapter proteins, indicating that cage formation is a thermodynamically favorable process. It has been proposed that the formation of clathrin cages mechanically bends the membrane to invaginate the budding vesicles. The way in which the curvature of the clathrin lattice changes to progressively form a cage out of a flat lattice is still poorly understood.

The presence of adapter proteins is necessary for the formation of clathrin lattices in vivo. Three types of adapter complexes have been reported, and more are likely to be discovered in the near future. AP1 is the TGN adapter complex, AP2 is recruited to the plasma membrane, and the localization of AP3 is not yet definitive, although it also appears to be involved in post-Golgi transport, perhaps to endosomes. Interestingly, despite significant sequence homology to AP1 and AP2, it is still not entirely clear whether the AP3 adapter recruits clathrin, since no coats can be observed on AP3-positive membranes by electron microscopy. Clathrin coats have also been observed on recycling endosomes, but the adapter complex involved in their formation has not yet been identified unambiguously.

Adapter complexes consist of two 100-kDa subunits, one 50-kDa medium chain, and one 20-kDa small chain. They are directly implicated in the sorting process that concentrates receptors in coated pits (see above). The small GTP-binding protein ADP ribosylation factor (ARF1) is believed to regulate adapter binding and clathrin coat assembly on the TGN. However, ARF (or a homologue) does not seem to be required for clathrin coat recruitment at the plasma membrane, indicating that the mechanism used to recruit clathrin coats varies depending on the pathway in which it is used.

Once the plasma membrane has been invaginated by the action of the clathrin coat, dynamin, the mammalian homologue of the *Drosophila shibire* gene product, constricts the neck of the forming vesicle, using GTP hydrolysis to generate the necessary force. After the vesicle has pinched off, the coat is rapidly depolymerized, perhaps via the action of the chaperone protein hsp70c, in conjunction with the coat protein auxilin, a DnaJ homologue. Interestingly, dynamin may not be involved in neck constriction at the TGN, suggesting once again that the formation of clathrin-coated vesicles from different donor membranes occurs by different mechanisms.

The involvement of COPI (coatomer) in vesicular transport at early steps of the biosynthetic pathway is well established. COPI exists as an equimolar heteroheptamer consisting of α (160-kDa), β (107-kDa), β' (102-kDa), γ (100-kDa), δ (60-kDa), ε (36-kDa), and ζ (20-kDa) subunits. These subunits are associated in a complex (the coatomer), which can be either cytosolic or membrane associated. COPI-coated vesicles can be readily distinguished from clathrin-coated vesicles, since their coat appears thinner and less regular by classical electron microscopy. Recruitment of the coat onto Golgi membranes and coat depolymerization, once the vesicle has formed, also depend on ARF1.

Recently, a COPI subcomplex, consisting of α, β, β', ζ, and ε COP, has been found on early endosomes. It is not clear whether the δ and γ subunits are missing altogether or whether their endosomal homologues have not yet been found. However, a morphologically distinct coat, similar to the biosynthetic COPI coat, has not been observed on endosomes, perhaps because of differences in protein composition. In vivo and in vitro experiments indicate that transport from early to late endosomes and formation of ECVs/MVBs from early endosomal membranes depend both on this COPI subcomplex and on endosome acidification. The two mechanisms are likely to be related functionally, since COP recruitment onto endosomal membranes is itself pH dependent. The precise mechanism of COP action in the endocytic pathway is still unclear, but it is tempting to speculate that COPI proteins participate in the selective incorporation of proteins into forming ECVs/MVBs.

Cytoskeleton

Movement from peripheral early endosomes to perinuclear late endosomes is facilitated by microtubules, as is the case for other transport steps over long distances. This is particularly well illustrated during axonal retrograde movement of ECV/MVB-like endosomal vesicles, where vesicles are moved from the early endosomes located in the presynaptic region through the length of the axon back to the cell body where late endosomes are located. Microtubules have precise orientations in the cell, radiating from the centrioles in the cell center to the periphery in a polarized manner, and they can thus provide tracks for vesicle movement, with directionality being provided by specific motor proteins such as cytoplasmic dynein in the endocytic pathway. Cytoplasmic dynein is part of a larger protein complex, dynactin, that can bind organelles and is proposed to mediate movement along microtubules. How the binding of this complex to the organelle is regulated or how force is generated is still not completely clear.

Actin microfilaments have been implicated in different transport steps, including in the endocytic pathway. In contrast to microtubules, actin filaments display no polarity in their orientation in the cell, making it hard to envision a role in directional transport. However, endocytic vesicles have to travel through the dense actin network of the cortical cytoskeleton, which defines an organelle exclusion zone underneath the plasma membrane. It is attractive to propose that selective remodeling of this actin network is involved at early steps of the endocytic pathway. Undoubtedly, major actin remodeling occurs during phagocytosis, where actin filaments play a central role in phagosome formation. Alternatively, actin filaments may contribute to tethering endosomal membranes and thereby facilitating endosome dynamics. This view is supported by the finding that unconventional myosins, which are actin-based motor proteins, are involved in more than one step of endocytic transport. Small GTP-binding proteins of the rho and rac families are currently believed to regulate organization and dynamics of the actin cytoskeleton. In fact, some rho proteins have been detected on endosomes, and rhoD has recently been functionally implicated in endosome dynamics. In addition, both rho and rac are believed to be involved in the formation of clathrin-coated vesicles at the plasma membrane. Recent studies also suggest that annexin II (see below) may serve as an interface for the association of elements of the cortical actin cytoskeleton to endosomal membranes.

rab Proteins

rab proteins are a family of small GTPases of the ras superfamily. Eukaryotic cells contain more than 40 different rab proteins, which exhibit a compartment-specific localization in the biosynthetic and endocytic pathways. For all members of the family which have been characterized, a direct involvement in membrane transport has been demonstrated. Their restricted localization and function make rab proteins potential candidates to mediate or regulate the specificity of docking and/or fusion reactions during vesicular transport.

Like all small GTP-binding proteins of the ras family, rab proteins are posttranslationally isoprenylated, which allows them to be anchored into membranes upon activation and to recruit cytosolic components to the membrane. Thus, rab proteins regulate membrane fusion by acting as molecular switches, cycling between an active, GTP-bound state, and an inactive, GDP-bound state. Since the intrinsic GTPase activity of most rab proteins is rather small, GTPase-activating proteins (GAPs) are required to promote rab deactivation by GTP hydrolysis. rab activation, on the other hand, is mediated by interaction with guanine nucleotide exchange factors (GEFs). GDP dissociation inhibitor (GDI) traps rab proteins in the GDP form during cytosolic recycling, avoiding inappropriate activation. Recent studies of rab5 function indicate that GTP hydrolysis is not required for docking and fusion but may serve to down-regulate the activity of the protein in order to provide sufficient but limited time to engage the membrane in docking and/or fusion. rab function is mediated by effector proteins, most of which still remain to be discovered. The rab effectors that have been described to date are clearly unrelated to each other, suggesting that rab proteins regulate different transport steps in different ways.

In the endocytic pathway, rab5 regulates clathrin-dependent internalization, as well as homotypic fusion of early endosomes. rab5 interacts with the cytosolic factor rabaptin 5, as well as with the phosphoinositol-3-phosphate (PI3P)-binding protein early endosomal antigen (EEA1), and both of these proteins are downstream effectors of rab5 function in early endosome fusion. Rabaptin 5 is part of a large protein complex that also contains the GEF rabex 5. Furthermore, rabaptin 5 can also interact with rab4, which has been implicated in the recycling of transferrin receptors to the plasma membrane, suggesting that early stages of endocytosis and recycling are regulated in a coordinated manner. rab7 has been proposed to act during transport from early to late endosomes, during homotypic fusion of late endosomes, and perhaps during transport from late endosomes to lysosomes. rab9 and its putative effector p40 regulate recycling of the mannose-6-phosphate receptor from late endosomes back to the TGN. rab11 is localized to recycling endosomes and is required for the normal transport through this compartment. In addition, rab11 has been found in the TGN, suggesting that it may act in a transport process between the biosynthetic and endocytic pathways.

SNARE Proteins, NSF, and SNAPs

Transmembrane proteins acting as soluble NSF attachment protein receptors (SNAREs) were identified at the nerve terminal both on synaptic vesicles (v-SNARE) and on the presynaptic target membrane (t-SNARE). Many homologues of both v- and t-SNAREs have since been found on other, mostly biosynthetic membranes. SNAREs are relatively small (15- to 40-kDa) transmembrane proteins. They are abundant and exhibit a certain compartment specificity, which makes them ideal candidates for the long-sought membrane receptors guiding vesicle docking to the target membrane. In the endocytic pathway, only a few v-SNAREs have been identified so far. Cellubrevin is the best-understood endosomal SNARE and is involved in transferrin receptor recycling.

SNAREs can interact with each other via their amino-terminal cytoplasmic coiled-coil domains. Parallel interactions can occur between SNAREs on the same membrane. When SNAREs on opposite membranes (such as vesicle and target membrane) bind to each other in a parallel manner, bending of their cytosolic domains may provide the mechanism for bringing the two membranes into close vicinity to promote fusion. The precise manner in which fusion occurs remains to be elucidated. An important challenge in unraveling SNARE function will be to understand the interplay between SNAREs and rab proteins in achieving specificity of docking and/or fusion.

NSF and soluble NSF associated protein (αSNAP) were originally identified as soluble factors required for intra-Golgi transport. Since then, these proteins were shown to be involved in docking and fusion during most transport steps in the cell. Karyogamy in yeast depends on cdc48, a member of the NSF protein family, and Golgi dynamics in mammalian cells depend on both NSF and p97, a cdc48 homologue. The precise role of NSF is still unclear. NSF and αSNAP were long believed to bind to the already formed v-SNARE/t-SNARE complex, but recent data suggest that they may in fact act at an earlier, predocking stage. One view is that NSF may separate complexes involving SNAREs of the same membrane, so that they can take part in interactions with SNAREs on the opposite membrane to engage in fusion.

Phosphoinositides

Over the last few years, phosphatidylinositols have emerged as regulators of membrane trafficking. Phosphorylation of their inositol ring at one or a combination of positions generates unique stereoisomers, some of which have been implicated in the regulation of vesicular transport, the dynamics of the cytoskeleton, and cell growth. In particular, early endosome fusion has been proposed to depend on phosphoinositide 3-OH kinase activity. ARF may be activated by phosphatidylinositol-4,5-bisphosphate, and several proteins known to take part in vesicle budding, such as adapters, dynamin, and coatomer, bind phosphatidylinositols, although the function of these interactions is still unknown. Thus, phosphatidylinositols have been implicated at virtually all steps of membrane trafficking and may provide, in combination with small GTPases and SNAREs, the temporal and spatial regulation necessary for organized vesicular traffic.

Membrane Microdomains

Recent studies have revealed that cholesterol/sphingolipid microdomains, including caveolae, are involved in protein sorting in the biosynthetic pathway and signal transduction at the plasma membrane. Relatively little is known about the role of specialized lipid domains in the endocytic pathway, although there are indications that cholesterol-containing domains may exist within early endosomal membranes. In addition, internal membranes of late endosomes exclude some lysosomal glycoproteins but are highly enriched in the phospholipid LBPA, down-regulated EGF receptors, as well as, in some cell types, the mannose-6-phosphate receptor. It is tempting to speculate that these proteins are restricted to internal membranes by preferential partitioning within specialized membrane microdomains.

Annexins

Annexins form a protein family of approximately 18 members, which have long been implicated in membrane dynamics, including secretion in chromaffin cells, apical transport in polarized cells, and endocytosis. These soluble proteins contain highly conserved Ca^{2+} binding repeats at the C terminus, which mediate Ca^{2+}-dependent binding to lipids. Annexin II, however, in addition to its Ca^{2+}-dependent binding to the plasma membrane, appears to be associated with early endosomes via a Ca^{2+}-independent mechanism regulated by membrane cholesterol. In contrast, annexin I binding to endosomal membranes depends solely on Ca^{2+}. The function of annexins is still unclear. In vivo and in vitro studies suggest that annexin II may be involved in early endosome dynamics, perhaps as an interface for binding of elements of the cortical actin cytoskeleton. Annexin I is phosphorylated via the EGF receptor and was proposed to play a role in multivesicular body biogenesis. It is unclear whether all annexins play a similar role at different steps of membrane transport or whether individual members of the family have different functions.

Selected Readings

Elsevier. 1995. Special issue on phagocytosis. *Trends Cell Biol.* **5**:85–142.

Gruenberg, J., and F. R. Maxfield. 1995. Membrane transport in the endocytic pathway. *Curr. Opin. Cell Biol.* **7**:552–563.

Harder, T., and K. Simons. 1997. Caveolae, DIGs, and the dynamics of sphingolipid-cholesterol microdomains. *Curr. Opin. Cell Biol.* **9**:534–542.

Hay, J. C., and R. H. Scheller. 1997. SNAREs and NSF in targeted membrane fusion. *Curr. Opin. Cell Biol.* **9**:505–512.

Kobayashi, T., F. Gu, and J. Gruenberg. 1998. Lipids, lipid domains and lipid-protein interactions in endocytic membrane traffic. *Semin. Cell Biol.* **9**:517–526.

Novick, P., and M. Zerial. 1997. The diversity of Rab proteins in vesicle transport. *Curr. Opin. Cell Biol.* **9**:496–504.

Riezman, H., P. G. Woodman, G. van Meer, and M. Marsh. 1997. Molecular mechanisms of endocytosis. *Cell* **91**:731–738.

Robinson, M. S. 1994. The role of clathrin, adaptors and dynamin in endocytosis. *Curr. Opin. Cell Biol.* **6**:538–544.

8

Where To Stay inside the Cell: a Homesteader's Guide to Intracellular Parasitism

Dᴀᴠɪᴅ G. Rᴜssᴇʟʟ

The adoption of an intracellular life-style confers several advantages to microbial pathogens; they become inaccessible to humoral and complement-mediated attack, they no longer require a specific adherence mechanism to maintain their site of infection, and they have ready access to a range of nutrients. However, aspiring intracellular pathogens must develop specific strategies to secure and maintain their life-style. Some characteristics, such as the lipid-rich cell wall of mycobacteria, are obvious preadaptations that confer a head start in the acquisition of an intracellular life-style. However, the majority of mechanisms exhibited by intracellular pathogens are the product of evolutionary selection in response to their intracellular existence.

Successful establishment and maintenance of an intracellular infection require the resolution of a series of interconnected problems, with the solution of each frequently having profound influence on subsequent "decisions." This chapter describes these problems for a range of bacterial, protozoal, and fungal pathogens and explores, in the order in which they are encountered by the pathogen, the consequences of each decision point in the establishment of an intracellular infection.

Routes of Invasion

To infect a cell, a microbe must first adhere to it. For some pathogens, this adherence phase determines the choice of the cell to be infected, while for others, although they are capable of binding to many cell types, development of the pathogen is restricted to only certain cell lineages. Because adherence and host cell entry are tightly associated phenomena, they are discussed together as functions of the route and mechanism of invasion. There are three basic mechanisms of invasion: (i) phagocytosis, i.e., entry into professional phagocytes such as macrophages, monocytes, and neutrophils via a process dependent on the host cell contractile system; (ii) induced endocytosis and phagocytosis, i.e., entry into nonprofessional phagocytes by the active induction of internalization through the activity of the host cell contractile system; and (iii) active invasion, i.e., active entry into a passive host cell without triggering any contractile event in the host

131

cell cytoskeleton. The different routes of host cell invasion are diagrammed in Figure 8.1.

Phagocytosis

Many microbial pathogens are capable of either transient or sustained infection of professional phagocytes. There are two related reasons for this seemingly anomolous phenomenon. First, macrophages and other phagocytes represent the frontline defense of the host against microbial invasion. These cells migrate through tissues, internalizing and degrading foreign particles; this behavior will obviously have maximized the frequency of interaction between phagocytes and microbes. Second, the macrophage has receptors which recognize a range of ligands including the serum opsonins antibody and complement. Pathogens that activate the alternate pathway of complement will accumulate C3b and iC3b on their surface, and because the macrophage is equipped with high-affinity receptors for these ligands, CR1 and CR3/CR4, respectively, these opsonized microbes will bind to and be internalized by the phagocyte (Box 8.1). The only specific mechanism required by the pathogen is to facilitate complement deposition while avoiding lysis by insertion of the terminal membrane attack components. Pathogens such as *Salmonella* and *Leishmania* achieve this through elongation of their surface lipopolysaccharide or lipidoglycans, respectively. These long carbohydrate chains activate complement but avoid lysis because the activating convertase is maintained some distance from the outer membrane. Complement receptors trigger phagocytosis without stimulating a strong superoxide burst from the macrophage.

Entry via phagocytic receptors activates the signaling pathways for maturation of the phagosome into an acidic, hydrolytically active compartment. Therefore, in the absence of mechanisms to subvert this process,

Figure 8.1 Diagrammatic representation of the three different routes of invasion of mammalian cells by intracellular pathogens. In each instance, the "active" cell or cells are labeled with a plus sign. In phagocytosis, the infecting pathogen is relatively passive in the process following ligation to host cell receptors capable of triggering internalization. This process requires little if any metabolic activity from the parasite. Examples include *Leishmania, Mycobacterium,* and *Histoplasma.* In induced endocytosis and phagocytosis, the pathogen induces a normally nonphagocytic cell to internalize the microbe. This is the least well understood route of entry and involves subversion of the host cell signaling pathways. Examples include *Salmonella* in nonprofessional phagocytes and *Trypanosoma cruzi.* In active invasion, the pathogen invades the host cell without the participation of the contractile apparatus of the host cell. In this process, the host cell is inert. Examples include all the apicomplexan parasites, *Plasmodium, Toxoplasma, Eimeria,* and microsporidia.

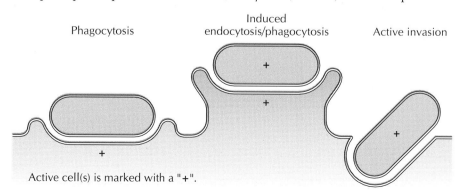

| Phagocytosis | Induced endocytosis/phagocytosis | Active invasion |

Active cell(s) is marked with a "+".

BOX 8.1

Complement opsonization

The complement system is one of the major defense barriers against microbial infection. It can be mobilized in both the presence (classical pathway) and absence (alternative pathway) of immunoglobulins specific to the infective agent. The direct consequence of complement activation and deposition is the polymerization of the latter components, C5 to C9, to form a membrane attack complex that creates a pore in the target membrane.

More relevant to this chapter, however, is the coating of microbes with fragments of C3, which are recognized by receptors expressed on the surface of phagocytes. C3 is the central component of both the classical and alter-

native pathways. It is a 185-Da αβ heterodimer that contains an internal thiolester bond which becomes unstable when the protein is cleaved into its C3b form by removal of the first 77 amino acids of the α chain. The thiolester bond will interact with water or with hydroxyl (carbohydrate) or amino (protein) groups on cell surfaces, forming covalent ester or amide linkages, respectively. The alternative pathway relies on the spontaneous production of serum C3b, which will deposit on surfaces and, together with factor B, will form a C3 convertase, C3bBb. In contrast, the classical pathway relies on antibody binding to allow C1q attachment and formation of a different C3 convertase consisting of C4bC2a.

Following formation of a C3 convertase, C3, which is present at more than 1 mg/ml in serum, is cleaved to form C3b, leading to massive deposition through a positive-feedback loop. C3b can be inactivated by hydrolysis or through the specific activity of factor I, which cleaves C3b into iC3b. Both C3b and iC3b are ligands for high-affinity receptors present on the surface of phagocytes and some other cell types. CR1 is expressed on erythrocytes, B lymphocytes, monocytes, granulocytes, and macrophages and recognizes both C3b and C4b. CR3 is present on granulocytes, monocytes, and macrophages and recognizes iC3b. Both these receptors, on professional phagocytes, will trigger phagocytosis under physiologic conditions.

---

all phagocytosed microbes will be delivered to the lysosomal system of the cell.

Induced Endocytosis and Phagocytosis

Other pathogens have evolved their own ligands for adherence to host cells, an event most frequently observed in pathogens that infect nonprofessional phagocytes. Host cell entry by many of these pathogens is a facilitated process that involves the induction of an internalization response in the cell by the adherent microbe. The signaling pathways activated during induced uptake are explored in depth in chapter 5. One of the better-studied examples is *Salmonella*, which induces the actin cytoskeleton to form extensive membrane ruffles, or "splash," during its entry into cells. This response is accompanied by the phosphorylation of several host cell proteins as a consequence of activation of host cell signaling cascades.

Active Invasion

In contrast to both preceding mechanisms, active invasion involves invasion of the host cell without triggering any contractile events such as phagocytosis. In fact, for *Toxoplasma*, the ability to block the maturation of the entry vacuole into an acidic, lysosomal compartment is determined at the time of host cell entry. All apicomplexan parasites, including *Plasmodium*, *Toxoplasma*, and *Eimeria*, have motile invasive stages called zoites. These stages show an actin-based motile system that mediates both gliding motility and host cell invasion. Recent work with mutant host cells and *Toxoplasma* lines resistant to the antimicrofilament agent cytochalasin D has shown that mutant parasite lines invaded host cells in the presence of the drug irrespective of the phenotype of the host cell. If this invasion process was subverted by opsonizing the parasites with immunoglobulin G (IgG), the Fc receptors on the host cell which were ligated during invasion prevented the parasites from maintaining their vacuoles outside the endosomal continuum. The vacuoles acidified, and the parasites died. Conversely, if *Toxoplasma* was allowed to invade and infect host cells and was subsequently killed by treatment with pyrimethamine, an inhibitor of the parasite's dihydrofolate reductase activity, the vacuoles persisted as isolated intracellular compartments, presumably because the host cell membrane fusion apparatus could not recognize them. Because *Toxoplasma* is promiscuous in its ability to invade a range of different host cells, the ligand on the host cell must be virtually ubiquitous. Although no ligand has been identified formally, experiments have implicated extracellular matrix proteins. It is also of interest that studies on sporozoites of the related parasite *Plasmodium* have demonstrated that these parasites employ a protein, thrombospondin-related adhesion protein (TRAP), to bind to glycosaminoglycans on host cells. Homologues of TRAP have been found in other coccidia including *Eimeria*, *Cryptosporidium*, and *Toxoplasma*.

Selection of an Intracellular Niche

After invasion, intracellular microbes use many different strategies to ensure the maintenance of an intracellular infection. The niches exploited by these intracellular pathogens fall readily into three different groupings, illustrated in Figure 8.2. The first is intralysosomal, in which pathogens persist in acidic, hydrolytic compartments that interact with the endosomal network of the host. The second is intravacuolar, in which pathogens persist in nonacidic vacuoles that exhibit modified or little interaction with the endosomal system of the host. The third is cytoplasmic, in which pathogens exit the phagosome and reside within the host cell cytosol.

Intralysosomal Pathogens

In terms of modulation of host cell function and the biogenesis of phagosomes and endosomes, the pathogens that remain within the "normal" differentiation cascade of the phagolysosome appear relatively passive. *Leishmania* parasites and *Coxiella* bacilli are phagocytosed by macrophages and reside in compartments that are fully acidic, achieving a pH of 4.7 to 5.2.

Leishmania.   *Leishmania* organisms are flagellated protozoans that are close relatives of the trypanosomes. The promastigote form found in the insect vector has a single, anteriorly orientated flagellum, which becomes vestigial in the amastigote form found within vertebrate macrophages. *Leishmania* species induce a range of diseases varying in severity from a simple lesion at the site of a fly bite to visceral leishmaniasis, where the parasite multiplies in the liver and spleen.

The vacuoles containing parasites of the *Leishmania mexicana* complex (Figure 8.3) interact freely with material internalized by the endosomal network of the host cell; in fact, the fusigenicity and access to these compartments increase as the infection continues. The vacuoles are also competent to fuse with other particle-containing phagosomes, facilitating transfer of the particles into the parasitophorous vacuoles. The lysosomal hydrolases within the parasitophorous vacuoles of *Leishmania*, most notably cathepsins B, L, and D, are fully active and hydrolytically competent; nonetheless, the pathogens appear impervious to degradation. The vacuoles have abundant LAMP1 and LAMP2 (lgp110 and lgp120), while transferrin, which traffics through the rapid recycling pathway of mammalian cells and does not

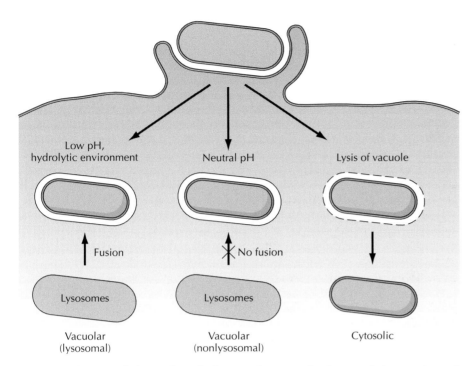

Figure 8.2   Intracellular niches. Pathogens have evolved to exploit a variety of intracellular locations, which fall readily into three different groups. The first includes those that reside in acidic, hydrolytically competent lysosomes and appear undeterred by the hostile nature of their compartment. Examples include *Leishmania*, *Coxiella*, and possibly *Salmonella* (in macrophages at least). The second includes those that remain vacuolar yet avoid the normal progression of their vacuole into a lysosomal compartment. This group of pathogens is the most diverse with respect to the nature of their intracellular vacuole. Examples include *Plasmodium*, *Toxoplasma*, *Legionella*, *Chlamydia*, and *Mycobacterium*. The third group includes those that avoid the consequence of remaining within a phagocytic vacuole by escaping into the cytoplasm. Examples include *Trypanosoma cruzi*, *Shigella*, *Rickettsia*, and *Listeria*.

normally access the late endosome or lysosome, is not detected in *Leishmania*-infected macrophages.

Leishmania parasites also have access to cytosolic material as potential sources of nutrients through fusion with autophagosomes. These parasites are purine auxotrophs and need to salvage purines from the host, since these are probably present at low concentration in the lysosome. Access to cytosolic components sequestered in autophagosomes may provide a source of these purines. The *Leishmania* amastigote has an abundant external nucleotidase that would allow the dephosphorylation of nucleoside triphosphates, enabling uptake and use of host purines.

Existence within vacuoles that remain integral to the endosomal network carries two severe problems. The first is immediate, i.e., resistance to hydrolytic attack, while the second is more insidious, i.e., the continued sampling of parasite-derived peptides and their presentation to the host immune system. With respect to the first problem, *Leishmania* appears to adopt a strategy of avoidance. The surface of the intracellular amastigote stage of the parasite is protein poor and is covered in a coat of short lipidoglycans (GIPLs). These GIPLs are relatively resistant to degradation by the host lysosomal hydrolases and thereby present minimal targets for at-

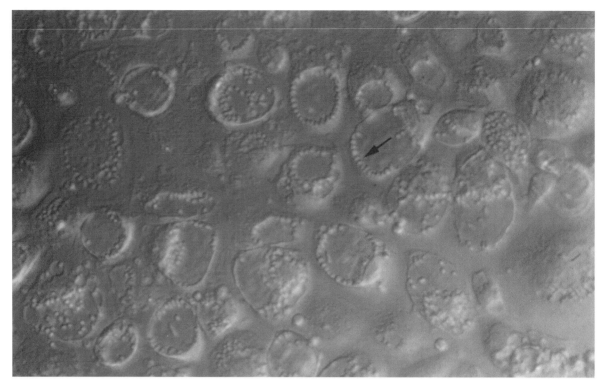

Figure 8.3 Hoffman modulation contrast micrograph of a monolayer of murine bone marrow-derived macrophages infected with *Leishmania mexicana*. This species of *Leishmania* tends to form large fluid-filled vacuoles that contain multiple parasites, which tend to line up along the periphery of the vacuoles (arrow). The vacuoles are acidic and contain active lysosomal hydrolases.

tack. The key to the survival of *Leishmania* within the lysosome is probably its flagellar pocket. The flagellar pocket of all trypanosomatids is the sole point of endocytosis and exocytosis in the cell, because the rest of the cell body is subtended by an extensive array of microtubules. Access to the pocket is partially regulated by a series of hemidesmosomal junctions that anchor the flagellum to the cell body at the mouth of the pocket, which may protect the endocytic receptors of the parasite from attack by host hydrolases.

The second problem relates to the long-term maintenance of the leishmanial infection within a host cell capable of presenting antigen via both class I and class II antigens (Box 8.2). Since it is known that both class I and class II major histocompatibility complex (MHC) molecules traffic through the parasitophorous vacuoles of *Leishmania mexicana*- and *L. amazonensis*-infected macrophages, it is possible that they bind parasite-derived peptides which could subsequently be presented at the surface of infected cells. The induction of a T-cell response leading to the release of macrophage-activating cytokines would be detrimental to the long-term success of the parasite. Leishmanial infections are therefore an interesting system to study strategies of immune system evasion for intracellular pathogens. Examination of the ability of infected macrophages to present exogenous antigen revealed a slight diminution in the T-cell responses induced; however, this reduction varied with both T-cell lines and epitopes, suggesting that the

BOX 8.2

Interface between intracellular pathogens and the immune system: the postcard version

Since intracellular pathogens are sequestered beyond the ready access of the humoral immune system, it is critical that infected hosts be able to identify infection foci or the infected cells themselves. This is achieved through the antigen-processing and presentation capabilities of mammalian cells. All nucleated cells possess major histocompatibility antigens capable of binding peptides acquired inside the cell and presenting these peptides to T lymphocytes, triggering these cells to respond. The major histocompatibility antigens fall into two groups: class I antigens sample peptides derived predominantly from cytoplasmic antigens and therefore play a major role in the presentation of antigens from cytosolic pathogens such as *Listeria*, *Shigella*, and *T. cruzi*, whereas class II antigens sample peptides generated within the endosomal-lysosomal system of the host cell and play a central role in the immune response to intravacuolar pathogens such as *Mycobacterium*, *Salmonella*, and *Leishmania*. Class I MHC antigens are expressed by all nucleated cell types. The class I molecules are trafficked through the cell as heterodimers complexed to β_2-microglobulin and acquire peptides in the ER. These peptides are translocated from the cytoplasm to the ER by specific transporters or TAP proteins. Class II MHC antigens are expressed by cells of the monocyte/macrophage lineage. Class II molecules are α/β heterodimers that complex with another protein, the invariant chain, which regulates intracellular trafficking and the point of peptide acquisition.

Effective presentation to T cells is achieved when the MHC molecule with bound peptide is recognized by an appropriate T-cell receptor on a T lymphocyte. Intracellular pathogens have evolved a spectrum of mechanisms to avoid or suppress the induction or consequences of such immune system responses. These strategies include sequestration of antigens, down-regulation of MHC molecule or costimulatory-molecule expression, suppression of T-cell proliferation, and blocking of the ability of the host cell to respond to activating cytokines. A full discussion of this interplay is provided in chapter 16.

effect was on the intracellular processing and loading rather than on the expression of class II MHC or costimulatory molecules.

Class II MHC molecules are synthesized as nascent chains that must bind to an invariant chain in order to be correctly processed and trafficked in cells. When the invariant chain is degraded, it facilitates the binding of appropriate peptides to the peptide-binding grove of the class II dimer. Recent analysis of *L. amazonensis*-infected macrophages showed the presence of both class II molecules and invariant chain within the parasitophorous vacuoles. The abundance of the invariant chain was enhanced by treatment of infected macrophages with inhibitors of cysteine proteinase activity. These inhibitors would block both the host cell lysosomal hydrolases cathepsins B and L and the parasite cysteine proteinases, and it was suggested that *Leishmania* degrades class II MHC molecules and thus suppresses antigen presentation by the host cell. However, this contrasts with data from another series of experiments in which the acid phosphatase gene of the parasite was engineered to express the protein in the cytosol, on the parasite surface, or secreted into the parasitophorous vacuole milieu. In live infections, the acid phosphatase was presented effectively to T cells when it was expressed as a parasite surface protein and when it was expressed as a secreted protein. Only after killing of intracellular parasites could effective presentation be detected with wild-type cells or cells overexpressing the intracellular form of the protein. These data indicate that the antigen presentation machinery of the host macrophage does interact functionally with the parasitophorous vacuole and, conversely, that degradation of class II MHC molecules was not the major route of immune system avoidance.

Coxiella. Although the intracellular compartment inhabited by *Coxiella burnetii* has not been the subject of such intensive study as that of leishmanial parasites, it appears to present a similar set of lysosomal characteristics. Early studies indicated that colloidal tracers endocytosed by infected cell lines were delivered to the bacterium-containing compartments. Moreover, macrophages coinfected with *Coxiella burnetii* and other pathogens or particles (*Leishmania, Mycobacterium avium,* or latex beads) all exhibited enhanced delivery of material to the bacterial vacuole, suggesting that this compartment fused readily with different endosomal stages. Immunolocalization studies with infected HeLa cells demonstrated that the vacuoles were strongly positive for the late endosomal/lysosomal proteins LAMP1 and LAMP2 and stained with antibodies against cathepsin D and the vacuolar protein-ATPase.

The apparent preference shown by *Coxiella* for the host cell lysosome is supported by data showing that the bacterium grows optimally at a pH below 5.0. Furthermore, treatment of infected cells with lysosomotropic bases which increase the lysosomal pH inhibits the growth of the bacteria markedly.

Salmonella *Salmonella* is frequently described as an intracellular pathogen; however, it is extremely cytopathic and does not form a stable or enduring relationship with its host cell. *Salmonella* invades both nonprofessional phagocytes and macrophages through the induction of a macropinocytosis-like event during which the bacilli are internalized into large "spacious" phagosomes (Figure 8.4). Although *Salmonella* is the sub-

Figure 8.4 *Salmonella* induces an extreme response in mammalian cells during entry. In contrast to tight, zippering phagocytosis through which many particles are internalized, these bacteria induce a membrane "splash" or ruffle that captures the bacteria along with an appreciable volume of fluid. This phenomenon is illustrated in a series of time-lapse video frames. The point of initial contact of the bacterium is marked with an arrow in the 30-s and all subsequent time frames. The macropinosome forms (120 s), and several fluid-filled vesicles coalesce (135 and 170 s), until, finally, the phagocytosed bacilli are translocated towards the cell body (250 s). The mechanism appears analogous to the formation of macropinosomes. Courtesy of Hiroshi Morisaki, Michelle Rathman, and John Heuser.

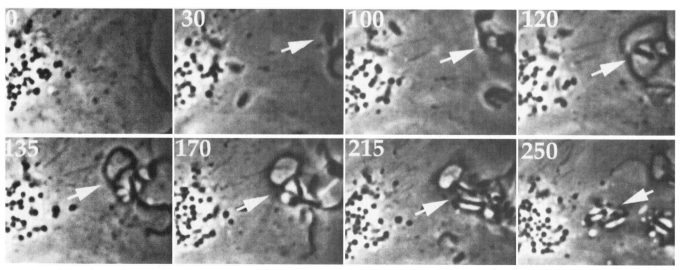

ject of study in several laboratories, there is considerable disagreement about the nature of the vacuole in which this pathogen survives. Recent work suggested that the intracellular population of infecting bacilli is extremely heterogeneous with respect to metabolic activity and contains a significant number of "dormant" bacteria. This heterogeneity or the different host cell types used in experiments may be the source of disparity among some of these studies.

Experiments on the infection of nonprofessional phagocytes, HeLa cells, revealed that bacterial vacuoles showed limited communication with lysosomal compartments preloaded with fluorescent dextrans or horseradish peroxidase. Although the vacuoles containing bacteria were positive for LAMP1, they were connected to LAMP1-containing compartments via a complex tubular network. In addition, these vacuoles were never positive for cation-independent mannose-6-phosphate receptor, although some of the vacuoles acquired cathepsin D. It was concluded that the vacuoles were a subpopulation of lysosome-like compartments that did not communicate with either secondary lysosomes or endosomes.

In contrast, analysis of *Salmonella*-containing vacuoles in phagocytic cells provides a picture more consistant with the vacuole being late endosomal or lysosomal in character. pH measurement following uptake indicated that acidification was partially reduced in the vacuoles containing live (pH > 5.0) versus dead (pH < 4.5) bacilli but that the pH of vacuoles containing live bacteria was still relatively acidic. Although the bacilli were internalized into unusual spacious phagosomes, these vacuoles showed unrestricted fusion with lysosomes, as demonstrated by the acquisition of lysosomal contents. Moreover, Texas red dextran internalized by macrophages infected with *Salmonella* entered the majority of bacterium-containing vacuoles. To address the possibility that the small number of "nonlysosomal" bacterial vacuoles observed in the infection contained bacilli that would multiply and form the basis of the intracellular infection, careful analysis of the number of liberated bacteria with respect to scoring of phagolysosome fusion was performed. It was concluded that the bacteria within phagolysosomes must constitute the bulk of the viable bacteria in the infection. In agreement with this conclusion, recent analysis confirmed that *Salmonella* vacuoles in macrophages acidify and, furthermore, that blocking the acidification of phagosomes with bafilomycin A, a vacuolar ATPase inhibitor, reduced the viability of the bacilli. This suggests that a drop in pH is required for full induction of intracellular survival strategies in *Salmonella*.

Pathogens Sequestered in Modified or Isolated Vacuoles

Of the three different groups of intracellular pathogens, the group of pathogens sequestered in modified or isolated vacuoles is the most disparate in terms of both the properties of their intracellular compartments and the degree to which the vacuole is modified or sequestered outside the normal endocytic continuum. Some pathogens, like *Histoplasma*, survive in lysosomal compartments that fuse with endosomes yet fail to acidify; others, like *Mycobacterium* and *Nocardia*, block the normal maturation procedure of their phagosome and fail to fuse with lysosomes; and finally, some microbes, like *Chlamydia, Legionella, Toxoplasma, Plasmodium,* and *Cryptosporidium,* form a compartment that appears sequestered completely outside the normal membrane trafficking pathways of their host cell.

Nonacidified Lysosomes

Histoplasma. The yeast *Histoplasma capsulatum* parasitizes macrophages in its vertebrate hosts. These vacuoles are readily accessible to endocytic tracers such as fluorescein isothiocyanate dextran and fuse with lysosomal compartments with an efficiency indistinguishable from that of *Saccharomyces*. Further examination revealed that although *Histoplasma* vacuoles were competent to fuse with lysosomes and endosomes, they did not acidify. Analysis of the vacuolar pH up to 20 min postinternalization revealed that while the vacuoles containing the yeast cell wall preparation zymosan acidified rapidly to pH 5.5, the vacuoles containing *Histoplasma* maintained a pH of 7.0.

Developmentally Arrested Phagosomes
Some pathogenic microbes enter their host macrophages by phagocytic uptake yet prevent the normal course of differentiation of their phagosome into a phagolysosomal compartment. This phenotype has been described for a range of pathogenic mycobacterial species capable of infecting mammals. These species include *Mycobacterium tuberculosis*, *M. leprae*, *M. microti* (a rodent pathogen), and the *M. avium* complex, which is an opportunistic pathogen of immunosuppressed individuals. Although the diseases induced by these species differ markedly, they show strong parallels with respect to the behavior of their intracellular compartments.

Mycobacterium **spp.** Early work by D'Arcy Hart and colleagues revealed that the vacuoles inhabited by pathogenic mycobacteria did not fuse with lysosomes preloaded with electron-dense colloids, in contrast to vacuoles containing nonpathogenic or dead mycobacteria. Recent analysis of *M. tuberculosis* and *M. avium* in both human monocyte-derived and murine bone marrow-derived macrophages has yielded a more complete appreciation of the properties of these vesicles. Vacuoles containing these bacteria did not acidify below pH 6.2 to 6.5 and showed a paucity of the vacuolar proton-ATPase responsible for acidification of endosomal and lysosomal compartments. The paucity of proton-ATPase complexes on mycobacterial vacuoles is clearly seen on platinum replicas from isolated phagosomes (Figure 8.5). Despite this, the vacuoles possess LAMP1 (although it is less abundant than in neighboring lysosomes), class I and II MHC molecules, cathepsin D, and transferrin receptor.

The classical view that mycobacterial vacuoles were "nonfusigenic" was shown to be oversimplified when experiments investigating the partitioning of plasmalemma-derived glycosphingolipids in infected cells were conducted. GM_1 ganglioside, complexed with cholera toxin B subunit, was delivered rapidly to the mycobacterial vacuoles, reaching steady state within 10 min. This was faster than delivery to IgG-bead phagolysosomes and suggested that the mycobacterial vacuoles may be accessible to early endocytic compartments. This was consistent with the observation that *M. tuberculosis*-containing vacuoles in human monocyte-derived macrophages were positive for transferrin receptor.

Cathepsin D had been detected by immunoelectron microscopy in *M. tuberculosis*-containing vacuoles. The lysosomal aspartic proteinase is synthesized as a 51- to 55-kDa precursor, or proenzyme, that is activated by removal of its propeptide, producing a 48-kDa protein which is subsequently cleaved into two chains of 31 and 17 kDa on delivery to the lyso-

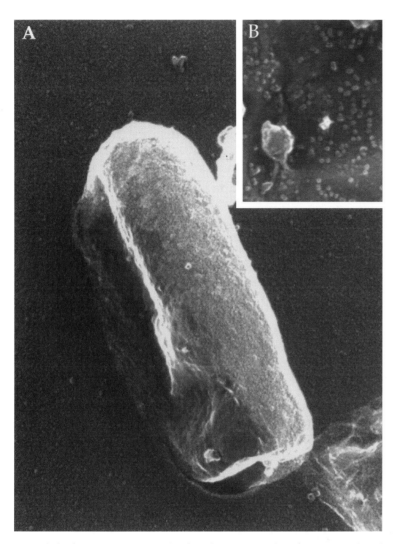

Figure 8.5 **(A)** Electron micrograph of a platinum replica from an isolated *Mycobacterium avium*-containing phagosome. The view is of the cytoplasmic face of the phagosomal membrane and reveals the atypical smooth texture of the phagosome. **(B)** Region of a *Leishmania mexicana*-containing phagosome viewed at comparable magnification. The stud-like structures represent proton-ATPase complexes, which are rare on mycobacterial vacuoles. Proton-ATPases are responsible for the normal acidification of phagosomes. Courtesy of David G. Russell and John Heuser.

some. This sequence of events provides an indication of both the hydrolytic capacity of an endosomal compartment and the route that cathepsin D has taken prior to its appearance within a vacuole. Characterization of isolated *M. avium*-containing phagosomes from macrophages infected up to 9 days previously revealed that the cathepsin D was present in an immature form. In contrast, the proteinase was found in its fully processed form in IgG-bead phagolysosomes isolated 60 min after internalization. This provides independent evidence of the limited hydrolytic capacity of the mycobacterial vacuole and, more importantly, suggests that endosomal constituents, such as cathepsin D and possibly LAMP1, may be delivered from the synthetic pathway rather than through fusion with existing, acidified endosomes. In this manner, endosomal constituents may be acquired prior to extensive accumulation of vacuolar ATPase. Analysis of IgG-bead phago-somes at very early time points indicates that this remodeling step with endosomal components delivered from the synthetic pathway is a phenom-enon common to "normal" phagosome biogenesis.

Experiments on both *M. tuberculosis* in human macrophages and *M. avium* in murine macrophages revealed that the transferrin receptor was functionally active and that its presence within mycobacterial vacuoles cor-related with delivery of iron-loaded transferrin. Pulse-chase experiments on established infections revealed a flux of transferrin through these vac-uoles, providing evidence that mycobacterial vacuoles were stabilized within the rapid recycling endosomal machinery of the host cell. Although these vacuoles would be highly fusigenic, the fusion capacity would be restricted to homotypic events, and without further differentiation, these vacuoles would not be capable of fusing directly with lysosomes. This view is supported by recent data revealing that the BCG strain of *M. bovis* main-tained the early endosomal GTP-binding protein rab5 even in established infections in the macrophage-like cell line J774.

The mechanism whereby mycobacteria retard the normal maturation of their phagosomes is obviously an area of considerable interest. Research-ers have suggested two different possibilities. First, *M. tuberculosis* and *M. bovis* BCG produce ammonia through urease activity, and it was proposed that this ammonia prevents acidification of the bacterial vacuoles. Initially, this suggestion did not appear to address the restricted fusion profile of the mycobacterial vacuoles; however, recent data on endosome trafficking and delivery of endocytosed markers to lysosomes have indicated that delivery to late endosomes and lysosomes is a pH-sensitive phenomenon and that agents that block vacuolar proton-ATPase function retard delivery to ly-sosomes. Second, mycobacterial lipids, most notably cord factor (α,α'-trehalose-6,6'-dimycolate), block membrane fusion in vitro, and the pres-ence of cord factor on *Nocardia* strains (see below) correlates with reduced fusion of vacuoles with lysosomes. There is, however, no direct confirma-tion of the role of either of these possibilities in restriction of vacuolar mat-uration.

Finally, Nramp, the product of the resistance gene *Bcg^r* which confers partial protection against BCG, has been localized to the membrane of phagosomes and endosomes and shown to have a strong influence on the acidification and maturation of BCG-containing phagosomes. These data indicate that failure of *Bcg^s* to limit the susceptibility to BCG infections lies at the level of the *Mycobacterium*-containing vacuole within the host mac-rophage.

Nocardia.   The actinomycete *Nocardia asteroides* also resides in the phagocytes of its host, and, like *Mycobacterium*, the vacuoles containing the infecting bacilli show minimal acidification and do not fuse readily with lysosomes. In broth culture, these bacteria do not tolerate low pH, and the survival of different isolates of *Nocardia* within macrophages correlates di-rectly with their ability to block acidification of their phagosome. The vir-ulence of the strains also correlates with the amount of cord factor that they synthesize. Recent data indicate that cord factor and other mycolic acid glycolipids can inhibit the fusion of liposomes in vitro. Although there is no direct evidence that *Nocardia* vacuoles remain within the endocytic path-way like those containing *Mycobacterium*, *Nocardia* is discussed in this section because the data are comparable to those for mycobacteria prior to biochemical and immunohistological characterization of the vacuoles.

Sequestered Compartments

The pathogens that reside in sequestered or isolated vacuoles fall into two main groups, (i) those that enter by phagocytosis but subvert the normal maturation pathway to establish a specialized compartment that no longer intersects with the endocytic network, such as *Legionella* and *Chlamydia*, and (ii) those that invade actively and build their own vacuole at the time of host cell entry, like *Toxoplasma* and *Plasmodium*. Because these compartments lie outside the normal endocytic pathway of the host cell, one of the obvious questions is how the pathogen obtains nutrients from the host. For this reason, there is particular interest in identifying pathogen-derived proteins that become constituents of the parasitophorous vacuole membrane.

Legionella. *Legionella pneumophila* is a facultative intracellular pathogen that parasitizes phagocytes. Its ability to survive within macrophages may represent a preadaptation for infection, because it is known to associate with free-living amoebae. Entry into the phagocyte can be mediated by a peculiar form of "coiling" phagocytosis, although it is unclear if this uptake pathway has any influence on the outcome of an infection. Within a few hours of entry, *Legionella*-containing vacuoles recruit intracellular organelles and become studded with ribosomes. The pH of these compartments has been calculated at pH 6.1, which is 0.8 pH unit higher than that of vacuoles formed around dead bacilli. Formalin-fixed or antibody-opsonized bacteria enter compartments that fuse with secondary lysosomes, whereas treatment of cells with antibiotics postinfection does not induce fusion of bacterial compartments with lysosomes. These data suggest that a "viable" bacterial compartment, once established, cannot revert to the endosomal network.

Immunofluorescence studies on *Legionella* vacuoles demonstrated the presence of the endoplasmic reticulum (ER) protein BiP, and it has been suggested that these vacuoles represent nascent autophagous vacuoles. Treatments which enhanced autophagy (amino acid starvation of the host cells) increased the association of ER membranes with these vacuoles and enhanced bacterial division. Recent mutational analyses have identified the *dot* or *icm* locus, which comprises a family of genes that encode an inducible secretion apparatus similar to the type III secretion systems described in chapter 13. *Legionella* organisms deficient in key members of this gene locus were unable to prevent delivery to lysosomes, culminating in the death of the bacilli. Although the substrates for this transport system have not been characterized with respect to function, it is reasonable to suppose that they play key roles in modulation of the differentiation of these vacuoles.

Chlamydia. The human pathogen *Chlamydia trachomatis* and the animal pathogen *C. psittaci* are both obligate intracellular parasites that invade nonprofessional phagocytes by activation of the host cell contractile system. Although the bacilli are actively internalized by the host cell, the vacuoles are rapidly remodeled by the pathogen to form parasitophorous vacuoles termed inclusion bodies (Figure 8.6). Early analysis of these compartments revealed that they did not possess acid phosphatase, nor did they fuse with ferritin-labeled secondary lysosomes. More detailed immunochemical analysis of the vacuole constituents revealed that they lacked cation-

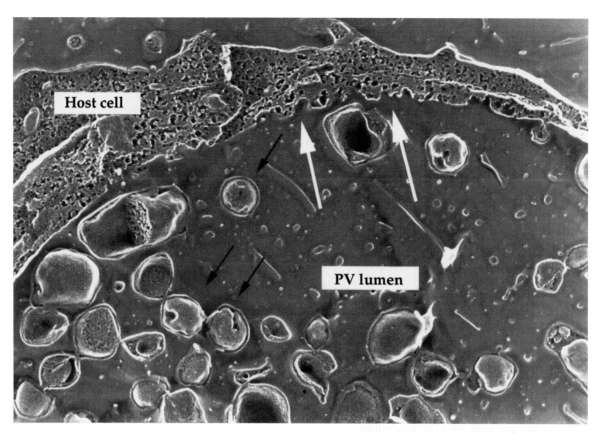

Figure 8.6 Electron micrograph of a freeze-etch preparation of a HeLa cell infected with *Chlamydia psittaci*. The bacteria (black arrows) form an inclusion body or parasitophorous vacuole (PV) that lies within the host cell cytosol (host cell) and is excluded from the normal endocytic routes of that cell. The vacuole membrane is smooth over most of its surface; however, it is ruffled with processes (white arrows) that extend into the host cell cytoplasm in the region that subtends the host cell endoplasmic reticulum. Courtesy of David G. Russell and Ted Hackstadt.

independent mannose 6-phosphate receptor (M6PR), transferrin receptor, both LAMP1 and LAMP2, cathepsin D, and vacuolar proton ATPase. Despite this clear lack of host cell proteins, the inclusion bodies showed a close association with annexins III, IV, and V. Annexins are implicated in Ca^{2+}-dependent, intracellular membrane fusion events. Indeed, although the chlamydial inclusion body did not fuse with endosomes or lysosomes, they were able to fuse with one another. Moreover, recent analysis with the fluorescent sphingomyelin precursor C_6-nitrobenzoxadiazolyl-ceramide demonstrated transport of a significant proportion of the label to inclusion body membranes and, subsequently, to the bacilli inside. These data suggest that there is at least some communication between the host cell Golgi apparatus and the parasitophorous vacuoles.

The successful formation of inclusion bodies is dependent on chlamydial protein synthesis early in the infection. Unlike *Toxoplasma* (discussed below), the route of entry appears irrelevant to bacterial survival. Antibody-opsonized bacilli internalized by Fc receptor-transfected HeLa cells were able to form viable inclusion bodies, unlike antibody-opsonized *Toxoplasma gondii*. These data suggest that bacterial products actively subvert the machinery of the host cell for normal endosomal differentiation irrespective of the receptor(s) ligated during entry.

Given that chlamydial vacuoles show limited communication with the host cell endosomal system, the routes of acquisition of nutrients are unclear. Experiments conducted to see if the inclusion membrane had pores for passage of molecules into the vacuole gave negative results for molecules down to 520 Da. However, inclusion bodies do contain pathogen-derived proteins in their membranes, and the identification and characterization of such proteins should elaborate the extent and nature of communication between *Chlamydia* and its host cell.

Toxoplasma. *Toxoplasma gondii* is a widespread apicomplexan parasite related to *Plasmodium.* The parasite is spread primarily through ingestion of cysts in undercooked meat or in food contaminated with cat feces. This can lead to the formation of cysts in the brain of the intermediate hosts, which include humans. Although in immunocompetent individuals this infection is almost invariably subclinical, in immunocompromised persons the parasite expansion is unrestricted and usually fatal. In some rural areas of the world where consumption of raw or rare meat is common, the sera of at least 70 to 90% of the population are positive for *Toxoplasma.*

The parasite is capable of invading and developing in an extremely diverse range of host cell types. Infection is initiated by ingestion of oocysts or tissue cysts, and the parasites form tachyzoites, which are highly motile and spread the infection. As discussed above, invasion is an active process that does not require any participation by the host cell cytoskeleton; on the contrary, ligation of receptors capable of triggering phagocytosis appears to direct the parasite into the lysosome.

Because the parasites are the sole mediators of this process, it is not surprising to find that invasion and the formation of parasitophorous vacuoles containing *Toxoplasma* require the triggered secretion of an array of proteins that are released into the nascent vacuole during the entry process (Figure 8.7). Early ultrastructural studies described an extensive tubular vesicular network in the vacuole that appeared to be contiguous with the vacuole membrane. More recent biochemical and immunolocalization studies have shown that the released proteins originate from three different secretory organelles: the micronemes, the rhoptries, and dense granules. The proteins were targeted differentially to the lumen of the parasitophorous vacuole, the tubular vesicular network, and the vacuolar membrane. One of these proteins, the nucleotide adenosine triphosphatase, has a known enzymatic function and may be involved in purine salvage, because *Toxoplasma* is a purine auxotroph. NTPase activity would yield AMP, which could be further dephosphorylated by the 5'-nucleotidase of the parasite facilitating transport of adenosine into the cell. In addition, the marked association between the membranes of the parasitophorous vacuoles and membranes of the host cell ER and mitochondria have added further significance to the localization of the rhoptry proteins ROP2, ROP3, ROP4, and ROP7 on the parasitophorous vacuole membrane. It has been suggested that these proteins may aid in the acquisition of nutrients by the parasite via two possible mechanisms: (i) through the sequestration of cellular organelles such as mitochondria which would provide a ready source of ATP and (ii) through the formation of pores in the vacuolar membrane, which have recently been shown to facilitate the passive transport of molecules up to 1,300 Da, facilitating access to nucleotides, amino acids, and small peptides from the host cell cytoplasm.

► *For Figure 8.7, see color insert.*

Plasmodium. *Plasmodium falciparum,* another apicomplexan parasite, is the cause of cerebral malaria, which is responsible for more than two million deaths annually in children under 5 years old in sub-Saharan Africa. In addition, the parasite induces severe morbidity in an extensive proportion of the population. Although it has no animal reservoir, it has been extremely resistant to control and, through the spread of drug resistance, represents an increasing threat.

The parasite has two intracellular stages in its human host. One, the exoerythrocytic schizont, appears similar to the parasitophorous vacuole of *Toxoplasma* (see above). The other is the intraerythrocytic trophozoite, and because of the metabolically inert nature of its host cell, the parasite has evolved some extremely interesting means of obtaining nutrients. The organism is delineated by two membranes, which are derived from the plasmalemma of the invading merozoite and the cell membrane of the erythrocyte, although this latter membrane most probably becomes parasite-derived once the infection is established.

The parasites appear to acquire macromolecular nutrients from their host via two discrete pathways. The first pathway is responsible for the defining characteristics of the disease, with its transient fevers that are induced when the parasites lyse their host erythrocyte and release a toxic pigment, hemozoin, that was generated during degradation of host cell hemoglobin. During the period of parasitization of the erythrocyte, which lasts only a matter of hours, trophozoites will degrade 50 to 90% of the host cell hemoglobin. The hemoglobin is internalized through a process similar to endocytosis, except that because parasites are bounded by two membranes, the vacuoles around the hemoglobin have a double membrane. These transport vesicles bud off from defined regions of the parasite surface known as cytostomes and then fuse with the digestive vacuole, or lysosome. This pathway is illustrated in the electron micrograph shown in Figure 8.8. Hemoglobin degradation appears to be an ordered process dependent on both aspartic and cysteine proteinases, which have acidic pH optima. The toxic heme group released during this process is sequestered in the inert hemozoin pigment stored in the vacuole.

In addition to this conventional pathway, the parasite is capable of obtaining macromolecular nutrients directly from the host serum. Early experiments indicated that small fluorescent tracers could access the intraerythrocytic parasite. It was suggested that this was the function of a continuous "duct" which maintained communication between the parasite and the external environment, a conclusion supported by additional studies in which fluorescent lipids were used to label both the parasite and the intraerythrocytic membranous labyrinth.

The contiguous or fragmented nature of this labyrinth has been the subject of debate. Recent analysis of the membranous labyrinth, which is also referred to as the tubovesicular membrane (TVM), demonstrated that it incorporated sphingomyelin, metabolized from a labelled ceramide analog, and possessed high levels of sphingomyelin synthase. It was therefore suggested that the TVM was highly dynamic, maintaining transient interactions with the erythrocyte plasmalemma and the external environment. In spite of the controversy surrounding the stability of this organelle and the selection of molecules capable of being transported, it has been shown that its maintenance is crucial for the acquisition of purines, orotic acid and glutamate. Inhibitors of the sphingomyelin synthase of the parasite caused breakdown of the TVM and had an immediate effect on nutrient acquisi-

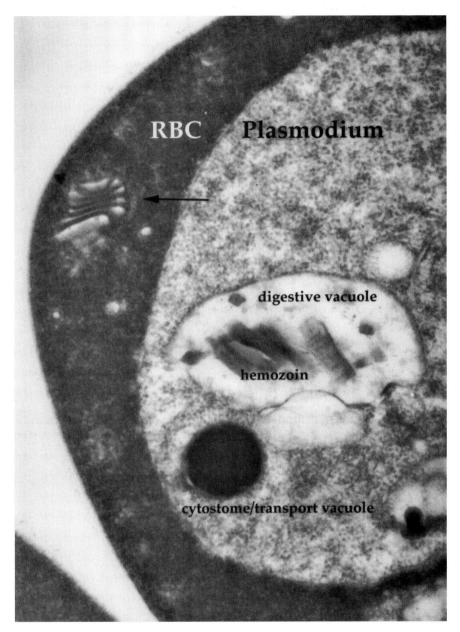

Figure 8.8   Electron micrograph of an erythrocyte (RBC) infected with *Plasmodium falciparum* (Plasmodium). The micrograph illustrates the degradative pathway followed by host cell hemoglobin. The protein is internalized through the parasite's cytostome into double-membrane transport vacuoles, which deliver the hemoglobin to the hydrolytic digestive vacuole. Following degradation of the protein, the released heme, which is toxic, is polymerized into the nonreactive polymer hemozoin. The hemozoin appears as crystalline structures in the digestive vacuole. Membranous processes known as Maurer's clefts are visible in the cytoplasm of the erythrocyte (arrow). Courtesy of David G. Russell and Daniel Goldberg.

tion, providing additional support for the notion that the TVM is relatively dynamic.

Another possible role for this extratrophozoite structure is the export and incorporation of parasite proteins into the plasmalemma of the erythrocyte. Parasitized erythrocytes show aggregates, or "knobs" of parasite-derived proteins on their surface. These proteins play a role in adherence and mediate the cerebral sequestration which is the main cause of pathology and death associated with these infections. Given the relatively inert nature of the erythrocyte, *Plasmodium* provides an interesting example of the strategies used by a pathogen that has to seriously remodel its host cell to build an inhabitable environment.

Pathogens Resident in Host Cell Cytosol

A few pathogens have evolved mechanisms that allow them to avoid the potentially hostile environment of the endosomal and lysosomal network of the host cell by escaping into the host cell cytosol. This group includes both bacteria (e.g., *Shigella, Listeria,* and *Rickettsia*) and a protozoan (*Trypanosoma cruzi*). All these pathogens exhibit membrane-disruptive activities that are optimal at low pH, suggesting that there is an "activation" step during the acidification of the vacuoles formed during parasite entry. Genetic characterization of the loci involved in vacuolar escape by the bacterial pathogens within this group has revealed a battery of secreted products, many with analogous functions, delivered by a secretory apparatus composed of homologous subunits. The acquisition of clustered arrays of genes involved in the maintenance of infection and virulence has obvious implications for the evolution of pathogenic mechanisms.

Although the adoption of a cytosolic location enables these pathogens to bypass the degradative pathway to the lysosome, exploitation of this niche carries several other problems, most specifically efficient cell-cell spread and avoidance of antigen presentation via class I MHC molecules.

Shigella. *Shigella flexneri,* one of the causative agents of bacillary dysentary, was the subject of some early experiments in the early and mid-1980s that marked the advent of the new era in microbial cell biology. It was observed that the ability of these bacilli to replicate inside cells in culture correlated not with the production of Shiga toxin but with the presence of a 140-MDa plasmid, pWR100. Moreover, introduction of the plasmid into *Escherichia coli* through transconjugation conferred the ability to survive inside HeLa cells. Ultrastructural analysis of infected cells revealed recombinant bacilli free in the cell cytosol. Coincubation of plasmid-positive and -negative strains of *Shigella* and *E. coli* together with erythrocytes demonstrated that the plasmid enabled the bacteria to induce lysis of erythrocytes.

More recent detailed analysis of the plasmid-borne genes identified a family of products, IpaA to IpaD. These proteins are secreted on contact with a host cell through a dedicated secretory system, encoded by the *mxi-spa* locus on the bacterial chromosome. The secreted proteins all perform functions designed to maximize the chances that the bacillus will infect the host successfully. IpaB associates with the interleukin-1β-converting enzyme and leads to the induction of apoptosis, and is discussed in depth in chapter 15. IpaC and IpaD are known to form polymers and to be required for binding and entry; however, it is still unclear if one or both of these proteins mediate vacuolar lysis. Binding of IpaC to liposomes was enhanced by acidic pH, suggesting that the intercollation of the protein into vesicles may be favored within the low pH of the endosome. *Shigella* mutants deficient in IpaC and IpaD could be complemented by the hemolysin of *E. coli,* suggesting that the molecules perform related functions. Intriguingly, however, recent characterization of the secreted proteins (Ssp) of *Sal-*

monella typhimurium revealed that the genes encoding SspB, SspC, SspD, and SspA are homologs of those encoding IpaB, IpaC, IpaD, and IpaA and also play a role in host cell entry; however, in contrast to *Shigella, Salmonella* remains within its phagosome.

Listeria.   *Listeria* is the causative agent of listeriosis and has emerged recently as the paradigm for defining the mechanisms of survival of an intracytosolic pathogen. Like *Shigella, Listeria monocytogenes* escapes from its vacuole into the cytoplasm shortly after entry into its host cell. However, in contrast to *Shigella,* where there is a membrane-disruptive activity but no defined molecule, *Listeria* possesses three well-characterized molecules with membrane-disrupting capabilities. In early experiments in 1990, it was shown that expression of the listeriolysin (LLO) gene in the nonpathogenic *Bacillus subtilis* converted this bacterium into one capable of survival in tissue culture cells through its acquired ability to escape from its phagosomes. LLO is a thiol-activated cytolysin that belongs to a family of homologous proteins present in several pathogenic bacilli, most notably *Clostridium* and *Streptococcus.* Experiments conducted on LLO-negative *Listeria* complemented with the closely related perfringolysin (PFO) from *Clostridium perfringens* demonstrated that minimal alterations in the amino acid sequence of PFO rendered it capable of fulfilling the same role as LLO in vacuolar escape.

Although LLO appears to be sufficient to confer vacuolar escape on *Listeria,* these bacteria also possess two different phospholipases which are released during bacterial invasion. The lipases include a broad-spectrum phospholipase C (PC-PLC) and a phosphatidylinositol-specific phospholipase C (PI-PLC). The PC-PLC is released from the bacilli in an active form, while the PI-PLC is secreted as an inactive proenzyme that requires cleavage for activation. Activation can be mediated by a bacterial metalloproteinase (Mpl) or by the activity of host cell cysteine proteinases, most probably cathepsins B and/or L. The activation of PI-PLC requires a low-pH environment, suggesting that activation is concomitant with maturation of the phagosomes into acidic, hydrolytic vacuoles. Experimental data suggest that the lipases play a role in escape from the double-membrane vacuoles formed during cell-cell spread.

To survive and move within and between cells, *Listeria* exploits the ability of ActA to promote binding and polymerization of the host cell actin cytoskeleton. This activity is fundamental to the intracytosolic life-style of this pathogen and is detailed extensively in chapter 10.

Rickettsia.   *Rickettsia prowazekii* is a louse-borne pathogen that is the causative agent of epidemic typhus; like both *Listeria* and *Shigella,* it is characterized by its ability to lyse its phagosome and escape into the host cell cytoplasm. Although it can independently express both phospholipase A_2 and hemolytic activities, it is not clear whether either or both of these activities are required for vacuolar escape. The hemolytic activity showed a mildly acidic pH optimum, which was enhanced by Ca^{2+}, implying the presence of an activity analogous to that of LLO.

Trypanosoma cruzi.   The one protozoan capable of gaining access to the host cell cytoplasm is the South American trypanosomatid *Trypanosoma cruzi. T. cruzi* is the etiologic agent of Chagas' disease, an infection transmitted by the bloodsucking reduvid *Rhodnius.* The infection is characterized by a transient, flu-like fever followed by a prolonged period of subclinical infection during which the parasite invades cardiac muscle cells and triggers immune system-mediated damage to the heart. In these later stages, infected individuals may suffer cardiac arrest prior to any other overt sign of disease.

Escape of *T. cruzi* trypomastigotes from their host vacuoles following internalization is pH dependent, because lysosomotropic bases such as chloroquine block entry into the cytoplasm. This requirement for an acidic, hydrolytic environment to trigger successful vacuolar escape is graphically illustrated by recent data demonstrating that, in contrast to many pathogens that try to avoid the rapid transition of their phagosome into a lysosome-like compartment, *T. cruzi* actively recruits lysosomes to its sites of host cell invasion. These lysosomal compartments were observed to fuse with nascent entry vacuoles prior to full internalization of the parasites. Entry by the parasites is also an active process that does not appear to require host cell actin, because antimicrofilament agents such as cytochalasin D actually enhance rather than suppress host cell invasion.

The ability to exit from phagosomes is a characteristic of both the trypomastigote and amastigote stages of the parasite, and fractionation of the activity assayed by hemolysis led to the identification of a protein, TcTox, which showed immunological cross-reactivity with the complement component C9. C9 is a component of the terminal complex, or membrane attack complex, polymerized to form a transmembrane channel or pore. It is hypothesized that TcTox has a function analogous to the pore-forming properties of LLO.

Another protein that has been implicated in facilitating the escape of *T. cruzi* from vacuoles is the transialidase of the parasite. The transialidase is encoded by a heterogeneous, multigene family whose products are glycosylated and range from 60 to 250 kDa. The enzyme is externally orientated on the surface of the trypomastigote stage and is retained in the parasite plasmalemma by a glycophosphatidylinositol anchor. The enzyme is capable of cleaving sialidase from host donor glycoproteins and transferring the sialic acid residues to its own proteins. This enzyme has been implicated in many phases of the host-parasite interplay, from host cell binding to immune evasion. However, mammalian lysosomal glycoproteins, such as LAMP1 (lgp120), are heavily sialylated, ostensibly to protect them from autocatalysis. Experimental infections of Lec2 cells, which are deficient in sialylation, demonstrated that vacuolar escape was significantly more rapid than in wild-type host cells, suggesting that the sialidase and transialadase activity of the parasite might fulfill an accessory role through facilitating easier access of TcTox to its target membranes.

Modulation of Intracellular Compartments by the Host Immune Response

The ability of the immune system to regulate intracellular pathogens is based on its capacity (i) to detect infection foci and (ii) to produce the appropriate response in either infected cells or bystander effector cells. Recently, much of the attention on killing of intracellular pathogens, in macrophages in particular, has focused on inducible nitric oxide synthase (iNOS). Expression of iNOS is turned on by gamma interferon and tumor necrosis factor alpha and leads to the production of NO from arginine. Inhibitors of NO production block the killing of many intracellular pathogens, and infections with *Listeria, Salmonella, Leishmania, Plasmodium,* and *Mycobacterium* species run uncontrolled in iNOS knockout mice. Obviously, iNOS fulfills a necessary function in regulation of these infections, but a full appreciation of its mode of action must take into account the cascade of other physiologic changes that occur during macrophage activation.

For intracellular pathogens that need to actively maintain their intracellular compartments, such as *Mycobacterium* species, the obvious question is which event comes first: the death or compromise of the infecting microbe or the differentiation of its compartment into an acidic, hydrolytically competent lysosome. If the latter is true, this translocation could drastically alter both the environment and the cofactors that would potentiate the efficacy of NO. Experiments on murine macrophages infected with *Mycobacterium avium* demonstrated that activation facilitated the acidification of mycobacterial vacuoles in both de novo and established infections. The functional translocation toward more lysosomal compartments preceded any marked drop in microbial viability, suggesting that it was the product of an alteration in macrophage physiology rather than a consequence of microbial death.

The lysosomal environment of activated macrophages could potentiate NO toxicity in several ways. Oxidation of NO to nitrite and nitrate is retarded at acidic pH. NO can combine with reactive oxygen intermediates, whose production is up-regulated in activated macrophages, to make peroxynitrite ($ONOO^-$). NO can release metal ions, such as Fe^{2+}, from metalloproteins, which can combine with H_2O_2 to produce ·OH and hypervalent Fe. Furthermore, the activity of lysosomal hydrolases on the microbial cell wall will probably expose more targets to oxidative attack. The microbicidal responses of activated macrophages are probably based on the complex interactions of several antimicrobial phenomena, and more work on the effects of activation on the regulation of intracellular fusion within the endosomal-lysosomal continuum is required before these interactions can be appreciated.

Cell-Cell Spread (Metastasis) of Intracellular Pathogens

To be successful, intracellular pathogens must evolve a strategy to enable them to spread to new host cells or ultimately to new hosts. Most pathogens cause the death of their host cell at this stage. For some pathogens, such as *Plasmodium, T. gondii,* and *T. cruzi,* this has led to development of a "swarm" response whereby the host cell is lysed and there is a coordinate release of infective forms of the microbe. Other pathogens, such as *Leishmania,* just induce a progressive deterioration of their host cell, and, because they exist as amastigotes (infective forms), they infect fresh macrophages on phagocytosis following release by their degenerate host cell. These strategies appear to require minimal participation by the host cell. In contrast, both *Shigella* and *Salmonella* are capable of triggering programmed cell death (apoptotic) responses in their host cells. For *Shigella,* at least, this occurs through a very specific route, and the mechanisms behind the induction of these cell destruction pathways are discussed in chapter 15.

Finally, one of the most intriguing means of cell-cell spread is through direct transfer, which is exploited by *Listeria, Shigella,* and *Rickettsia* and is a direct product of the ability of these bacteria to subvert the actin component of the host cell cytoskeletal network to their own devices (this is discussed in chapter 10).

Conclusion

Intracellular parasitism covers the diverse life-styles of a broad phylogenetic spectrum of pathogens. This chapter has attempted to present the major points in the biology of these pathogens within a thematic framework from the time of initial infection, through the choice of intracellular niche, avoidance or exploitation of the immune response, and culminating in the metastasis or spread of the infection. These processes function as continua that have critical points of decision which determine the fate of the microbe, and this chapter has described these decision points for different pathogens and explored the consequences of the "wrong" decision. Since the infection process is a continuum, it is equally important to read the chapters that deal with issues bordering on the establishment and maintenance of an intracellular infection, and these points of intersection have been noted in the text.

Selected Readings

Bielecki, J., P. Youngman, P. Connelly, and D. Portnoy. 1990. *Bacillus subtilis* expressing a haemolysin gene from *Listeria monocytogenes* can grow in mammalian cells. *Nature* **345:**175–176.

Burleigh, B. A., and N. W. Andrews. 1995. The mechanisms of *Trypanosoma cruzi* invasion of mammalian cells. *Annu. Rev. Microbiol.* **49:**175–200.

Clemens, D. L., and M. A. Horwitz. 1996. The *Mycobacterium tuberculosis* phagosome interacts with early endosomes and is accessible to exogenously administered transferrin. *J. Exp. Med.* **184:**1349–1355.

Doborowski, J. M., and L. D. Sibley. 1996. *Toxoplasma* invasion of mammalian cells is powered by the actin cytoskeleton. *Cell* **84:**933–939.

Finlay, B. B., and S. Falkow. 1997. Common themes in microbial pathogenicity revisited. *Microbiol. Mol. Biol. Rev.* **61:**136–169.

Lauer, S. A., P. K. Rathod, N. Ghori, and K. Haldar. 1997. A membrane network for nutrient import in Red cells infected with the malaria parasite. *Science* **276:**1122–1125.

Sansonetti, P. J., A. Ryter, P. Clerc, A. T. Maurelli, and J. Mounier. 1986. Multiplication of *Shigella flexneri* within HeLa cells: lysis of the phagocytic vacuole and plasmid-mediated contact hemolysis. *Infect. Immun.* **51:**461–469.

Sturgill-Koszycki, S., P. Schlesinger, P. Chakraborty, P. Haddix, H. Collins, A. Fok, R. Allen, S. Gluck, J. Heuser, and D. G. Russell. 1994. Lack of acidification in *Mycobacterium* phagosomes produced by exclusion of the vesicular proton-ATPase. *Science* **263:**678–681.

Sturgill-Koszycki, S., U. Schaible, and D. G. Russell. 1996. *Mycobacterium*-containing phagosomes are accessible to sorting endosomes and reflect a transitional state in normal phagosome biogenesis. *EMBO J.* **15:**6960–6968.

Vogel, J. P., H. L. Andrews, S. K. Wong, and R. R. Isberg. 1998. Conjugative transfer by the virulence system of *Legionella pneumophila*. *Science* **279:**873–876.

9

Actin Filaments: Self-Assembly and Regulatory Interactions

FREDERICK S. SOUTHWICK AND DANIEL L. PURICH

The cytoskeleton is composed of three topologically distinct protein polymer networks, each of which is assembled from monomers consisting of actin, tubulin, or intermediate-filament proteins. Current evidence suggests that the actin cytoskeleton is involved mainly in microbial pathogenesis. Actin has the capacity to hydrolyze the terminal P-O-P bonds of nucleoside 5'-triphosphates, thereby providing the thermodynamic impetus for both polymer assembly/disassembly and force generation. Thus, in addition to serving as passive scaffolds for myosin motors by nucleotide hydrolysis, actin filaments can actively provide energy for cell motility.

Not surprisingly, the precision required for cell crawling, phagocytosis, and other vital motile processes necessitates the involvement of many regulatory proteins. Taking pseudopod and lamellipodium formation as an example, the viscoelasticity and structural organization of the peripheral cytoplasm must undergo rapid and extensive remodeling. This is accomplished at spatially defined sites for filament assembly and disassembly in response to specific intracellular and external stimuli. For the leading edge of the cell to advance, facile addition of actin to the ends of actin filaments at or near the junction of the cytoplasm and peripheral membrane is required; disassembly at the opposite ends of these filaments then serves to replenish the actin monomer pool needed for continual assembly.

Researchers generally agree that actin polymerization per se can serve as a force-generating mechanism. This process appears to function during the rapid expansion of the leading edge of a host cell or in the intracellular locomotion of a pathogen. The major challenge faced by host cells and pathogens alike is how to devise suitable mechanisms for harnessing the forces of actin filament elongation to achieve vectorial locomotion. Molecular adaptability is the essence of a pathogen's capacity to survive within host cells and to adopt unique infective strategies for spreading from cell to cell. The pathogens *Listeria, Shigella, Rickettsia,* and vaccinia virus have developed their own signature mechanisms for usurping elements of the host cell contractile apparatus. Because dynamic changes in the actin cytoskeleton appear to be an essential feature, this chapter focuses on actin and actin regulatory proteins. Future work may demonstrate that bacterial pathogens can also utilize tubulin and possibly intermediate filaments. In fact, there is already evidence that tubulin is essential for the synthesis of both Sendai virus and vesicular stomatitis virus RNAs. Therefore, in the future, microtubules and intermediate filaments may emerge as important factors in microbial cell biology.

Basic Properties of Actin Monomers and Filaments

Actin exists in two principal forms, the 43-kDa globular monomer (designated G-actin) and filamentous actin (hence the term F-actin). During polymerization, actin-bound ATP is hydrolyzed. After actin polymerizes, some fraction of the energy freed during ATP hydrolysis can be stored within the filament, most probably as a conformationally altered actin structure. This stored energy provides a driving force to ensure that actin filaments can disassemble later in response to the appropriate regulatory cues. There is also emerging recognition that the actin-nucleotide complex "matures" during the polymerization process—first binding to the filament end as actin-ATP, then hydrolyzing to filament-bound actin-ADP-P_i, and finally releasing phosphate (P_1) to form the actin-ADP complex. There may also be additional intermediates, such as a conformationally energetic [actin*-ADP-P_1] species, that lie at higher potential energy and form transiently during the F-actin maturation process. By analogy to the GTP-regulatory proteins, the ability of actin to assume these various conformational and/or phosphorylation states may represent a timing mechanism that controls the strength of regulatory-protein interactions with actin subunits in distinct regions running along the assembled actin filaments. For example, various filament-severing processes are likely to sense the maturation state of assembled filaments by binding to ADP-rich regions rather than those with a higher ATP content (see below).

Atomic-level structures have been solved by X-ray analysis for the ATP and ADP complexes at effective resolutions of 2.8 and 3Å, respectively. The actin molecule has two easily discernible domains that can be further subdivided into two subdomains (Figure 9.1). Accordingly, actin has a deep central groove or cleft serving as the adenine nucleotide-binding site. ATP or ADP can bind in this cleft, and a calcium ion is also bound to the β- and γ-phosphoryls of ATP or the β-phosphoryl of ADP. The overall tertiary structure of the monomer closely resembles that of the well-known glycolytic enzyme, hexokinase, and both are characterized by their own five-stranded β-sheet motif, consisting of a so-called β-meander and a right-handed β-α-β unit in each domain. As with hexokinase, the overall structure suggests that gene duplication probably occurred early in the course of evolution. Actin also exhibits extensive structural similarity to the amino-terminal domain of the heat shock proteins, and although there is no obvious similarity of their amino acid sequences, both actin and HSP70 have an ATP-binding site that closely resembles hexokinase.

Although the nucleotide-binding site appears to be readily accessible to the solvent, actin-bound nucleotide is relatively firmly held within the nucleotide-binding pocket. In the absence of any exchange-promoting factor, reconversion of actin-ADP to actin-ATP is slow. ATP cannot directly phosphorylate actin-bound ADP; instead, ADP must dissociate from the actin-ADP complex to form nucleotide-free actin, and only then can ATP bind and become incorporated. ADP release is apt to be the slowest or rate-

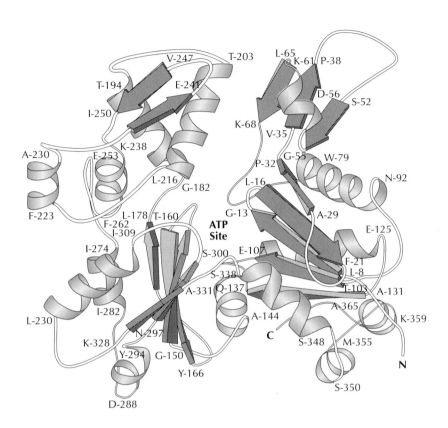

Figure 9.1 Atomic-level structure of an actin monomer showing the ATP-binding site. Based on the work of W. Kabsch, H. G. Mannherz, D. Suck, E. F. Pai, and K. C. Holmes, *Nature* **347**:37–44, 1990. The vertical axis of the monomer (as depicted in this figure) runs parallel to the long axis of the filament. The right-hand side of the molecule is exposed to the outside of the actin filament, whereas the left-hand side is nearest the long axis of the filament. Residues 262 to 274 are thought to reach across this axis and interact with the adjacent actin monomer of the double-stranded helix. As oriented here, the polarity of the filament would correspond to the pointed end at the top and the barbed end at the bottom.

limiting step in the exchange process. The rate of this dissociative process is markedly accelerated in the presence of the actin-regulatory protein known as profilin, and the significance of this facilitated exchange reaction is considered later in this chapter.

The atomic-level structure of the polymerized actin filament has been inferred by computer-based modeling and construction. This process requires one to fit the atomic structure of the actin monomer with fiber structure data obtained from independent X-ray studies of actin filaments. The derived orientation of actin subunits within the filament is completely consistent with what is known about the head-to-tail polymerization process, confirming that the two ends of an assembled actin filament are structurally different. Actin monomers assemble into a double helix, allowing parallel monomer subunits to interact and stabilize the filament. Interstrand stability also appears to benefit from an extended β-hairpin or "hydrophobic loop" (residues 262 to 274) that reach across the filament axis to a site on the adjacent strand of the double-stranded actin filament. The polarity of actin filaments can be readily demonstrated by electron micrographs of actin filaments that have been decorated by the attachment of myosin head fragments. These subfragments bind tightly to the actin filament, assuming a 45° angular tilt toward the fast-growing end, and define a barbed and pointed filament end. As described below, the two ends of an actin filament differ greatly in their ability to interact with actin monomers.

Actin Assembly Kinetics

The four principal stages of actin polymerization are nucleation, elongation, attainment of monomer-polymer equilibrium, and treadmilling. It is worth emphasizing that each step depends on the state of phosphorylation of actin-bound adenine nucleotide in monomers and filaments.

Nucleation is thought to be the most unfavorable step in the actin assembly process. In vitro experiments suggest that three actin-ATP molecules must come together to form a polymerization nucleus. The thermodynamically unstable nuclei rapidly dissociate back into monomers, and unless the concentration of actin-ATP complex is sufficiently high, nuclei cannot persist at sufficient concentrations to promote efficient polymerization. Note also that in cells any spontaneous nucleation is probably efficiently suppressed at all regions other than those actively undergoing actin assembly; otherwise, indiscriminate actin filament assembly would result in unregulated changes in the consistency and shape of the cell. Unregulated nucleation would also deplete the stores of actin monomer needed for regulated filament assembly. Actin monomer-sequestering proteins (see below) probably limit nucleation by decreasing the available concentration of free actin-ATP monomers. ADP-actin complex also appears to be a potent inhibitor of nucleation, so conversion of ADP-actin to ATP-actin via nucleotide exchange can be expected to promote nucleation. The process of nucleation can persist only when actin-ATP is maintained at a high concentration. During actin filament elongation in vitro, the concentration of available actin-ATP begins to decrease as the polymer accumulates. For example, if [actin-ATP$_{total}$] is the total monomer concentration at the beginning of polymerization, then the rate of spontaneous nucleation will be proportional to the third power of A_{total} (i.e., rate$_{nucleation}$ = k_N [actin-ATP$_{total}$]^3), and a decrease to 0.8 [actin-ATP$_{total}$] would reduce the rate of nucleation by more than a factor of 2.

Elongation occurs as an indefinitely repeated series of monomer addition reactions. Actin-ATP first binds to a growing filament end; the bound ATP then undergoes hydrolysis during or slightly after the next monomer addition step occurs (Figure 9.2A). This ATP hydrolysis reaction forms actin-ADP-P$_1$ within the polymer lattice, and P$_i$ is released later. From a macroscopic point of view, actin polymerization can be measured by a number of techniques. (i) Capillary viscometry is a method based on the higher viscosity of F-actin than of G-actin. The time required for a solution to pass through the narrow orifice of a glass capillary viscometer is a direct reflection of the mean filament length of the actin solution; the lower the flow rate, the longer the filament length distribution. (ii) Fluorescence, the most widely applied technique, uses an extrinsic chromophore (typically NBD or pyrene) that is covalently attached to actin monomers at cysteine-374. The increase in fluorescence intensity directly correlates with the actin filament concentration. (iii) Light scattering takes advantage of the fact that assembled filaments scatter light with much more intensity than individual actin subunits do. (iv) Electron microscopy allows direct assessment of polymer length after suitable fixation and contrast staining. (v) DNase I inhibition capitalizes on the ability of DNase I to bind preferentially to actin monomers with sufficient affinity to block the ability of this enzyme to cleave DNA. (vi) Ultracentrifugal pelleting relies on the much higher sedimentation coefficient of actin filaments than of monomeric actin. (vii) Filtration assays allow one to rapidly separate monomeric and polymeric actin by using filter disks with 0.45-μm pores. Depending on the experiment, each technique has inherent advantages and limitations.

The observed kinetics of actin elongation conform to that predicted by the following rate law:

$$dC_p/dt = k_+[\text{F}][\text{actin-ATP}_{\text{monomer}}] - k_-[\text{F}]$$

where C_p is the concentration of polymerized actin, k_+ is the macroscopic rate constant for monomer association, [F] is the number concentration of polymer ends that react with actin monomers, $\text{ATP}_{\text{monomer}}$ is the monomer concentration, and k_- is the macroscopic first-order rate constant for monomer dissociation. If [F] is constant during the elongation phase (i.e., if nucleation occurs only at the outset of in vitro polymerization), then the observed rate process will fit a simple first-order decay curve. The polymer elongation phase will continue until the rate of monomer addition (i.e., $k_+[\text{F}][\text{actin-ATP}_{\text{monomer}}]$) is exactly balanced by the rate of monomer release ($k_-[\text{F}]$). This condition defines the critical actin concentration ([actin]$_{\text{critical}}$ $= k_-/k_+$), a parameter representing the concentration of monomeric actin that coexists with assembled filaments. In a plot of polymerized actin versus actin monomer concentration (Figure 9.3, left), no polymer will be observed until the monomer concentration exceeds [actin]$_{\text{critical}}$. (This behavior is precisely the same as that of the critical concentration for micelle formation observed with lipids and other amphiphilic solutes.)

In most cases, both actin filament ends are free to self-assemble, and the macroscopic on-rate constant equals the sum of the microscopic on-rate constants k_{+b} and k_{+p} describing actin monomer addition to the barbed and pointed ends of an actin filament, respectively; likewise, the macroscopic off-rate constant equals the sum of the corresponding microscopic off-rate constants k_{-b} and k_{-p} for the two ends of each polymer.

Two additional dynamic processes can continue, even after the elongation phase has terminated. First, the rapid onset of monomer-polymer equilibrium often occurs before the filaments achieve their own equilibrium concentration behavior. Thus, assembled filaments will undergo polymer length redistribution until the system reaches an equilibrium distribution of polymer lengths. This very slow process resembles crystallization in many respects (i.e., initially formed, small imperfect crystals will tend to dissolve away as larger, well-structured crystals accumulate). The second and more significant dynamic property of assembled actin filaments is termed "treadmilling." This process was first described by Wegner (1982) to account for the effects of ATP hydrolysis on actin polymer dynamics. As noted above, the barbed end and the pointed end have different rate and equilibrium constants: $K_b = k_{-b}/k_{+b}$ and $K_p = k_{-p}/k_{+p}$, such that [actin]$_{\text{critical}} = \Sigma k_{\text{off}}/\Sigma k_{\text{on}} = (k_{-b} + k_{-p})/(k_{+b} + k_{+p})$. K_b has been determined experimentally to be 0.2 μM, and K_p has been determined to be 0.5 to 0.6 μM (Figure 9.3, right). The macroscopic actin critical concentration lies between K_b and K_p and is 0.25 μM. Because the barbed end has a higher exchange rate than the pointed end, this end makes a greater contribution

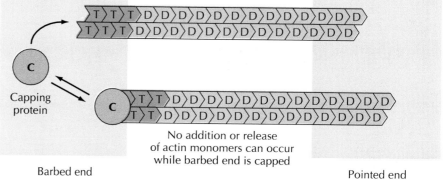

Figure 9.2 **(A)** Model of actin filament elongation in vitro showing how actin-ATP complexes preferentially add to the barbed end. **(B)** Schematic diagram of the action of monomer-sequestering agents, such as Tβ4. **(C)** Mechanism of a barbed-end capping protein binding to the barbed (or plus) end of an actin filament. Bound capping proteins prevent both association and dissociation of monomers; under such conditions, only the pointed end can interact with the actin monomer pool. See panel A. **(D)** Model for the action filament-severing proteins. The severing protein first binds along the side of the actin filament, next interposes itself between neighboring actin subunits within the filament, and then remains tightly bound to the barbed end of one of the severed filaments. **(E)** Schematic diagram showing the bundling protein α-actinin, which cross-links actin filaments into parallel arrays. α-Actinin molecules form an antiparallel dimer, and each subunit contains an actin filament-binding site. **(F)** Action of a depolymerizing factor in enhancing disassembly from the pointed end of the actin filament. The dark square with a D represents an ADF or cofilin molecule binding alongside an actin filament at a site near the pointed end. Upon binding, this protein enhances the rate of the pointed end, thereby accelerating the treadmilling rate of uncapped filaments.

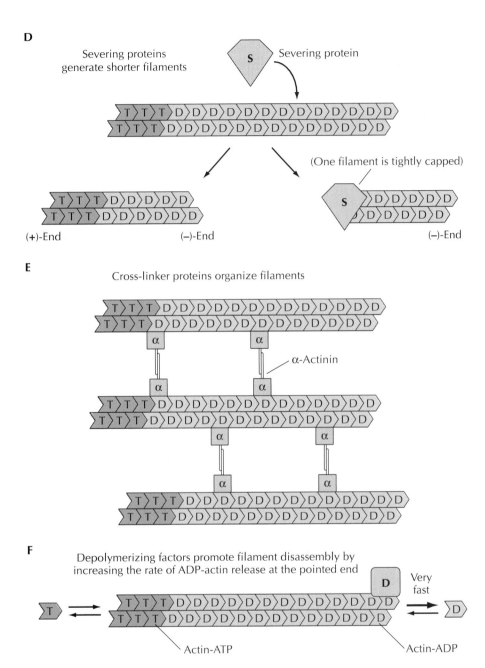

Figure 9.2 *continued*

to the macroscopic critical concentration. These characteristics result in an important set of inequalities that explain treadmilling.

$$(K_b \text{ or } k_{-b}/k_{+b}) < [\text{actin}]_{\text{critical}} < (k_{-p}/k_{+p} \text{ or } K_p)$$

When ATP is present, actin-ATP monomers are taken up at the more stable, higher-affinity barbed ends. Over time these monomers treadmill through the filament and are eventually released from the less stable, lower-affinity pointed ends. As the name implies, actin treadmilling does not involve any net increase in the amount of polymerized actin, since the addition of actin

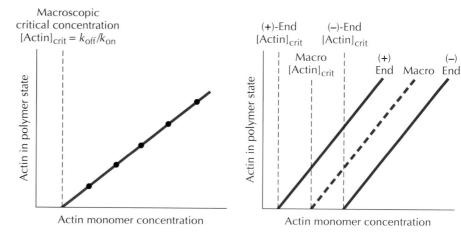

Figure 9.3 Critical-concentration behavior in actin polymerization. (Left) Plot of the steady-state actin filament concentration as a function of monomer concentration. This macroscopic behavior is often measured by the increase in fluorescence when pyrenyl-actin is incorporated into filaments. (Right) When both filament ends are uncapped, the macroscopic critical concentration lies between the microscopic critical concentrations for actin monomer interactions at the barbed end [or (+) end] and the pointed end [or (−) end]. At steady state, monomers will naturally dissociate from the pointed ends and will associate with the more stable barbed ends. This phenomenon is known as treadmilling.

monomers to the barbed end is equal to dissociation of actin monomers from the pointed end.

Although investigations of the in vitro polymerization have served as a theoretical underpinning for understanding actin dynamics, there are obvious voids in our comprehension of the exact events occurring in vivo. Intracellular actin-based motility in many cell types attains elongation rates of 0.5 to 1 μm/s, corresponding to the addition of 50 to 100 monomers per filament per second. This is much faster than can be observed in any in vitro kinetic study, and the mechanism of assembly may differ substantially from the model studies. For example, although exchange of ATP for ADP on actin is not a rate-limiting reaction in vitro, this may not be true in the cell. Intracellular F-actin formation is also highly localized, occurring within discrete polymerization zones established by assembly of actin-based motility (ABM) complexes assembled from a battery of regulatory proteins. Moreover, actin assembly is controlled by extracellular cues transmitted to the interior of the cell by focal adhesion and adherens junction proteins.

Interactions of Regulatory Proteins with Actin Monomers and Filaments

Changes in actin filament concentration and length underlie the shape changes required for leukocytes to migrate to sites of infection, for fibroblasts and endothelial cells to migrate to areas of wound healing, and for platelets to plug leaking vessels. When these processes become unregulated or deranged, there are serious breaches in the ability of the host to ward off pathogens. To understand the ways in which pathogens may subvert actin-based motility, one must consider the properties of the large repertoire of actin-regulatory proteins found in nonmuscle cells. These proteins can be classified into five categories: (i) actin monomer-binding proteins, (ii) actin filament-capping proteins, (iii) actin filament-severing proteins, (iv) actin filament-bundling and cross-linking proteins, and (v) actin-depolymerizing proteins (Table 9.1).

Monomer-Binding Proteins

One mechanism for regulating actin filament assembly is by binding to and altering the ability of this large population of actin monomers to polymerize. Although purified actin displays a low critical concentration of 0.25 µM (see above), the cytoplasm of nonmuscle cells contains monomeric actin at concentrations of 100 to 400 µM, corresponding to 50 to 60% of the total cell actin. Two actin monomer-binding proteins work in concert to maintain the high concentrations of unpolymerized actin and help to regulate actin filament assembly in nonmuscle cells.

The factor primarily responsible for preventing actin monomer assembly into filaments is the 5-kDa polypeptide known as thymosin β4 (often abbreviated as Tβ4). This small protein binds to an actin monomer to form a 1:1, or binary, complex. When bound to actin, Tβ4 sterically hinders spontaneous actin filament assembly (Figure 9.2B). Tβ4 concentrations in cells are also in the 100 to 400 µM range, matching well with the concentrations of polymerization-incompetent monomeric actin in resting nonmuscle cells. Tβ4 has a higher affinity for ATP-actin monomers (the form of actin that most readily forms actin filaments) than for ADP-actin, and the exchange of actin-bound adenine nucleotide appears to be blocked by the interaction of Tβ4 with G-actin. Because the affinity of the free barbed filament end for actin monomers is greater than that of Tβ4 (K_D for ATP actin monomer barbed end = 0.20 µM versus 1 to 2 µM for ATP actin monomer-Tβ4), the barbed end of an actin filament is capable of removing actin monomers from the Tβ4 complex. Addition of stable actin nuclei results in the rapid formation of filaments from the Tβ4-actin monomer. Thus, while Tβ4 is acting as an actin monomer-sequestering protein to prevent indiscriminate assembly of new filaments, its biochemical properties permit regulated growth at the barbed ends of preformed actin filaments.

The other actin monomer-binding component, a 15-kDa protein known as profilin, was the first actin-regulatory protein discovered, originally being isolated from spleen extracts. The intracellular concentration of profilin is approximately one-fourth that of Tβ4 and is too low to account for the high concentrations of monomeric actin in nonmuscle cells, suggesting that profilin could be serving other functions. Like Tβ4, profilin binds actin monomers to form a 1:1 complex, but the versatility of profilin interactions with G- and F-actin go far beyond its ability to bind actin monomers. In this respect, profilin has a complex set of functions in actin assembly that remain to be completely clarified. Profilin clearly reduces the final steady-state F-actin concentration, and while it was initially thought to act primarily as a monomer-sequestering protein, its primary inhibitory effect is on nucleation of actin assembly. This action prevents spontaneous nucleation, which could result in the uncontrolled formation of new actin filaments. Unlike Tβ4, which inhibits nucleotide exchange, profilin binds to actin monomers and markedly enhances the rate of nucleotide exchange (Figure 9.4). Low concentrations of profilin can catalyze nucleotide exchange, even in the presence of high Tβ4 concentrations. Therefore, profilin would be expected to readily convert ADP-actin to ATP-actin, the form of

Figure 9.4 Schematic diagram of how profilin may facilitate the exchange of ATP for ADP on an actin monomer. When profilin binds to an actin monomer, the central cleft of actin opens, making the nucleotide-binding site more accessible for release and exchange. Because the ATP concentration far exceeds the ADP concentration in living cells, ATP will readily replace ADP from the actin monomer.

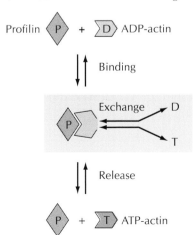

Profilin facilitates nucleotide exchange

Table 9.1 Actin-regulatory proteins

Protein	Interactions
Monomer-binding proteins	
Tβ4	Preferentially interacts with actin-ATP; weaker binding to actin-ADP
Profilin	Accelerates ATP-ADP nucleotide exchange; binds to oligoproline sequences in VASP and other related proteins; binds PIP_2
Barbed-end-capping proteins	
Profilin	May bind as profilin or profilin-actin complex
CapG	Calcium-sensitive binding interaction; binding of PIP_2 inhibits capping function
CapZ	Calcium-independent binding interaction; binding of PIP_2 inhibits capping function
Gelsolin	High-affinity binding interaction associated with actin filament severing; requires calcium; binding of PIP_2 inhibits capping function
Filament-severing proteins	
Gelsolin	Initial side binding to filaments, followed by filament severing (also see above)
Cross-linking/bundling proteins	
α-Actinin	Creates colinear arrays of actin filaments; interacts with vinculin and phospholipids
ABP-280	Creates orthogonal networks
Actin-depolymerizing proteins	
Cofilin/ actin-depolymerizing factor	Accelerates disassembly at the pointed end of actin filaments; thought to selectively interact with ADP-rich regions in actin filaments; probably weakly severs actin filaments
Other proteins	
Actin-related proteins (ARPs)	Bind to the sides of actin filaments; caps pointed ends; binds profilin; nucleates actin assembly
VASP	Binds to zyxin and vinculin; has multiple polyproline-binding sites for profilin
Mena and N-Mena	Similar in primary structure to VASP; bind profilin
Vinculin	Found in focal contacts; binds VASP; α-actinin, and actin filaments
Zyxin	Found in focal contacts; binds VASP; may directly bind profilin

monomeric actin that favors actin assembly. Because the concentration of ATP-actin is unlikely to be rate limiting in unstimulated cells, this capacity to stimulate nucleotide exchange function may not be critical in the resting cell. However, in cellular regions undergoing rapid actin filament turnover, the concentrations of ADP-actin would be expected to increase locally and the supply of ATP-actin might limit the actin polymerization rate if profilin were absent. Thus, profilin may assist rapid actin filament assembly by continuously maintaining the supply of ATP-actin monomers. Finally, profilin also interacts with the barbed ends of actin filaments, and in the presence of Tβ4, profilin may act to lower the critical concentration of the barbed filament ends. Thus, in addition to increasing the supply of ATP-actin monomers, profilin may usher actin monomers onto the barbed ends of actin filaments. To summarize our present understanding of profilin function, this protein blocks the formation of new filaments by inhibiting nucleation, whereas at the same time it can stimulate the elongation of preformed actin filaments by enhancing ATP-ADP exchange on actin monomers and enhancing actin monomer addition to the barbed ends of actin filaments.

Profilin concentrates in regions where actin filaments are assembling, an observation that is consistent with its involvement in localized assembly of actin. How can profilin concentrate in regions where actin filaments are assembling? Profilin is the only actin-regulatory protein known to bind to poly-L-proline, a property that has been widely exploited in its purification. The ability to bind to proline stretches is likely to play an important role in localizing profilin to sites of new actin filament assembly. A number of proteins have recently been shown to contain oligoproline repeats that are capable of binding profilin. The best studied of these profilin-docking proteins is vasodilator-stimulated phosphoprotein (VASP) (see below). Profilin also associates with the plasma membrane and likewise binds the phosphoinositide phosphatidylinositol-4,5-bisphosphate (PIP_2). When micelles containing this phospholipid bind to a profilin-actin complex, actin is released from the complex. Profilin-PIP_2 complexes inhibit phospholipase C, and therefore profilin may also serve as a regulator of the phosphoinositide pathway.

Actin Filament-Capping Proteins

Proteins that bind to the barbed ends of actin filaments have a profound effect on filament growth (Figure 9.2C). The barbed end has a high affinity for actin monomers and can readily compete with Tβ4 for sequestered ATP-actin monomers. When the barbed ends of actin filaments are capped, the critical concentration of the actin filament increases to that of the pointed end, 0.6 μM (Figure 9.5, left). Barbed-end capping also lowers the depolymerization rate of actin filaments because the K_- for the free pointed ends is considerably slower than that for the barbed end (Figure 9.5, right). In the presence of Tβ4 and profilin, the lower-affinity pointed end is unable to efficiently compete for actin monomers, and significant growth of actin filaments is unlikely to occur. Therefore, in the cell, the barbed filament end is the primary site for actin filament assembly. For unstimulated nonmuscle cells to maintain the high concentrations of unpolymerized actin, nearly all of the barbed ends must be blocked from competing with Tβ4 for actin monomers. Given the importance of regulating the barbed end to maintain high concentrations of monomeric actin in resting cells and to initiate new actin filament assembly during cell movement, it is not surprising that nonmuscle cells contain multiple proteins that are capable of binding the barbed end and blocking monomer exchange. These proteins are called barbed-end-capping proteins.

CapZ is a heterodimer consisting of a 36-kDa α subunit and a 32-kDa β subunit. The protein derives its name from the observation that it localizes to the Z line in skeletal muscle. This protein has also been called β-actinin and capping protein. CapZ is found in all nonmuscle cells and is particu-

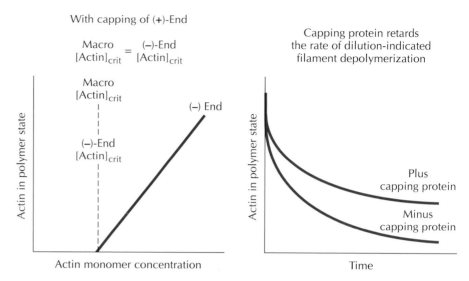

With capping of (+)-End

$$\frac{\text{Macro}}{[\text{Actin}]_{\text{crit}}} = \frac{(-)\text{-End}}{[\text{Actin}]_{\text{crit}}}$$

Figure 9.5 Effects of a barbed-end capping protein on the critical concentration and rate of depolymerization of actin. (Left) Graph of steady-state actin filament concentration as a function of actin monomer concentration. As shown in Figure 9.2C, capping blocks all exchange at the barbed end. The free pointed end has a lower affinity for actin monomers, and this lower affinity is reflected as an increase in the critical concentration (see Figure 9.3 for comparison). (Right) Plot of the decrease in filamentous actin versus time after diluting actin filaments to below their critical concentration in the absence and presence of a barbed-end capping protein. Capping of the barbed end retards the depolymerization, because dissociation of actin monomers occurs only at the pointed end.

larly abundant in neutrophils, where it represents 1% of the total cytoplasmic protein. This protein binds the barbed ends of actin filaments and prevents the release or addition of actin monomers. The dissociation constant for the CapZ filament end is 0.5 to 3 nM. The ability of CapZ to cap filaments is not affected by ionized calcium concentration but is blocked by PIP_2. In the submicromolar Ca^{2+} concentrations found in unstimulated cells, CapZ would be expected to cap the barbed ends, preventing them from competing with Tβ4 for actin monomers and allowing the cell to maintain a high concentration of unpolymerized actin (see above). Studies of CapZ in platelets reveal that stimuli that induce platelet actin assembly result in the uncapping of the ends of actin filaments by CapZ.

Another capping protein that differs structurally from CapZ is CapG, a 38-kDa protein that is closely related to the actin filament-severing proteins known as gelsolin and villin. While most abundant in macrophages and neutrophils, CapG is also found in most other cell types, with the notable exception of platelets. The affinity of CapG ($K_D = 1$ nM) for the barbed end is comparable to that of CapZ. Unlike other members of the gelsolin-villin family, CapG caps only the barbed ends of actin filaments but demonstrates no actin filament-severing activity. Like other members of this family, CapG is calcium sensitive and requires 1 μM Ca^{2+} for half-maximal capping of actin filaments. The phosphoinositide PIP_2 inhibits barbed-end capping by CapG, suggesting that CapG responds to two intracellular signals, PIP_2 and ionized calcium. Its calcium sensitivity may permit CapG to cap and uncap filaments in response to the brief fluctuations in ionized calcium observable in the periphery of motile phagocytes.

Cycles of filament capping and uncapping may represent an essential feature of peripheral membrane ruffling.

Actin Filament-Severing Proteins

The most abundant severing protein in nonmuscle cells is gelsolin, an 82-kDa monomeric protein found in the cytoplasm of all nonmuscle cells. In the presence of micromolar concentrations of ionized calcium, gelsolin binds to the sides of actin filaments and interferes with monomer-monomer binding within the filament. These actions result in the severing of actin filaments (Figure 9.2D). Gelsolin is closely related to CapG. Other members of the gelsolin family that also sever actin filaments include villin, a 95-kDa protein found in the intestinal brush border, and adseverin (also named scinderin), found in the adrenal medulla. Like CapG and the other members of the gelsolin/villin family, gelsolin also caps the barbed ends of actin filaments. Once gelsolin caps an actin filament, the protein binds the filament end with high affinity (K_D in the subpicomolar range) and does not dissociate from the barbed end when the ionized calcium concentration is lowered. However, as observed with both CapG and CapZ, binding to actin filaments can be inhibited by PIP_2. Experiments with increasing and decreasing gelsolin levels in cells prove that in vivo, gelsolin enhances the recycling of actin filaments and facilitates actin-based motility in nonmuscle cells. The severing of actin filaments reduces the viscosity of the peripheral cytoplasm and allows the actin cytoskeleton to be rapidly remodeled.

Actin Filament-Bundling and Cross-Linking Proteins

Electron micrographs of nonmuscle cells reveal that actin filaments are organized into a network in which many filaments appear to cross each other at right angles, forming an orthogonal mesh. The cross-linking protein ABP-280, or filamin, is responsible for organizing these networks. This spatially extended homodimeric protein consists of two 280-kDa subunits linked at a flexible hinge region. Each subunit possesses a single actin-binding site, thereby allowing the dimer to link two actin filaments. In addition to this cross-linking protein, nonmuscle cells possess smaller actin filament-bundling proteins, the most prominent member of this class being α-actinin (105 kDa). This protein links actin filaments into bundles of filaments in a parallel array (Figure 9.2E). Another member of this class is plastin, a 65-kDa protein (also known as fimbrin), which occurs in the so-called T and L forms.

Actin-Depolymerizing Proteins

The rate of actin filament treadmilling is much higher in vivo than in vitro. This increase in actin monomer turnover can be accounted for by a family of proteins called actin depolymerization factor (ADF) or cofilins. These are low-molecular-weight proteins (19 to 20 kDa) that increase the dissociation rate of ADP-actin monomers from the pointed ends of actin filaments (Figure 9.2F). These proteins can bind to ADP actin monomers with considerably higher affinity (0.1–0.2 μM K_D) than ATP actin monomers (1.3 μM K_D) and enhance the dissociation rate of ADP-, but not ATP-actin.

Other Actin-Regulatory Proteins

Actin-Related Proteins

The actin-related proteins (ARPs) possess homology to actin but are incapable of forming filaments. Recently, the ARP2-ARP3 complex was shown to bind to profilin as well as to the sides of actin filaments. In addition the ARP2-ARP3 complex caps the pointed ends of actin filaments with high affinity (K_D = 5 to 20 nM). At high concentrations, ARP2-ARP3 may also form nuclei to initiate actin assembly from the barbed ends. This complex is found at the leading edge of motile cells and therefore is likely to play a role in generating new filaments for actin-based motility.

Tropomyosin and Myosins

Other proteins capable of binding to actin include tropomyosin and the myosins. There are multiple subtypes of tropomyosin, and all appear to bind in the groove of the actin filament in a fashion which prevents myosin binding and also blocks gelsolin side binding and severing. Myosins are another large family of proteins that produce force and movement by binding to actin filaments through the myosin head regions. The ATPase activity of the head region is activated by binding to the actin filament. The energy of hydrolysis is transduced into a structural change in the binding angle of the myosin head from 90° to 45°. This change in angle advances the filament toward its pointed end. There are two major classes of myosins. While myosin II is most abundant in muscle, this force-producing motor is also found in nonmuscle cells. The tail regions of myosin II self-associate and form filament bundles in which the myosin heads arrange at opposing ends with respect to each other. These heads can pull actin filaments toward the center of the bundle. Myosin I is a more recently described class of proteins possessing a shorter tail, which fails to self-associate. Myosin I tails can bind to actin filaments or membranes and may be responsible for moving these structures. Furthermore, myosin I proteins tend to localize in the leading edge of moving cells whereas myosin II tends to localize toward the posterior region.

Proteins Found in Focal Contacts

When nonmuscle cells adhere to a surface, specific regions of the cell form close contacts. These regions are called focal contacts and contain high concentrations of actin filaments. In addition, talin (a 270-kDa protein) is found in these regions and binds vinculin (120 kDa). Vinculin in turn can bind α-actinin and actin filaments, and through the amino acid sequence 840-PDFPPPPPDL-849, called an ABM-1 sequence (see below), it binds VASP. Vinculin contains a 90-kDa head region and a 30-kDa tail region. The ABM-1-binding site is located near the carboxy terminus of the head domain, while the F-actin-binding site is found in the tail region. The tail folds over the head, masking both the F-actin- and ABM-1-binding sites. The folding and unfolding of vinculin acts as a molecular switch. Conditions such as proteolysis, phosphorylation, and binding of PIP_2 can unfold the molecule and allow the protein to bind VASP and actin filaments. Another protein found in focal contacts is zyxin. This 84-kDa protein contains two ABM-1-binding sites and binds VASP with high affinity. VASP, a 45-kDa protein with multiple phosphorylation sites, is the central adapter protein responsible for concentrating profilin in regions where actin filaments are assembling. Each VASP monomer contains four amino acid sequences of the type XPPPP, where X is glycine, alanine, lysine, or serine and defines an ABM-2 sequence, and these sequences bind profilin. VASP exists as a tetramer in solution; therefore, one VASP tetramer has 16 potential profilin-binding sites. Mena and N-mena, two other proteins that are structurally related to VASP, also contain ABM-2 sites and are capable of concentrating profilin.

Polymerization Zone Model

How might cells use actin polymerization to generate force during locomotion? We favor models that stress functional parallelisms between bacterial rocket-tail assembly and host cell processes. In both systems, a surface-bound component creates a local site for concentrating the profilin-actin complex into the polymerization zone between the membrane surface and the capped barbed ends of actin filaments. Delivery of polymerization-competent monomers probably coincides with the uncapping of barbed filament ends, and surface-bound profilin is likely to usher actin-ATP monomers onto the growing filament ends. Because actin filaments are likely to become lodged within the cytomatrix by the action of cross-linking and bundling proteins, expansion of the actin filament network would provide thrust to the host cell peripheral membrane or the bacterium. The mechanisms controlling uncapping and regulating of filament length remain to be determined, but phosphatidylinositides and profilin are likely to play key roles.

What factors are responsible for attracting profilin-actin complex into the polymerization zone? Two consensus docking sequences appear to be centrally involved in the assembly of an ABM complex. These sequences were originally defined by the oligoproline modules in *Listeria monocytogenes* ActA surface protein and human platelet VASP. Analysis of the known actin regulatory proteins led to the identification of two distinct ABM homology sequences: ABM-1, (D/E)FPPPPX(D/E) (where X is P or T), and ABM-2, XPPPP (where X is G, A, L, P, or S). To generate movement in a discrete area of the cell, a complex of actin-regulatory proteins must be organized in the region of actin filament assembly. A fully assembled ABM complex can readily attract high concentrations of profilin in the region where actin polymerization is required. In *Listeria* (see above), the bacterial surface protein ActA contains a series of four ABM-1 sequences of the type EFPPPPTDE. Each of these sequences may attract a VASP tetramer. Because each VASP tetramer contains 16 to 20 ABM-2 profilin-binding sites, one ActA molecule can potentially attract 64 to 80 profilin molecules. By using this protein-binding amplification cascade, extremely high concentrations of profilin can be attracted to the surface of *Listeria*. In a similar fashion, vinculin and zyxin both contain ABM-1 sequences that can attract VASP and profilin to form an ABM complex at the leading edge of motile cells (Figure 9.6). For actin-based *Shigella* motility, proteolysis of vinculin unmasks its ABM-1 site. Unmasking of this binding site serves as a molecular switch that initiates the assembly of an actin-based motility complex containing VASP and profilin.

Intracellular pathogen motility occurs at rates up to 1 μm/s, corresponding to an apparent first-order rate of about 400 monomers added per second. Published values for the bimolecular rate constant for actin-ATP addition to barbed ends allowed us to estimate that about 200 μM actin-ATP must be immediately available so that microbial pathogens can sustain such active assembly. This high local actin-ATP concentration greatly exceeds the G-actin critical concentration (0.25 μM) by a factor of 800, raising the compelling question: What is the basis of this great discrepancy between intracellular and in vitro polymerization behavior? From work conducted in our laboratory, we now favor the concept of a polymerization zone—an activated region that recruits actin-ATP and permits the explosive filament growth needed to propel the pathogen (Figure 9.7). The presence of profilin

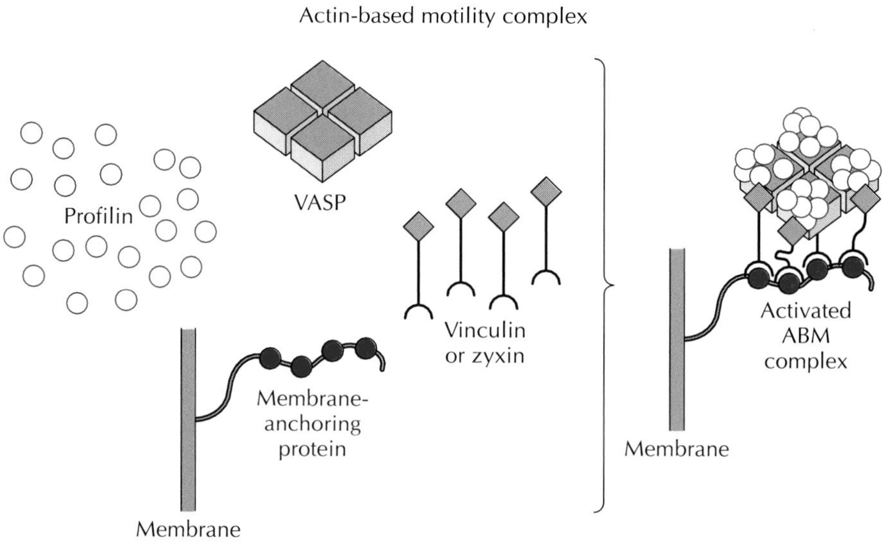

Actin-based motility complex

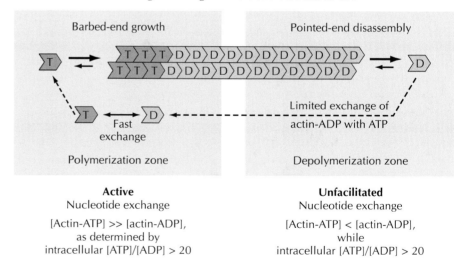

Figure 9.6 Assembly of an ABM complex. Changes in the structure of vinculin or zyxin act as a molecular switch that exposes ABM-1-binding sequences for attracting and tethering VASP in the polymerization zone. Bound VASP then binds numerous profilin molecules, which increase the local concentrations of actin-ATP within the polymerization zone to stimulate filament assembly.

Exchange-limited growth of actin filaments in vivo

Barbed-end growth	Pointed-end disassembly

Active	**Unfacilitated**
Nucleotide exchange	Nucleotide exchange
[Actin-ATP] >> [actin-ADP], as determined by intracellular [ATP]/[ADP] > 20	[Actin-ATP] < [actin-ADP], while intracellular [ATP]/[ADP] > 20

Figure 9.7 Schematic diagram of discrete polymerization and depolymerization zones in a cell. Within the polymerization zone, the ATP-actin concentration is high and rapid filament growth occurs. Nucleotide exchange of ATP for ADP on actin monomers is very active as a result of the high concentrations of profilin in this region. Thus, the ADP-actin concentration is expected to be low in the polymerization zone. By contrast, the depolymerization zone is a region where actin filaments disassemble at their pointed ends. The profilin concentrations are much lower in the depolymerization zone, and lower rates of nucleotide exchange should increase ADP-actin levels and prevent filament assembly.

within this zone also ensures that the weakly polymerizing actin-ADP complex will undergo facilitated exchange with ATP to produce a strongly polymerizing actin-ATP complex. Other agents that promote filament growth, including α-actinin and ARPs, may likewise be recruited to the polymerization zone to control the rate of filament assembly in the vicinity of the microbe. In our model, pointed-end filament depolymerization occurs exclusively within more distal cell regions that contain much lower profilin concentrations. Such an arrangement would greatly limit the nucleotide exchange rate, and the presence of more actin-ADP is likely to prohibit filament assembly at all regions outside the active polymerization zone. Depolymerizing factors (e.g., ADF and cofilin) are apt to be selectively localized within the regions of filament depolymerization.

Conclusion

Formation of the ABM complex probably constitutes one of the principal underlying mechanisms for initiating and choreographing the explosive growth of cortical actin needed in cellular motility. The term "actin filament rocket-tail" is frequently used to emphasize that these structures are responsible for the forward propulsion of cellular components and microbial pathogens. Peskin et al. (1993) proposed that the rocket-tail behaves as a "Brownian ratchet" in which the insertion of new monomers occurs at the interface of the bacterium or activated membrane whenever the microorganism or membrane moves even slightly in advance of the rocket-tail. ATP-actin addition to barbed filament ends is thermodynamically favorable, and hydrolysis of filament-bound ATP-actin during filament elongation can provide the useful work needed for movement. Microbial pathogen motility can be viewed as "point-rocketing" (i.e., a process whereby a small point or object is propelled by actin filament assembly).

The precise mechanisms regulating the formation of new actin filaments in the motile cell remain to be determined. Actin filament growth can be regulated by the capping and uncapping of the barbed end. In addition, filaments can be severed by gelsolin. When combined, uncapping and severing should greatly increase the number of free barbed-filament ends available for rapid actin assembly. Another filament assembly-promoting mechanism may require the localization of nucleating proteins (such as the ARP2-ARP3 complex) in the polymerization zone. Through the action of ABP-280 and α-actinin, newly formed filaments should be stabilized into orthogonal networks and bundles. These proteins increase the rigidity of the actin filaments, giving the peripheral cytoplasm the structure and mechanical properties required for shape change and movement. Osmotic forces and myosins may also play a role in advancing the peripheral membrane. Given the complexity of amoeboid movement, the task of determining the primary mechanisms has proved difficult.

As described in the next chapter, bacterial pathogens that utilize actin and actin-regulatory proteins for their intracellular locomotion have become model systems for identifying the minimal pathway(s) for actin-based motility.

Selected Readings
Haffner, C., T. Jarchau, M. Reinhard, J. Hoppe, S. M. Lohmann, and U. Walter. 1995. Molecular cloning, structural analysis and functional expression of the proline-rich focal adhesion and microfilament-associated protein VASP. *EMBO J.* **14**:19–27.

Holmes, K. C., D. Popp, W. Gebhard, and W. Kabsch. 1990. Atomic model of the actin filament. *Nature* **347**:44–49.

Peskin, C. S., G. M. Odell, and G. F. Oster. 1993. Cellular motions and thermal fluctuations: the Brownian ratchet. *Biophys. J.* **65**:316–324.

Southwick, F. S., and D. L. Purich. 1996. Mechanisms of disease: intracellular pathogenesis of listeriosis. *N. Engl. J. Med.* **334**:770–776.

Stossel, T. P. 1993. On the crawling of animal cells. *Science* **260**:1086–1094.

Wegner A. 1982. Treadmilling of actin at physiological salt concentrations. An analysis of the critical concentrations of actin filaments. *J. Mol. Biol.* **161**:607–615.

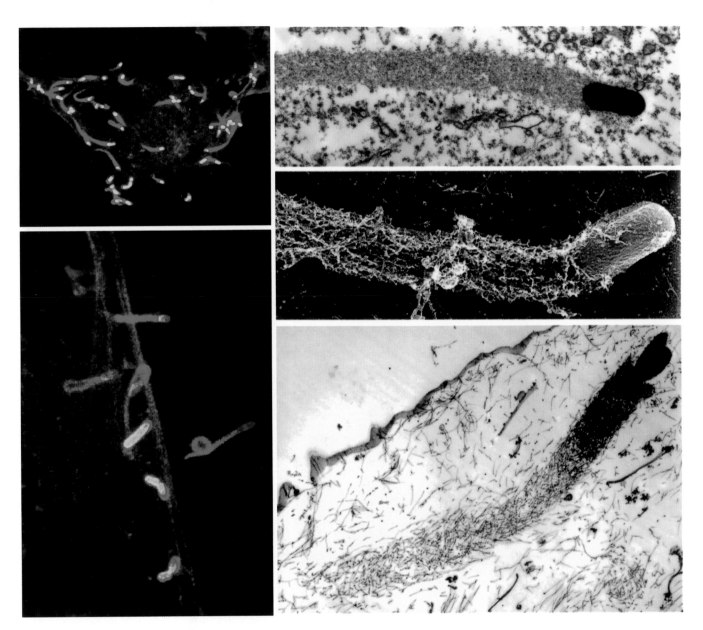

Figure 1.3 *L. monocytogenes* actin tails. (Left top and bottom) actin tails visualized by immunofluorescence. *L. monocytogenes*-infected cells were labeled with fluorescein isothiocyanate-conjugated phalloidin (to label F-actin) and anti-*Listeria* antibodies. (Right) actin tails visualized by electron microscopy. (Top) Thin-section electron micrograph from an infected tissue culture cell showing a moving bacterium associated with an F-actin comet tail (Tilney technique). (Middle) Three-dimensional visualization of the actin comet tail by the quick-freeze/deep-etch technique. (Courtesy of J. Heuser.) (Bottom) Thin section through an *L. monocytogenes* cell with a tail whose actin filaments have been decorated with subfragment 1 (S1) of myosin.

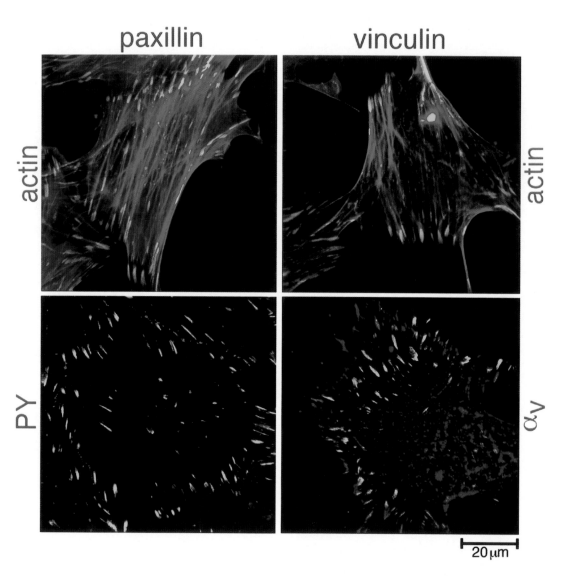

Figure 5.3 (Top) Double fluorescent labeling of fibroblasts for actin (green) and vinculin or paxillin (red). (Bottom) Double immunofluorescent labeling (right) for vinculin (red) and integrin (green) or (left) for paxillin (red) and phosphotyrosine (PY) (green). Both vinculin and paxillin are associated with the termini of actin-containing stress fibers. Vinculin and phosphotyrosine are also associated with cell-cell adherens junctions, while paxillin and integrin are present only in matrix adhesions.

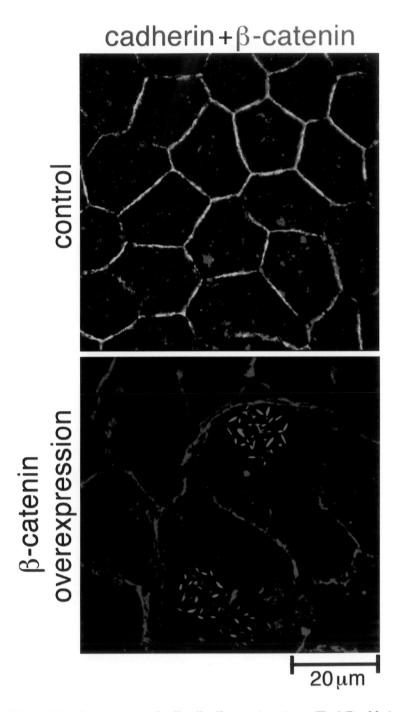

Figure 5.4 Components of cell-cell adherens junctions. (Top) Double immunofluorescent labeling of epithelial MDCK cells for cadherin (red) and β-catenin (green). Cell-cell junctions are seen as yellow lines due to superposition of red and green colors, showing nearly complete overlap between cadherin and β-catenin at cell-cell adhesion. (Bottom) Overexpression of a chimeric molecule consisting of β-catenin and green fluorescent protein (which allows visualization of the molecule) in MDCK cells by transient transfection followed by immunolabeling for cadherin (red). The fluorescent β-catenin appears green. Note that when overexpressed, β-catenin accumulates in the nucleus, forming aggregates of different shapes.

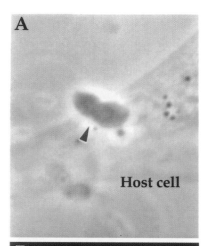

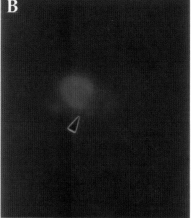

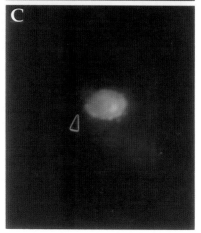

Figure 8.7 Phase-contrast **(A)** and fluorescence **(B and C)** micrographs of a tachyzoite of *Toxoplasma gondii* caught in the act of invading a mammalian cell. The preparation was reacted with antibody against the parasite surface protein SAG1 (B), permeabilized, and incubated with antibody against the microneme protein MIC2 (C). The fluorescence shows the capping of the surface protein SAG1 concomitant with the secretion of the microneme protein MIC2 into the nascent parasitophorous vacuole during the entry process. The junction point between the parasite and the host cell is indicated by an arrow. The parasite is known to remodel this vacuole extensively during establishment of intracellular infection. Courtesy of L. David Sibley.

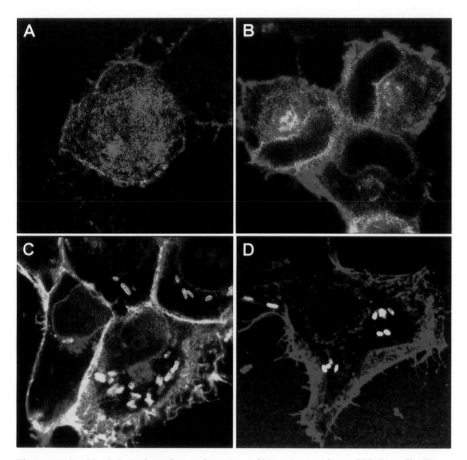

Figure 13.5 *Yersinia*-induced translocation of Yops into cultured HeLa cells. Translocation of Yops is visualized by confocal laser-scanning microscopy. **(A to C)** Yops are represented by the green fluorescence: YopH (A), YopE (B), and YpkA (C). HeLa cell plasma membranes are illustrated by the red fluorescence. Note that the individual Yops have different locations within the eukaryotic cell: YopH is widely distributed in both the cytosolic and nuclear compartments, YopE is enriched in the perinuclear region, and YpkA is localized at the inner surface of the plasma membrane. The yellow color is indicative of colocalization of YpkA and the plasma membrane. **(D)** A *yopB yopD* mutant is translocation deficient.

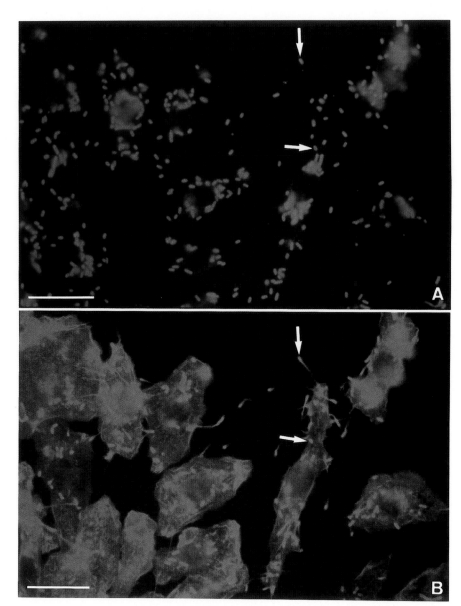

Figure 17.1 Localization of actin filaments by conventional fluorescence microscopy. Bone marrow-derived mouse macrophages were infected for 15 min with live *L. monocytogenes* LO-28 (hly⁺). After a 4-h chase in bacterium-free medium, the cells were fixed with 3% paraformaldehyde and permeabilized with Triton X-100. The cells were sequentially incubated with a specific rabbit polyclonal antibody raised against *L. monocytogenes* followed by goat anti-rabbit IgG coupled to Cy3 **(A)** and then with β-phalloidin coupled to fluorescein isothiocyanate to localize actin filaments **(B)**. Arrows indicate cytoplasmic bacteria surrounded by a network of actin filaments or displaying a comet of actin filaments. Bar, 5 μm.

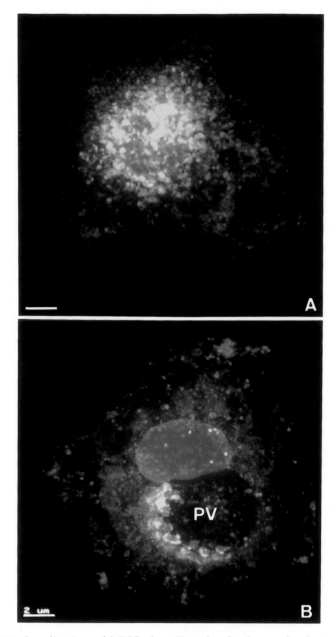

Figure 17.2 Localization of MHC class II molecules by confocal microscopy in uninfected macrophages (A) and in macrophages infected with *Leishmania mexicana* (B). Cells were treated for 16 h with recombinant gamma interferon and fixed either before or after a 24-h infection with the parasites. The cells were permeabilized and sequentially incubated with rat monoclonal antibodies (M5/114) and with rabbit anti-rat antibodies coupled to fluorescein isothiocyanate. Parasite and cell nuclei were stained with propidium iodide (red stain). These figures represent a three-dimensional reconstruction of the cell. **(A)** In uninfected macrophages, MHC class II molecules are located at the plasma membrane and in vesicles concentrated in the perinuclear region. **(B)** In infected cells, the label is found at the plasma membrane, in perinuclear vesicles, and associated with the parasitophorous vacuole (PV), where it is concentrated at the site of attachment of parasites. However, exact location of class II molecules in the parasitophorous vacuole cannot be resolved by this method. Bar, 2 μm. Prints kindly provided by Jean-Claude Antoine, Institut Pasteur, Paris, France.

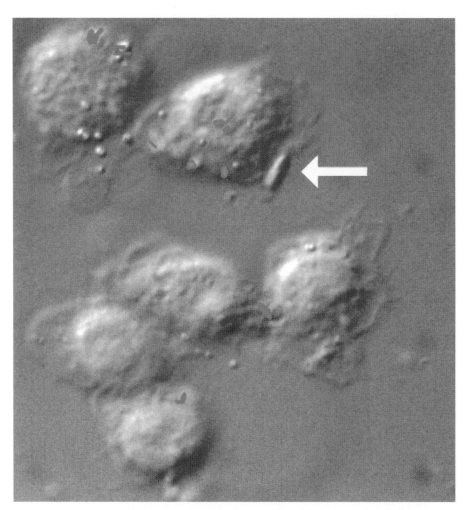

Figure 18.6 GFP as a tool to detect gene expression of intracellular bacteria. Human macrophages were infected with *S. typhimurium* containing a *gfp* reporter gene fused to a promoter expressed only when the bacteria reside within host cells. In this confocal micrograph, bacteria expressing GFP appear as bright green rods. The signal is always seen within the boundaries of the macrophages (which resemble fried eggs), consistent with the intracellular expression of the promoter. The arrow points to an extracellular bacterium, whose promoter is inactive and therefore does not drive GFP expression.

10

Bacterial Manipulation of the Host Cell Cytoskeleton

SANDRA J. MCCALLUM AND JULIE A. THERIOT

Throughout the coevolution of humans and pathogenic bacteria, each has sought new ways to overcome the other. Human physiology has evolved physical barriers and complex and specific immune responses to kill the pathogenic bacteria with which we come into contact. Bacteria have evolved to circumvent these defenses in a variety of different ways, using our own biochemistry and cellular makeup to their advantage to find niches through which to attack us. One of the host cell systems that has been exploited by a number of different bacterial species is the actin cytoskeleton.

The actin cytoskeleton is a meshwork of protein filaments that extend throughout the cytosol and form a dense web underlying the plasma membrane. In most vertebrate cells, the actin cytoskeleton is the primary determinant of cell shape and provides the machinery for whole-cell movement. It acts as both a framework for cellular morphology and a scaffold for intracellular processes including signal transduction and membrane organization. The actin cytoskeleton is involved in many signaling mechanisms in response to both intracellular and extracellular signals and can reorganize quickly to change local morphology or global cell shape. Different cell types use their actin cytoskeletons for different purposes; for example, macrophages and neutrophils use actin-dependent movement to crawl through tissues and engulf bacterial invaders, while epithelial cells use actin structures to maintain strong adhesive connections with their neighbors and with their underlying extracellular matrix. Actin microfilaments are polymers made up of repeating units of the protein actin. Actin spontaneously forms filaments 10 nm in diameter under the proper chemical conditions. Tight regulation of these conditions by a cell provides a system that is both structurally sound and extremely dynamic and that interacts with a wide variety of other proteins and cellular systems. More detailed information about the actin cytoskeleton and actin-binding proteins is given in chapter 9.

The importance of the actin cytoskeleton to eukaryotic cells and its responsiveness to a wide variety of signals have provided pathogenic bacteria with a plethora of chances for opportunistic exploitation (Figure 10.1).

A Invasion

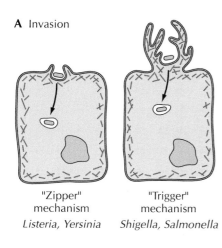

"Zipper"
mechanism
Listeria, Yersinia

"Trigger"
mechanism
Shigella, Salmonella

B Prevention of uptake

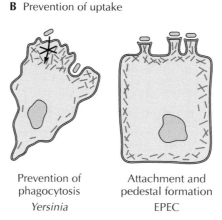

Prevention of
phagocytosis
Yersinia

Attachment and
pedestal formation
EPEC

C Actin-based motility

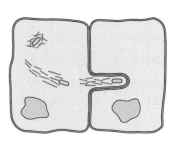

Intracellular motility
and intercellular spread
Listeria, Shigella

Figure 10.1   Schematic diagrams of host cell actin rearrangements caused by bacterial pathogens. **(A)** Some pathogens induce their own uptake by nonphagocytic epithelial cells, initiating signal transduction cascades that result in polymerization of actin filaments under the host cell plasma membrane. For some pathogens, including *Listeria* and *Yersinia,* tight adhesion of the bacterium to the cell surface is followed by modest actin polymerization only in the immediate neighborhood of bacterial attachment. This is the zipper form of induced phagocytosis, allowing the bacterium to be taken up by cells. For others, such as *Shigella* and *Salmonella,* the host cell throws up a large ruffling membrane that folds over and traps the bacteria in a membrane-bound compartment. This is the trigger form of induced phagocytosis. **(B)** Macrophages normally phagocytose foreign particles, including bacterial invaders, but *Yersinia* species are able to prevent phagocytosis by injecting virulence factors into the host cell cytoplasm that disrupt actin cytoskeletal structures. In contrast, EPEC colonizes the surface of epithelial cells without being internalized, building a dense pedestal of actin filaments that raises the bacteria above the cell surface. **(C)** *Listeria* and *Shigella* are able to escape from their vacuoles and nucleate the polymerization of actin filaments at their surfaces while growing in the host cell cytoplasm. This results in the formation of a dense tail structure that pushes the bacteria through the cytoplasm. At the interface between neighboring epithelial cells, the actin-rich tail pushes the bacteria out into a membrane-bound protrusion, which can be taken up by the second cell, allowing direct cytoplasm-to-cytoplasm transmission.

There are a number of ways that different pathogens have evolved to attack eukaryotic cells through manipulation of their cytoskeletons. Many pathogens have developed mechanisms for gaining access to parts of the human body that are normally sterile, where they can grow in a nutrient-rich environment without competition from other microorganisms. To do this, a clever pathogen may breach the epithelia that line the gastrointestinal tract, the lungs, or the urogenital tract. From the outside of epithelial cells, bacteria can interact with the host cell plasma membrane to mimic signal transduction mechanisms that induce local actin filament polymerization and thereby trick nonphagocytic epithelial cells into engulfing them and allowing them access to the intracellular milieu. This strategy shields the bacterium from the humoral component of the host immune surveillance system. Some bacterial species have developed elaborate secretion systems to deliver factors to the inside of host cells to manipulate the cytoskeleton, while others appear to express factors on their cell surface, which mimic eukaryotic signaling molecules, to transmit inappropriate signals across the host cell plasma membrane. Once inside, some bacteria can utilize the host actin cytoskeleton to provide their own motility and facilitate their intercellular spread. Others have developed systems to allow their growth in protected vacuoles and then spread through cell lysis.

While pathogens may benefit from being engulfed by epithelial cells, engulfment by immune system cells such as macrophages and neutrophils is not so desirable, since it generally results in lysosomal digestion and destruction of the bacterium. Therefore, some pathogens have developed ways of inhibiting immune-system cell phagocytosis by blocking key cytoskeletal functions and thus evading an important host cell defense. Others preempt macrophage phagocytosis by inducing engulfment by an alternative pathway, so that they can actually survive and thrive within the cellular body.

In addition to pathogens that influence cytoskeletal dynamics at close range in order to invade or avoid engulfment by epithelial cells and macrophages, some bacteria have developed secreted toxins with specific effects on actin filaments or the proteins that regulate the actin cytoskeleton. These toxins can act at a distance from the bacteria themselves. They are particularly interesting to cellular microbiologists because they provide important tools with which to study the actin cytoskeleton and yield information about microbial pathogenesis. These toxins are covered in chapters 11 and 12.

In this chapter, we will illustrate several modes of bacterial manipulation of the host cell cytoskeleton by using a few well-studied examples and explore interactions of various pathogenic bacteria with the actin cytoskeleton of both phagocytic and nonphagocytic host cells. These interactions are generally the result of specific bacterial gene products, i.e., virulence factors, whose sole function is to mimic or interfere with normal host cell signals to regulate actin filament dynamics. Judicious exploitation of preexisting regulatory pathways has enabled the pathogenic bacteria to induce their eukaryotic hosts to perform energetically demanding and ultimately self-destructive behaviors. Successful pathogens are frequently expert cell biologists and demonstrate a remarkably sophisticated practical understanding of highly complex multifactorial cellular systems.

Mechanisms of Bacterial Invasion

The human body is not a single organism but, rather, a complex ecological community, most of whose members are bacteria. When all the bacteria found on the skin, in the nose and mouth, in the lower gastrointestinal tract, etc., are taken into account, there are about 10 times more bacterial cells than human cells in the body. However, the immune system keeps them corralled in appropriate locations, and most internal tissues are normally sterile. Opportunistic pathogens must await an injury or other breach in the barriers before they can reach these niches, but some primary pathogens have developed mechanisms to sneak across them. With respect to manipulation of the host cell cytoskeleton, several interesting examples are found among pathogens that invade epithelial cells.

Two epithelia—the lining of the intestines and the lungs—are particularly vulnerable to attack by bacterial pathogens. Since the function of these epithelial is to absorb digested nutrients and oxygen, respectively, and pass them into the bloodstream, the barrier is a cellular monolayer. Bacteria that can invade this monolayer are one step away from access to the bloodstream and the rest of the body. Epithelial cells take up proteins and other factors from the outside, using pinocytosis and receptor-mediated endocytosis, but they are generally not capable of engulfing large objects such as bacteria. To be taken up by epithelial cells, the bacteria must induce their own phagocytosis.

There are two recognized mechanisms by which pathogenic bacteria induce phagocytosis by epithelial and other nonphagocytic cells; these are termed the zipper and trigger mechanisms. Both of these mechanisms are actin dependent and require that the extracellular bacteria induce intracellular polymerization and reorganization of actin filaments at the plasma membrane. In both cases, the bacteria manipulate the host cell signal transduction apparatus to induce cytoskeletal responses similar to those that occur in response to natural stimuli. The zipper mechanism exploits pathways normally involved in cellular adhesion, while the trigger mechanism resembles the membrane ruffling response to eukaryotic cell growth factors. Examples of both are described below.

Zipper Mechanism of Bacterial Uptake by Nonphagocytic Cells

Listeria monocytogenes

One bacterial pathogen that is a recognized champion at manipulating the host cell actin cytoskeleton is *Listeria monocytogenes*. *L. monocytogenes* is a gram-positive food-borne pathogen that can cause serious infections in pregnant women and immunocompromised people, resulting in a relatively high rate of mortality among infected adults, newborns, and fetuses. In healthy adults, the infection can be asymptomatic or result in mild, flu-like symptoms. *Listeria* is usually found in soil, in water, and on plants but can be isolated from animals as well, and infections can be transmitted in foods such as unpasteurized dairy products. In addition, *Listeria* can contaminate stored refrigerated foods due to its unusual ability to grow at 4°C.

Once ingested, some *Listeria* organisms pass through the upper gastrointestinal tract to the small intestine, where they are thought to enter the intestinal epithelium. In tissue culture models of infection, *Listeria* can invade a wide variety of cell types, including epithelial, endothelial, and fibroblastic cells, by using a zipper mechanism. This term refers to the ability of the bacterium to induce very local changes in actin organization which result in a close apposition of the bacterium to the host plasma membrane. The bacterium appears to sink into the cell, or to pull the plasma membrane up around it, in a process that does not involve formation of membrane ruffles as in the trigger internalization process (discussed below) and apparently does not require more than a very local rearrangement of the actin cytoskeleton.

Two bacterial factors called internalins have been identified that allow *Listeria* to zipper into eukaryotic cells. Internalin A (InlA) is an 88-kDa bacterial surface protein that contains leucine-rich repeats that are thought to participate in protein-protein interactions. InlA is necessary for invasion of cultured intestinal epithelial cells, and transposon insertion into the *inlA* gene abolishes the invasive ability of *Listeria*. Addition of the *inlA* gene on a plasmid restores functionality to this mutant, confirming that the phenotype is not due to disruption of nearby genes. During invasion of epithelial cells, InlA interacts with a host receptor, the E-cadherin molecule. E-cadherin is an adhesion protein responsible for homotypic association of epithelial cells. It is a transmembrane protein whose cytoplasmic side is linked to the actin cytoskeleton through a complex of proteins which includes α-, β-, and γ-catenin. It is not known whether these interactions are important for *Listeria* invasion, but they could represent a link between InlA and the cytoskeletal changes that result in *Listeria* internalization.

In other cell types, InlA is not required for *Listeria* invasion; it has been shown that the 65-kDa protein InlB is responsible for the internalization of bacteria into cultured hepatocytes, Chinese hamster ovary (CHO) cells, and HeLa cells. The host cell receptor for InlB has not been identified. *Listeria* harbors at least four other genes containing leucine-rich repeats that may function as internalins for other host cell types. Perhaps the diversity of the internalin multigene family reflects the diversity of their potential cellular targets; cadherins, too, are a diverse multigene family.

Internalization of *Listeria* by the zipper mechanism involves a dynamic actin cytoskeleton in the host cell as well as the activation of tyrosine kinases and lipid kinases. Inhibitors of tyrosine kinases and the lipid kinase phosphatidylinositol 3-kinase (PI3-K) block entry of *Listeria* into epithelial cells in culture. Dominant negative forms of PI3-K also block *Listeria* entry, and PI3-K is activated upon entry of *Listeria* into Vero cells. Cytochalasin D, an inhibitor of actin polymerization, also blocks bacterial entry. Interestingly, cytochalasin D does not block the activation of PI3-K, suggesting that PI3-K may act upstream of the signals to the cytoskeleton and may be involved in the activation of that signal. The precise steps of PI3-K activation and those leading to actin rearrangement in *Listeria* invasion have not yet been uncovered. However, it has been demonstrated in other systems that the lipid products of PI3-K have effects on actin polymerization, uncapping filament barbed ends to make them available for growth. In addition, PI3-K activates the Rac GTPase, a low-molecular-weight GTP-binding protein of the Ras family, that regulates actin rearrangements in a number of different cell types.

Yersinia Species

An analogous zipper-type entry mechanism may be used by some species of the gram-negative pathogen *Yersinia. Y. enterocolitica, Y. pseudotuberculosis,* and *Y. pestis* are the causative agents of mild to severe bouts of gastroenteritis, animal disease similar to that seen in human patients, and bubonic plague, respectively. (Bubonic plague is also called the "black death" due to the damage to the peripheral blood vessels that gives the skin its blackish appearance in *Y. pestis* victims.) *Y. enterocolitica* and *Y. pseudotuberculosis* enter the body by the fecal-oral route and invade the mesenteric lymph nodes around the intestines through the Peyer's patches. In tissue culture models of infection, these species are capable of invading epithelial cells. A *Yersinia* outer membrane protein, invasin, binds tightly to the subset of integrins that harbor a β_1 subunit. Integrins, like cadherins, are adhesion proteins, but they generally mediate adhesive interactions between cells and the extracellular matrix. As with *Listeria*, inhibitors of tyrosine kinases and actin polymerization inhibit *Yersinia* invasion. However, some evidence suggests that clathrin may also play a role in *Yersinia* uptake.

Both these examples of zipper-type invasion demonstrate bacterial exploitation of mechanisms normally involved in host cell adhesion. Cadherins (the receptors for *Listeria* internalin) and integrins (the receptors for *Yersinia* invasin) mediate attachment of epithelial cells to and spreading on their neighbors and the underlying extracellular matrix, respectively. Engagement of these receptors by their normal ligands results in receptor immobilization and clustering. These events induce signaling cascades that can result in strengthening of the cell-cell and cell-matrix contacts and in

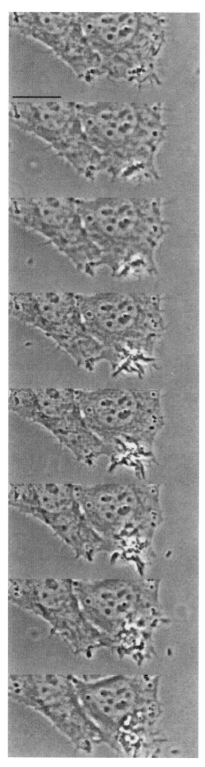

Figure 10.2 The trigger mechanism of bacterial invasion. Eight frames from a video sequence, recorded at 10-s intervals, show the large actin-rich ruffle formed when *Salmonella typhimurium* encounters a Henle epithelial cell in culture. Folding in of the ruffle forms numerous spacious vacuoles, visible as white circles in the later frames. Bar, 10 μm.

cellular differentiation. When a bacterial surface protein engages the adhesion proteins on a host cell, the host cell responds as it normally would, recruiting cytoskeletal elements to the location of the attachment and attempting to strengthen the attachment. However, since the bacterium is small compared to the responding cell, the attempt of the cell to spread against the bacterial surface quickly results in engulfment of the bacterium. This striking example of functional convergence suggests that it may be relatively easy for a bacterial pathogen to develop the ability to invade epithelial cells by a zipper mechanism. The only requirement appears to be that the bacterium express a surface protein that binds to a cell surface adhesion receptor with an appropriate affinity.

Trigger Mechanism of Bacterial Uptake by Nonphagocytic Cells

Shigella Species

The second mechanism of bacterially induced phagocytosis, the trigger mechanism, is dramatically different from the zipper mechanism in morphological terms and requires a much more complicated type of bacterial machinery. In these cases, brief contact between a bacterium and the surface of a host cell results in a rapid, large-scale cytoskeletal response, where explosive actin filament polymerization under the plasma membrane pushes out huge sheets or ruffles. The extending ruffles fold over and fuse back to the cell surface, trapping large membrane-bound pockets of extracellular medium within the cell, a process termed macropinocytosis (Figure 10.2). The nearby bacteria are trapped by the infolding membrane ruffles, and the bacteria then find themselves within the membrane-bound compartments. Where zippering invasion appears to be a modification of cellular adhesion, triggering invasion bears much more resemblance to the response of cells to growth factors.

Shigella is a gram-negative pathogen that is the causative agent of bacillary dysentery (bloody diarrhea). There are four species of *Shigella* that cause dysentery: *S. dysenteriae*, *S. sonnei*, *S. boydii*, and *S. flexneri*. They have similar mechanisms of virulence, and since *S. flexneri* is the best studied, the molecular descriptions presented here are for *S. flexneri* but presumably also apply to the other species. *Shigella*-mediated dysentery is usually a self-limiting although unpleasant diarrheal disease in healthy adults, but kills about 500,000 children every year in developing countries, where contaminated water is all that is available. Compounding this problem is the fact that shigellae are developing resistance to the antibiotics that are safe to administer to children. *Shigella* infection can occur at a very low dose of bacteria, with a 50% infectious dose of 100 to 200 organisms, probably because the bacteria are not killed by stomach acid and find their way unharmed to the colon. During *Shigella* infection, the bacteria can invade a variety of cell types including macrophages, M cells, and intestinal epithelia, and after escaping from the phagocytic vacuole, they can move within and between cells by actin-based motility. This motile ability is discussed below. In this section, we focus on the ability of *Shigella* to invade epithelial cells by a process that triggers global changes in the cellular actin cytoskeleton that result in membrane ruffles and macropinocytosis.

The trigger mechanism of *Shigella* invasion involves proteins whose genes are carried on a 220-kb virulence plasmid. Bacteria that have been cured of this plasmid lose invasive ability. The main system involved in this process is a type III secretion system, discussed in chapter 13, that is activated when the bacteria come into contact with the host cells. This is not a process of transcriptional activation but, rather, mobilization of proteins already existing in the cytosol of the bacterial cell. The genes involved in this process are called the *ipa* (invasion plasmid antigens) and the *Mxi-spa* (membrane expression of *ipa*-surface presentation of antigens) genes. The *Mxi-spa* genes encode the secretory apparatus, while the *ipa* genes encode the secreted proteins. The protein products, IpaB, IpaC, and IpaD, are all necessary for *Shigella* invasion.

IpaB and IpaC are secreted into the extracellular milieu and are part of a complex that is sufficient for invasion of the host cells. This "Ipa complex" does not contain IpaD but may contain other factors. Mutants in which secretion of the Ipa factors is impeded are not internalized, and soluble Ipa factors secreted by the wild-type *Shigella* can rescue these mutants, suggesting that the Ipa proteins act as a soluble complex from the outside of the cell. Beads that are coated with the Ipa complex can be internalized into HeLa cells in a process that resembles *Shigella* invasion morphologically. These results suggest that the Ipa complex interacts with the outside of host cells to activate the signal transduction pathways that are responsible for the generation of membrane ruffles on the host cells. However, this picture may be incomplete, and more work is required to determine whether the Ipa complex is delivered as part of a larger structure or whether other factors are involved in the generation of the signaling events resulting in invasion.

The host cell factors involved in *Shigella* invasion are those that are normally involved in the formation of membrane ruffles in response to growth factors and other signals that cause morphologically similar actin-rich structures to form. One study has suggested that the $\alpha_5\beta_1$-integrin may be the host cell receptor for the Ipa complex, but this has not been definitively proven. It remains possible that the Ipa complex acts by creating a pore through which it inserts effector proteins or that some other interaction with the cell membrane is necessary. The ruffles that form upon *Shigella* invasion extend actin-rich membrane projections near the site of the bacterial contact that contain the actin-bundling protein fimbrin (plastin). Cells that express a dominant-negative form of fimbrin are defective for invasion of *Shigella*, supporting a role for this protein in this process. Also localized to these structures are the focal adhesion proteins vinculin, paxillin, α-actinin, talin, and the focal adhesion kinase (FAK). The precise roles of these proteins in *Shigella* invasion are unknown, but they are assumed to be responsible for the organization of actin in these structures in response to regulatory signals.

The regulatory signals responsible for the actin rearrangements that occur in *Shigella* invasion include activation of the low-molecular-weight GTPase Rho. Members of the Rho family of small GTPases (including Cdc42, Rac, and Rho itself) mediate actin cytoskeletal reorganizations in response to extracellular growth factors and act as molecular switches in a variety of signal transduction pathways. In fibroblasts, Rho induces the formation of stress fibers and focal adhesions in response to growth factors. Inhibitors of the activity of Rho block *Shigella* invasion. Another signaling molecule that may be involved in this process is the tyrosine kinase $pp60^{c\text{-}src}$. Upon invasion by *Shigella*, c-Src phosphorylates one of its major substrates, cortactin, and cortactin is recruited to the actin-rich projections that are induced. Cortactin is an actin-binding protein of the cortical actin cytoskeleton and may be the connection between the activation of the c-Src tyrosine kinase and the actin cytoskeleton. How these events are connected to the activation of Rho is still unclear.

Salmonella Species

Salmonella species are responsible for food-borne general gastroenteritis (*S. typhimurium*) as well as more specific ailments such as typhoid fever (*S. typhi*). Like *Shigella*, *Salmonella* bacteria are gram-negative pathogens closely related to *Escherichia coli*. One of the important aspects of the virulence of *Salmonella* is its ability to pass through the intestinal epithelia and invade underlying cells including enterocytes and macrophages. Interaction of *Salmonella* with host cells causes the generation of membrane ruffles and macropinocytosis, resulting in trigger-type internalization. Invasion of epithelial cells allows *Salmonella* to colonize and (in some cases) cross the barrier monolayer in the intestine. In more serious infections, *Salmonella* also invades and grows inside macrophages. Apparently, induction of triggered uptake by the macrophages preempts their normal phagocytic activity. Whereas *Salmonella* could not survive in a normal macrophage phagosome, since the phagosome is quickly targeted to fuse with lysosomal degradative compartments, the spacious membrane-bound compartment that *Salmonella* generates for itself by triggering ruffling and macropinocytosis can support bacterial replication.

Researchers have identified a number of host cell signaling pathways that may be responsible for the rearrangement of the actin cytoskeleton that occurs upon interaction of *Salmonella* with the host cell membrane. In Henle 407 cells (a human epithelial tissue culture line), interactions of *Salmonella* with the cells in culture cause the activation of the epidermal growth factor receptor (EGFR) tyrosine kinase. Activation of this receptor triggers the mitogen-activated protein kinase, and this results in activation of phospholipase A_2 (PLA$_2$). Lipid products of PLA$_2$, such as arachidonic acid, can have effects on actin regulatory proteins, perhaps by interacting with the regulators of the Rho subfamily of GTP-binding proteins. Consistent with this hypothesis, the Rho subfamily member Cdc42 is crucial for *Salmonella* invasion. Cdc42 regulates the formation of the actin-rich filopodia that form on some cell types in response to extracellular stimulation. It is also responsible for the activation of a cascade of GTPases which causes the formation of membrane ruffles in Swiss 3T3 fibroblasts. Transfection with mutant forms of Cdc42 that are inactive for signaling do not allow *Salmonella* invasion, and mutants of Cdc42 that are constitutively activated for signaling allow internalization of invasion-defective *Salmonella* mutants. Recall that *Shigella*-triggered invasion requires the activity of Rho but not Cdc42. This is an excellent example of how bacterial exploitation of different specific host cell signaling pathways can result in the same net outcome, in this case induction of host cell ruffling and bacterial uptake.

The eicosenoid leukotriene D$_4$ (LTD$_4$), which is generated from arachidonic acid by PLA$_2$, may also play a role in ruffle formation by *Salmonella*. Generation of LTD$_4$ results in a rise in the intracellular calcium concentration that is common to many pathways that activate changes in the actin cytoskeleton. As discussed in chapter 9, many actin regulators are sensitive to changes in intracellular calcium levels. Actin-severing proteins such as gelsolin, villin, and fimbrin are regulated by calcium, and their activation could expose the uncapped ends of actin filaments and thus make them available for the rapid actin polymerization at the plasma membrane that accompanies *Salmonella* invasion. Invasion-defective *Salmonella* can be rescued by the addition of LTD_4 to the outside of cells, supporting the hypothesis that LTD_4 plays a role in invasion.

In some cell types, such as HeLa and B82 fibroblast cells, other pathways have been identified as important for *Salmonella* invasion. The B82 fibroblasts have no EGFRs on their cell surface. In these cell types, activation of PLC-γ results in the formation of inositol-1,4,5-triphosphate (IP_3) and release of calcium from intracellular stores. This is distinct from the situation in Henle cells, discussed above, which activates the influx of extracellular calcium. As in that circumstance, these changes in intracellular calcium concentrations could have profound effects on the actin cytoskeleton.

The bacterial proteins that are responsible for the activation of these host cell pathways have not been definitively identified, but some clues have been provided by the demonstration that *Salmonella* has a type III secretion system similar to that identified in *Yersinia* and *Shigella* species (type III secretion systems are discussed in detail in chapter 13). A number of gene products that are indispensable for *Salmonella* invasion have been identified. Some of these, like InvJ and SpaO, appear to be a part of the secretory apparatus, while others are likely to be the effectors responsible for the rapid cytoskeletal changes that occur when *Salmonella* interacts with the host cell. The factor that is responsible for the activation of the EGFR has not been identified but could be something that interacts specifically with the receptor or, more likely (since other cell types can be activated that do not have EGFR), is a factor that interacts with a common cellular moiety such as a carbohydrate that is involved in the activation of the EGFR signaling pathway in those cells.

One bacterial effector protein secreted upon *Salmonella* invasion that has been studied is the SptP tyrosine phosphatase. The N terminus of this protein has homology to the YopE effector protein from *Yersinia*, while the C terminus of SptP has homology to the YopH tyrosine phosphatase from *Yersinia*. (These *Yersinia* effectors are discussed in more detail below.) When the SptP tyrosine phosphatase is microinjected into cells, it results in a disruption of the actin cytoskeleton. Curiously, both halves of the protein elicit the same result, suggesting that both the YopE and the YopH homologies may contribute to its action. In addition, this experiment suggests that the tyrosine phosphatase activity of SptP is not necessary for its role in cytoskeletal disruption, since one of the halves does not have the activity but still elicits the cytoskeletal effects. This is in contrast to the YopH tyrosine phosphatase, which is not cytotoxic in its inactive form.

SptP acts with other secreted effectors to elicit the disruption and reorganization of the cytoskeleton that occurs upon *Salmonella* invasion. Other potential effector molecules delivered by the *Salmonella* type III secretion system include SipA, SipB, and SipC. Their functions in the *Salmonella* invasion process are not yet known, but they have homology to their *Shigella* counterparts—IpaA, IpaB, and IpaC—and are predicted to act in a similar manner.

Despite the morphological and mechanistic differences between the trigger and zipper mechanisms for bacterial invasion of nonphagocytic cells, they have certain important traits in common. Both require that extracellular bacteria send signals across the host cell plasma membrane to induce local actin polymerization. Both involve bacterial activation of signal transduction cascades that are already present in the host cell although used for other purposes. Both are mediated by bacterial expression of specific virulence factors, whose sole function is to induce these particular responses during host cell invasion. Importantly, both zipper and trigger uptake proceed by using energy derived from the host cell. The cytoskeletal and membrane rearrangements that are responsible for bacterial invasion require no energetic input from the bacterium; they invade by persuasion rather than by force. This is in contrast to invasion of some eukaryotic parasites, such as *Toxoplasma gondii.* In that case, the energy for invasion comes from the parasite, which actively pushes its way into the host cell, which is a passive victim.

Preventing Phagocytosis

Yersinia Species

Thus far, we have only considered cases where it is to the advantage of the bacterium to be taken up by a host cell. Clearly, this is not always the case. In particular, bacteria that have gained access to normally sterile host tissues must avoid being eaten and digested by patrolling macrophages. Macrophages are highly motile and use an actin-dependent mechanism of phagocytosis to engulf any foreign particles they encounter. As we saw above, *Salmonella* avoids this fate by preemptively invading macrophages through a trigger mechanism and delivering itself into an intracellular compartment that is not targeted to fuse with the macrophage lysosomes. Other bacteria go to the opposite extreme and prevent the macrophages from engulfing them at all. This is done by blocking the signaling mechanisms that control the cytoskeletal rearrangements involved in immune cell phagocytosis. *Yersinia* species, for example, disrupt the macrophage cytoskeleton by injecting factors into the host cell cytosol to block the normal rearrangements required for phagocytosis.

Y. pestis enters the body by transmission through the skin from an infected flea bite or by aerosol transmission from an infected individual. They are then transmitted through the bloodstream to the nearest lymph nodes, where they multiply. *Y. enterocolitica* and *Y. pseudotuberculosis* enter the body by the fecal-oral route and invade the mesenteric lymph nodes around the intestine through the Peyer's patches. *Yersinia* species have developed a complex mechanism to evade the host immune response in lymphatic tissues by injecting factors into host cell macrophages that inhibit their ability to reorganize their cytoskeletons and engulf and kill the invaders. Injection of these factors into the host cell macrophages is mediated through a type III secretion system (discussed in chapter 13 and also described above in the context of triggered invasion). These injected proteins are called Yops. Initially, the Yops were named after "*Yersinia* outer membrane proteins," but some of these factors are now thought to be soluble. However, the name had already caught on, and so they are still called Yops. The Yops are part of a virulon encoded by a 70-kilobase virulence plasmid. Activation of these genes has been shown to require association of the bacterium with the host cell plasma membrane. The factors that control the initial steps of association, YopB, YopD, YopN, and LcrG, are the delivery apparatus of the Yop system, while YopH, YopM, YpkA, and YopE are the factors injected into the host cell cytoplasm that have effects on the actin cytoskeleton and the signal transduction apparatus of the cell.

YopH, YpkA, YopM, and YopE are indispensable for virulence, as shown by site-directed mutagenesis. YopH is a 51-kDa protein that has homology to mammalian protein tyrosine phosphatases in its C-terminal half, a proline-rich region in its central domain that has been reported to bind to the SH3 domain of the c-Src tyrosine kinase, and N-terminal signals for bacterial secretion and translocation into cells. YopH is one of the most active tyrosine phosphatases ever isolated, even though there is no evidence for the presence of phosphotyrosine in bacteria. It is suspected that the gene for this tyrosine phosphatase was acquired by the bacterium in a lateral transfer from a eukaryotic host. YopH acts synergistically with YopE to inhibit phagocytosis and cause detachment of cells from the extracellular matrix. Two groups have recently reported that the focal adhesion proteins FAK and p130^{cas} are substrates for the action of YopH and hypothesize that the inappropriate dephosphorylation of FAK and p130^{cas} causes disruption of focal adhesions. This may result in release of the cell from the extracellular matrix, thus abolishing the ability of the cell to engulf the bacteria. Experiments involving an inactive form of the YopH phosphatase show that this protein binds to FAK and p130^{cas} in a phosphotyrosine-dependent manner and that the inactive YopH colocalizes with these proteins in focal adhesions. This work is the first step in delineating the pathway that leads from bacterial attachment at the host cell surface to the disruption of the actin-rich structures in adhesive contacts important for inhibition of bacterial uptake.

YpkA is an 81-kDa protein with homology to mammalian serine/threonine kinases and, like YopH, is thought to interfere with bacterial uptake by macrophages and the oxidative burst that is used to kill the bacteria. Its targets are still unknown. YopE is a cytotoxin that causes the disruption of actin microfilaments. It does not disrupt pure actin filaments in vitro, however, and thus is thought to exert its effects on actin through an unidentified factor(s). The function of YopM is still unknown. Future study of these proteins will help to elucidate both the mechanism of bacterial infection and the complex protein interactions involved in the actin cytoskeletal rearrangements blocked by these bacterial interactions.

Bacterial Attachment without Invasion

Enteropathogenic *Escherichia coli*

A completely different type of host cell cytoskeletal manipulation has been developed by enteropathogenic *Escherichia coli* (EPEC). Like *Listeria*, *Salmonella*, and *Shigella*, EPEC enters the human host orally and its initial site of host colonization is the intestinal epithelium. However, EPEC typically remains extracellular and forms colonies that are strongly attached to the surface of the epithelial cells. This strategy allows the bacteria to remain in the nutrient-rich environment of the intestinal lumen while other microbial competitors are washed out by the diarrhea caused by EPEC infection. One would expect that attachment of bacteria to the epithelial cell surface via epithelial cell adhesion proteins would result in zippering and bacterial invasion, as described above for *Listeria* and *Yersinia*. EPEC has developed a mechanism to generate very strong adhesive contacts with epithelial cells, but these contacts do not result in uptake. Paradoxically, this anti-invasive mechanism also involves recruitment of cytoskeletal elements and induction of actin filament polymerization at the site of attachment to the host cell.

EPEC and related gram-negative pathogens are important causative agents of diarrheal disease in children worldwide. In cell culture and in biopsy specimens of infected intestinal tissue, epithelial cell brush borders contain microcolonies of bacteria at sites where the microvilli have been destroyed and replaced by actin-rich structures termed pedestals. These areas of infection are called attaching and effacing lesions due to the intimate attachment of the bacteria and consequent loss of microvilli (effacement) associated with the disease.

The initial adherence of EPEC to the intestinal epithelia is mediated by the association of the bacterial protein intimin and its bacterially coded receptor translocated intimin receptor (Tir, previously called Hp90). The bundle-forming pilus of EPEC causes the subsequent clumping of the bacteria at that site. The bundle-forming pilus is a protein extension that is crucial for the formation of EPEC microcolonies by allowing for the clustering of the microbes through association of their pili. Once intimate attachment has been established, EPEC activates epithelial cells by injecting proteins into their cytosol through a recently identified type III secretion system (described in chapter 13). The intestinal epithelial cells, in addition to being induced to retract their microvilli, are induced to create a pedestal of actin filaments underneath the bacteria. The cytoskeletal proteins α-actinin, talin, and ezrin have been localized to this pedestal area and may be involved in the formation of this structure. The radical change in cytoskeletal organization is mediated by virulence genes in the bacterium which mimic host cell signal transduction mechanisms to direct the construction of the pedestal. On some tissue culture cell types, the pedestals themselves move laterally over the cell surface, a dramatic and unusual form of actin-based motility.

Almost all of the EPEC factors known to be important for the formation of the pedestal in epithelial cells are encoded by genes on the bacterial chromosome. The *eaeA* gene, which encodes intimin, is thought to be responsible for the organized formation of the actin-rich pedestal. Mutations in the *eaeA* gene cause signaling and cytoskeletal rearrangements but in a general, unorganized manner. The formation of the pedestal structure is blocked. Intimate attachment of EPEC to cells is followed by the injection of bacterial factors EspA and EspB into the host cell. These factors initiate the host signal transduction pathways that lead to pedestal formation. It is unknown how they interact with intimin. Mutations in the *espB* gene do not cause signal transduction or cytoskeletal rearrangement. The *espB* gene product is a 37-kDa protein that is secreted into the supernatant when EPEC is grown in tissue culture medium but not in bacterial medium. Another secreted protein, EspA, is thought to be secreted along with EspB by the type III system. The molecular basis of the effects of these proteins on the actin cytoskeleton has yet to be elucidated.

Intracellular Motility and Intercellular Spread

Pathogenic bacteria that have chosen to exploit an intracellular growth niche in the host are quickly faced with a unique problem. To continue to replicate, they must find some mechanism for moving from one host cell to another. In most cases, including *Salmonella* and other well-known intracellular pathogens such as *Mycobacterium*, the bacteria replicate within a membrane-bound compartment until the host cell lyses and the replicated bacteria can repeat the cycle of invasion. A small number of pathogens have taken an alternative approach and have developed mechanisms for undergoing intercellular spread prior to the lysis of the originally infected host cell. These bacteria leave the phagocytic vacuole by secreting membrane-degrading enzymes and then, in the cytosol, exploit the host cell actin cytoskeleton for their own motility. This allows them to divide and move within the cell they have invaded as well as to pass into neighboring cells via the engulfment of membrane extensions containing bacteria. This unique mode of spread allows these crafty pathogens to move from cell to cell in a host epithelium without ever leaving the cytoplasmic environment, and they are therefore not exposed to soluble antibodies. Three bacterial genera with this ability have been identified so far: *Listeria*, *Shigella*, and *Rickettsia*. Since *Rickettsia* species are obligate intracellular pathogens with limited molecular genetics, relatively little is known about them. For this reason, we focus on *Listeria* and *Shigella* species.

Listeria monocytogenes

As discussed above, *Listeria* invades epithelial cells through a zipper mechanism. Shortly after entering the host cell, the microbe secretes a pore-forming hemolysin, listeriolysin O, which degrades the phagocytic vacuole, releasing the bacterium free into the host cell cytoplasm. After entering the host cell cytosol, the *Listeria* organism is quickly surrounded by a "cloud" of short actin filaments which localizes to one end of the bacterium upon cell division. After reorientation of the actin associated with the bacterium, an actin-rich tail forms and the bacteria can be observed to move at speeds ranging from 6 to 90 μm per min depending on the host cell type. The tail, which can be tens of micrometers in length, is made up of many thousands of short actin filaments cross-linked into a dense meshwork. Electron microscopic analysis of the actin-binding portion of myosin bound to the actin tail of *Listeria* demonstrates that the actin filaments are oriented with their barbed (rapidly growing) ends toward the surface of the bacterium. The actin filaments within the tail remain stationary in the host cell cytoplasm. Videomicroscopic studies of the incorporation of fluorescently labeled actin into the actin tail of *Listeria* have shown that the filaments behind the bacteria are growing at the bacterial surface and depolymerizing uniformly throughout the tail, suggesting that the propulsive force is derived from the actin polymerization at the bacterial cell wall. As we saw above for cellular invasion, the energy for bacterial movement is provided entirely by the host cell.

One bacterial protein, ActA, has been identified as being necessary and sufficient for the subversion of the host cell actin cytoskeleton for motility by *Listeria*. ActA is a 67-kDa protein that is expressed on the surface of the bacterium. Expression of the ActA gene in a nonpathogenic *Listeria* species, *Listeria innocua*, confers motility on this species. The ability to move by actin-based motility was also observed in a different gram-positive bacterium, *Streptococcus pneumoniae*, that had been coated with an ActA-derived fusion protein. Interestingly, motility was observed only after bacterial division. This is consistent with the observation for *Listeria* that ActA must be polarized to one end of the bacterium to support motility. Polarization of ActA by *Listeria* also appears to be regulated by bacterial division.

ActA is not homologous to any known genes but has a number of interesting features that are important for its function. It is a surface-bound protein that has an N-terminal signal sequence and a single C-terminal membrane-spanning domain. Actin filament nucleation at the bacterial surface requires the presence of the N-terminal one-third of the protein. This N-terminal domain also mediates ActA dimerization, which might be important in the establishment of polarized localization. The central one-third of the protein includes four proline-rich repeats. Mutational analysis has shown that the proline-rich repeats contribute to *Listeria* motility by affecting its rate. Removal of any or all of the proline-rich repeats causes the mutant bacteria to move more slowly, with about a 20% decrease in speed for each repeat removed. Thus, ActA mediates three separable steps in *Listeria* actin-based motility. First, it catalyzes the nucleation of actin filament polymerization at the bacterial surface. Second, it mediates the rearrangement of these actin filaments into a polarized tail structure that can support unidirectional movement. Third, it uses specific sequences to accelerate the movement of the bacteria. Interestingly, purified ActA has no effect on actin polymerization or dynamics in vitro. All of its functions are mediated through association with factors supplied by the host cell.

In recent years, much experimental effort has been devoted to determining what host cell factors are important in the generation of *Listeria* actin-based motility. A number of host proteins are localized to the tail of the bacterium, while others interact directly with ActA. These studies have been greatly facilitated by cell-free systems that have been developed to study *Listeria* motility in vitro. Extracts from the oocytes of the clawed toad *Xenopus laevis*, as well as cytosolic and cytoskeletal extracts from human platelets, provide the host cell factors that are necessary for motility. When these extracts are supplemented with energy in the form of ATP and fluorescently labeled actin monomer, *Listeria* can be observed by videomicroscopy to move in a manner that is very similar to that seen in cultured cells (Figure 10.3). This provides a system to study both the biochemical requirements and the biophysical aspects of bacterial actin-based motility.

Biochemical studies have implicated a number of different host cell proteins in the process of actin-based motility by *L. monocytogenes*. Some of these, i.e., tropomyosin, fimbrin, vinculin, villin, ezrin, talin, and gelsolin, have been localized to the actin-rich tail but have not been shown to play a functional role in motility. It is likely that the high concentration of actin in the tail nonspecifically attracts some of these actin-binding proteins, and so their importance cannot be determined until further study verifies a requirement for them in the tail. The proteins whose functional role in actin-based motility has been demonstrated include the Arp2/3 complex, VASP, profilin, and α-actinin. The evidence for the involvement of each of these factors is discussed below, and their putative roles are summarized in Figure 10.4.

Recently, the purification of a complex from platelets that supports actin nucleation at the surface of *Listeria* has provided more information about this process. The Arp2/3 complex is named for the actin-related proteins (Arps) that are two of the seven polypeptides of the complex. This complex is homologous to one identified in *Acanthamoeba castellani* and appears to play an evolutionarily conserved role in regulation of the actin cytoskeleton in eukaryotes. The complex has actin-nucleating activity when tested in a fluorescence assay that measures the polymerization of actin filaments through incorporation of pyrenated actin monomers. This nucleating activity is synergistically activated by the presence of ActA. The Arp2/3 complex acts with the N-terminal domain of ActA (in the region upstream of the proline-rich repeats) to accelerate actin filament polymerization at the fast-growing, barbed end. This activity may be modulated by the interactions of VASP and profilin with the proline-rich and long repeats of the central domain of ActA. The Arp2/3 complex is found throughout the en-

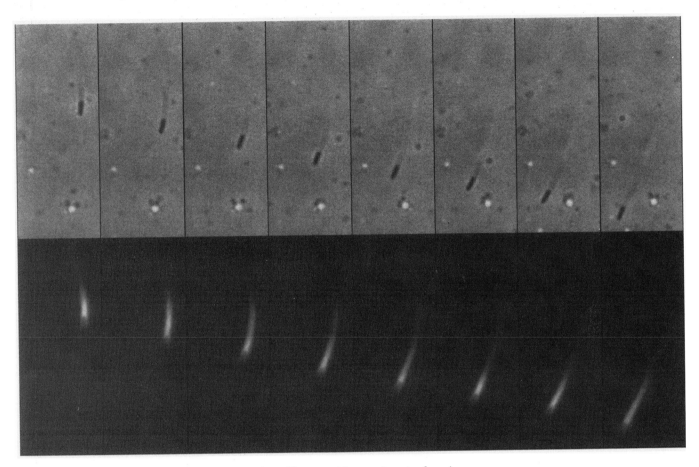

Figure 10.3   Actin tail formation and movement of *L. monocytogenes* in cytoplasmic extract. Eight frames from a video sequence are shown, recorded at 10-s intervals with phase-contrast images on the top and direct actin fluorescence on the bottom.

tire extent of the *Listeria* tail, suggesting that the complex may remain associated with one end of the filament that it nucleates after the filament is released from the bacterial surface. This work has influenced the study of actin polymerization in such processes as formation of lamellipodia and chemotaxis that also involve rapid polymerization of actin filament barbed ends because the Arp2/3 complex has been localized to lamellipodia in locomoting cells.

Profilin is a 14-kDa protein that binds to actin monomers. Profilin acts as a nucleotide exchange factor for actin, enhancing the dissociation of ADP and the binding of ATP to monomeric actin. In vitro, profilin can accelerate the elongation of actin filaments by acting as a chaperone to shuttle actin monomers from the thymosin-β4-bound monomer pool to the barbed ends of preexisting filaments. Its association with actin can be disrupted by the inositol lipid phosphatidylinositol-4,5-bisphosphate (PIP$_2$), and this interaction is thought to be one way that actin monomers are activated for rapid polymerization of the actin cytoskeleton in response to stimuli that generate PIP$_2$, such as growth factors. Profilin localizes to moving bacteria in a manner dependent on the presence of the proline-rich repeat domain of ActA on the bacterial surface. However, profilin does not bind directly to ActA, which suggests that another factor is bridging the gap.

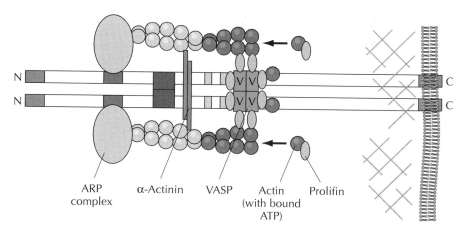

Figure 10.4 Schematic diagram of the interactions between *L. monocytogenes* ActA protein and host cell cytoskeletal factors. The C terminus is anchored in the bacterial membrane. The N-terminal third of the protein mediates dimerization and also mediates nucleation of actin filaments by the Arp2/3 complex. Filaments nucleated close together can be efficiently cross-linked by α-actinin. A second host protein complex that includes VASP tetramers and profilin associates with the proline-rich repeats in the central region of ActA. This interaction serves to accelerate the rate of actin filament elongation, since profilin can act as a chaperone to deliver ATP-bound actin monomers to filament ends.

The adapter factor that links profilin to the proline-rich region of ActA has been identified as VASP. VASP, like profilin, is localized to the end of motile bacteria associated with the actin-rich tail, and this association has been reported to occur before the formation of the actin clouds around the bacteria. This suggests that VASP plays a role in actin filament formation. VASP binds directly to ActA in vitro in a blot overlay assay by using radiolabeled VASP as a probe, and its consensus binding site is the oligopeptide FPPPP found within the proline-rich domain of ActA. Mutation of the proline-rich repeats of ActA abolishes VASP binding. In addition, VASP binds to profilin and has several other putative binding partners that may be involved in cytoskeletal regulation. A VASP-related protein called Mena (the mammalian homolog of the *Drosophila* enabled gene) also can bind both ActA and profilin. The observation that deletion of the proline-rich repeats results in a decrease of bacterial motility suggests that the acceleration function of ActA is mediated through its association with VASP or Mena and profilin. The known function of profilin in accelerating the rate of actin filament elongation is the most likely cause of this change in bacterial speed. However, neither VASP nor profilin is required for actin filament nucleation at the bacterial surface.

α-Actinin is a 100-kDa actin-cross-linking protein that acts by forming a homodimer that can associate with an actin filament at each end. α-Actinin has been localized to focal adhesions and filopodia, where it is thought to be important in maintaining the bundling of the microfilaments in those structures. Microinjection of a 53-kDa piece of the α-actinin molecule that contains the actin-binding domain but not the dimerization domain causes disruption of actin bundles because they are not cross-linked. This treatment disrupts the tail of *L. monocytogenes* and abrogates motility, indicating that α-actinin is important for cross-linking actin microfilaments in the actin-rich tail of *Listeria*. A cloud of actin filaments remains associated with the bacteria after this treatment, suggesting that the cross-linking function of α-actinin is required primarily for establishing the mechanical integrity of the tail that enables it to push the bacterium.

Thus, nucleation and elongation of actin filaments at the bacterial surface are mediated by the bacterial surface protein, but the construction and dynamics of the tail itself are governed by the host cell. First, the filaments are cross-linked by α-actinin to form a sound structure that can push the bacterium, and then the host also directs depolymerization of old filaments in the tail (catalyzed by the actin-binding protein cofilin/ADF) to regenerate the monomer pool, so that elongation can continue at steady state at the bacterial surface, resulting in a constant flux of actin subunits through the tail. Small local biochemical interactions within a few nanometers of the bacterial surface initiate a cascade of cytoskeletal rearrangements resulting in formation of the tail, which can be many micrometers in length. This thousandfold enhancement of scale is reminiscent of the host cell responses induced during bacterial invasion. Once again, the bacteria harness the power of the host cell cytoskeleton by small-scale signals, turning the host cell itself into a powerful and energetic ally.

The process of intercellular spread of *Listeria* is poorly understood. The bacteria appear to leave the host cell in a plasma membrane protrusion that forms when the motile bacterium comes into contact with the membrane. *Listeria* moves at the same rate in the membrane protrusions as in the cytosol and is either returned to the same cell or taken up by a neighboring cell. In the current hypothesis, the neighboring cell phagocytoses the membrane protrusion and then contains the bacterium in a double-membrane-bound vacuole. The bacterium escapes that vacuole by secreting the pore-forming hemolysin listeriolysin O, as well as two phospholipases, and then resumes its movement within the cytoplasm of the second cell. However, the mechanism that allows an adjacent cell to "bite off" a piece of its neighbor has not been studied. In addition, it is not known whether specific host or bacterial proteins are involved in the formation of the membrane protrusions and the subsequent phagocytic process. No mutants have been described that are capable of forming intracytoplasmic actin tails but are not capable of forming protrusions, implying that this aspect of behavior may be mediated primarily or solely by host factors. The development of new techniques that can distinguish between intra- and intercellular spread is required to dissect this important aspect of infection by these bacteria.

Shigella flexneri

In contrast to the zipper-type invasion of *Listeria*, *Shigella* invades host epithelial cells by a trigger mechanism and thereby ends up in a different type of intracellular membrane-bound compartment. However, after this point, the intracellular life-styles of *Shigella* and *Listeria* are quite similar. *Shigella* also escapes from its membrane-bound compartment, in a process that apparently requires IpaB. Once it has left the phagocytic vacuole, *Shigella* can usurp the host cell actin cytoskeleton for its own motility and forms an actin-rich tail that is very similar to that formed by *L. monocytogenes* but is due to the expression of a different bacterial protein, called IcsA (for "intercellular spread A") or VirG. Interestingly, IcsA has no sequence similarity to the ActA protein. IcsA is necessary and sufficient for generation of motility by expression of IcsA in nonmotile *E. coli*; *E. coli* expressing IcsA moves by actin-based motility in cytoplasmic extracts.

IcsA is a 120-kDa protein that has an N-terminal signal sequence for delivery to the periplasmic space and a C-terminal sequence of 344 amino acids (known as the β-domain) that delivers and anchors the protein to the outer membrane of the cell. The remaining 706 amino acids are known as the α-domain and are oriented with the N terminus of the protein extended away from the cell. This domain is predicted to be important for interactions with host cell proteins. The α-domain contains five glycine-rich repeats that may be important for binding to proteins which are responsible for the nucleation of actin filaments to form the actin-rich tail. IcsA is proteolytically cleaved at position 758 (between the α and β domains) by the SopA (IcsP) protease, and this yields a soluble 95-kDa fragment that has been localized to the actin-rich tail of *Shigella* but does not appear to be important for motility, because a *Shigella* mutant that does not express the protease and does not secrete the cleaved IcsA is still motile. IcsA appears to be expressed on the entire bacterial surface but is localized in a polar manner on locomoting bacteria. It has been hypothesized that the proteolytic cleavage of IcsA is responsible for maintaining the polar localization of this protein by clipping away the protein everywhere except at one end of the bacterium. This hypothesis is supported by the fact that SopA-defective mutants do not show a polar localization of IcsA. Interestingly, these mutants have increased motility. This is in contrast to the situation with *Listeria*, where polarization of ActA is critical for the generation of motility.

Two proteins that bind to IcsA have been identified: the actin-binding protein vinculin and the neuronal Wiskott-Aldrich syndrome protein (N-WASP). Two studies have shown that vinculin localizes to the "rear" end of motile *Shigella* in cells and one of these studies reported that vinculin accumulated at the same location as IcsA before the formation of actin clouds. This is similar to the results reported for VASP in *Listeria* motility and implicates vinculin in the formation of the *Shigella* actin-rich tail. Vinculin is a 120-kDa protein that has a cryptic site in its C terminus for actin binding that is revealed by proteolysis of vinculin into a 90-kDa N-terminal head portion and a 30-kDa C-terminal tail portion under certain conditions. One of these conditions appears to be *Shigella* invasion. In experiments with *Shigella* invasion of PtK2 epithelial cells, only the 120-kDa vinculin was detected by Western blotting of whole-cell extracts prior to invasion; however, after invasion, the 90-kDa head portion could be detected in large amounts on the surface of the bacteria. An in vitro association of the 90-kDa head portion of vinculin with the α-domain of purified IcsA has been observed. In further support of this work, only microinjection of the head fragment and not the full-length vinculin accelerates the motility of *Shigella*. Interestingly, vinculin contains conserved proline-rich motifs that may allow it to bind to VASP. This has led to the generation of a model where the acceleration of *Shigella* motility is the same as that of *Listeria* motility except for the addition of the vinculin head fragment to bridge the gap between IcsA and VASP, where VASP binds to ActA directly. This model remains to be proven but is supported by the observation that microinjection of profilin also accelerates *Shigella* motility.

One recent study has reported the association of IcsA with the N-WASP actin-associated protein. N-WASP is the neuronal relative of the hematopoietic WASP protein implicated in Wiskott-Aldrich syndrome. Wiskott-Aldrich syndrome is a condition where blood cells have improperly formed cytoskeletons and is characterized by thrombocytopenia. Both of these

WASP proteins are thought to be regulated by the small GTP-binding protein Cdc42. However, only N-WASP is important for the formation of actin-rich filopodial membrane projections. N-WASP binds to the glycine-rich repeat region of IcsA in vitro and localizes to the front of the actin-rich tail in locomoting bacteria in cultured cells. Depletion experiments showed that removal of N-WASP from *Xenopus* egg extracts abrogated the ability of *Shigella* to move and that addition of N-WASP restored motile ability. N-WASP does not seem to have any effect on the motility of *Listeria*. It is possible that N-WASP is capable of nucleating actin filament polymerization in association with IcsA. Alternatively, N-WASP might be acting as an adapter factor that mediates the localization of a nucleating factor (possibly the Arp2/3 complex). Intercellular spread in *Shigella* infection has been only slightly better studied than that in *Listeria* infection. The hypothesis is the same. The bacteria appear to protrude from the host cell in a plasma membrane projection that is taken up by a neighboring cell. The bacteria escape the double-membrane vacuole in the adjacent cell and start their motility again. For *Shigella* infection, the host cell protein E-cadherin has been implicated in the ability of the bacteria to spread to neighboring cells. Host cells with E-cadherin deleted show less intercellular spread than wild-type cells do. However, the basis for this has not been elucidated and could be due to global structural defects in the E-cadherin null cells. Further study is still required in this area.

The ActA-mediated actin-based motility of *Listeria* and the IcsA-mediated actin-based motility of *Shigella* have apparently arisen independently through convergent evolution, since there is no detectable similarity between the two bacterial proteins. Comparison of the two systems is thus particularly informative. In both cases, there appear to be several steps involved in actin-based motility. First, the bacterial surface proteins associate with host factors that cause actin filament nucleation (the Arp2/3 complex in *Listeria,* and N-WASP plus possibly other unidentified factors in *Shigella*). Second, the nucleated clouds are rearranged to form the polarized tail, a step critical in the generation of unidirectional movement that apparently involves both bacterial polarity and actin-cross-linking proteins provided by the host cell. Third, actin filament elongation at the bacterial surface is accelerated by yet another set of host factors, resulting in acceleration of overall bacterial speed. The dynamics and behavior of the tail at a distance from the bacterial surface are governed solely by host cell homeostatic mechanisms. In both cases, the bacterium provides no energy for its movement but moves through host cell cytoplasm by using host cell energy. The observation that this ability has appeared two and possibly three times independently in the evolution of bacterial pathogens (and at least once among the viruses [Box 10.1]) suggests that this evolutionary step is not particularly demanding.

The main utility of actin-based motility for the bacteria appears to be so that they can reach the host cell surface and spread intercellularly. Objects the size of bacteria cannot diffuse appreciable distances in the viscous eukaryotic cytoplasm and so must use some form of active movement. The cytoplasm is also too viscous for bacterial flagella to rotate, and so the bacteria must look elsewhere for a motile system. In the eukaryotic cytoplasm, they are surrounded by the robust and adaptable actin cytoskeleton, which is preadapted and poised to generate large-scale protrusive structures in response to small local signals. Apparently, all the bacteria must do is provide a binding site for host factors that catalyze actin filament

BOX 10.1

Everybody in the pool, viruses can swim too!

Bacteria are not the only pathogens that can exploit the host cell cytoskeleton. Recent evidence suggests that at least one virus can utilize the actin cytoskeleton for its own intracytoplasmic motility as well. Vaccinia virus is a complex virus related to the smallpox viruses and has a large genome of about 190 kb. Initial observations of vaccinia virus particles by electron microscopy revealed virus-containing microvillus-like structures at the surface of infected cells, and these structures were later shown to contain actin. Videomicroscopy has directly demonstrated that viruses associated with these actin structures are motile and that the movement of tail-associated viral particles is very similar to the actin-based movement of *Listeria* and *Shigella.*

Vaccinia virus may take a number of different forms during the course of infection. One, the intracellular mature virus (IMV), is released from the cell only upon cell lysis. A second form, the intracellular enveloped virus (IEV), consists of IMV (about 5 to 15% of the total IMV) that wraps itself in intracellular membranes derived from the *trans*-Golgi network. The third form, extracellular enveloped virus (EEV), is released when IEV fuses with the plasma membrane. The IEV form of the virus appears to be the only one that can move by actin-based motility. Even though infection is not sensitive to actin filament-disrupting drugs such as cytochalasin D, this drug does prevent the release of IEV particles from cells, suggesting that actin polymerization is involved in the release process. The approximately 260 open reading frames of vaccinia virus do not encode any homologs to the ActA and IcsA proteins involved in *Listeria* and *Shigella* motility, respectively, and so it is not obvious how IEV achieves motility. Recently, there have been a number of reports implicating IEV-specific factors in the generation of the vaccinia virus actin tail. The specific mechanisms of action of these gene products have yet to be elucidated.

The microvilli that are formed by the viruses appear to be morphologically distinct from normal cellular microvilli and can be observed extending into neighboring cells. This may mean that vaccinia virus, like *Shigella* and *Listeria,* can spread to adjacent cells in membrane protrusions as well as by more classical budding.

It may seeem surprising that these viruses have developed a pathogenic mechanism that bears such striking similarity to a pathogenic mechanism of bacteria. However, the selective pressures driving the evolution of this dramatic type of host-pathogen interaction may be similar in the two cases. Most budding viruses assemble into particles from their constituent components as the host cell plasma membrane; the unassembled components can diffuse rapidly through the cytoplasm to the site of assembly at the surface. Vaccinia virus, however, replicates in viral "factories" near the nucleus, and it is the intact (and relatively large) viral particles that must make their way to the plasma membrane. Harnessing the protrusive behavior of the host cell actin cytoskeleton appears to be an energetically cheap and evolutionarily nondemanding mechanism to generate intracellular motility, for these large viruses as well as for bacteria.

nucleation and elongation at the bacterial surface, and the later steps of construction of the comet tail and generation of propulsive force are all taken care of by the obliging host cell.

Selected Readings

Black, D. S., and J. B. Bliska. 1997. Identification of p130Cas as a substrate of *Yersinia* YopH (Yop51), a bacterial protein tyrosine phosphatase that translocates into mammalian cells and targets focal adhesions. *EMBO J.* **16:**2730–2744.

Cossart, P. 1997. Host/pathogen interactions. Subversion of the mammalian cell cytoskeleton by invasive bacteria. *J. Clin. Invest.* **99:**2307–2311.

Cudmore, S., P. Cossart, G. Griffiths, and M. Way. 1995. Actin-based motility of vaccinia virus. *Nature* **378:**636–638.

Finlay, B. B., and P. Cossart. 1997. Exploitation of mammalian host cell functions by bacterial pathogens. *Science* **276:**718–725.

Fu, Y., and J. E. Galan. 1998. The *Salmonella typhimurium* tyrosine phosphatase SptP is translocated into host cells and disrupts the actin cytoskeleton. *Mol. Microbiol.* **27:**359–368.

Galan, J. E., and J. B. Bliska. 1996. Cross-talk between bacterial pathogens and their host cells. *Annu. Rev. Cell Dev. Biol.* **12:**221–255.

Ireton, K., and P. Cossart. 1998. Interaction of invasive bacteria with host signaling pathways. *Curr. Opin. Cell Biol.* **10:**276–283.

Kuhn, M., and W. Goebel. 1998. Host cell signalling during *Listeria monocytogenes* infection. *Trends Microbiol.* **6:**11–15.

Suzuki, T., H. Miki, T. Takenawa, and C. Sasakawa. 1998. Neural Wiskott-Aldrich syndrome protein is implicated in the actin-based motility of *Shigella flexneri*. *EMBO J.* **17:**2767–2776.

Welch, M. D., A. Iwamatsu, and T. J. Mitchison. 1997. Actin polymerization is induced by Arp2/3 protein complex at the surface of *Listeria monocytogenes*. *Nature* **385:**265–269.

11

Bacterial Toxins

RINO RAPPUOLI AND MARIAGRAZIA PIZZA

Toxins were the first bacterial virulence factors to be identified and were also the first link between bacteria and cell biology. Cellular microbiology was, in fact, naturally born a long time ago with the study of toxins, and only recently, thanks to the sophisticated new technologies, has it expanded to include the study of many other aspects of bacterium-cell interactions. This chapter covers mostly the molecules that have been classically known as toxins; however, the last section also mentions some recently identified molecules that cause cell intoxication and have many but not all of the properties of classical toxins. These belong to a rapidly expanding field and are perhaps among the most interesting molecules being studied today in cellular microbiology. Table 11.1 shows the known properties of all bacterial toxins described in this chapter, while Figure 11.1 shows the subunit composition and the spatial organization of toxins whose structures have been solved either by X-ray crystallography or by quick-freeze deep-etch electron microscopy.

The observation that culture supernatants free of bacteria fully reproduced the symptoms of deadly diseases such as diphtheria, tetanus, cholera, and botulinum made it obvious that in these instances bacterial toxins were the only factors needed by bacteria to cause disease, making the study of their pathogenesis straightforward. In many cases, immunity against the toxins is enough to prevent the disease caused by the bacterium that produces them, and therefore they have been often used as vaccines after being subjected to chemical treatment to remove their toxicity. The powerful effects of toxins, such as cell death, and the striking morphology changes that they caused in vivo in animal models and in vitro on cells (Figure 11.2) made it easy to study their interactions with eukaryotic cells. For this reason, toxins were the first link between bacteria and cell biology. The link became stronger when it was found that the discovery of any toxin target led to the discovery of a new, important pathway in cell biology. This is because toxins need targets consisting of molecules that play a key role in the most essential and vital processes of living organisms. Two very important pathways of cell biology were discovered thanks to bacterial toxins, and they have been extensively investigated. The first is G-protein-

193

Table 11.1 Bacterial toxins and their targets

Category	Target	Toxin	Activity	Consequence	Three-dimensional structure known
Extracellular toxins acting on the cell surface	Immune system	Superantigens, (SEA–SEH, TSST-1, SPEA, SPEC, SSA)	Binding to MHC class II molecules and to Vβ or Vγ of T-cell receptor	T-cell activation and cytokine secretion	SEB, SED, SEC, TSST-1, SPEC, SPEA
		SPEB	Cysteine protease	Alteration in immunoglobulin-binding properties	−
	Cell membrane (pore-forming toxin)	Streptolysin O	Cell membrane permeabilization	Cell death	−
		Perfringolysin O			+
		Alpha-toxin			+
		Leukotoxins			−
		Aerolysin			+
		Delta hemolysin			−
		E. coli hemolysin			−
	Surface molecules	Bacteroides fragilis enterotoxin	Cleavage of E-cadherin	Alteration of epithelial permeability	−
Extracellular toxins acting on intracellular targets	Protein synthesis	DT	ADP-ribosylation of EF2	Cell death	+
		Pseudomonas aeruginosa exotoxin A			+
		Shiga toxin	N-Glycosidase activity on 28S RNA	Cell death	+
	Signal transduction	PT	ADP-ribosylation of G$_i$	cAMP increase	+
		CT	ADP-ribosylation of G$_s$	cAMP increase	+
		LT			+
		Pertussis adenylate cyclase	Binding to calmodulin, ATP → cAMP conversion	cAMP increase	−
		Anthrax EF	Binding to calmodulin, ATP → cAMP conversion	cAMP increase	−(PA+)
		Anthrax LF	Cleavage of MAPKK1 and MAPKK2	Cell death	−(PA+)
		CDT	Cell cycle arrest	Cytotoxicity	+
		C. perfringens alpha-toxin	Phospholipase C	Gas gangrene	+

Category	Toxin	Mechanism/Target	Cellular effect	
	C. difficile toxin A and B	Monoglucosylation of Rho, Rac, and Cdc42	Breakdown of cellular actin stress fibers	–
	CNF1 and CNF2	Deamidation of Rho, Rac, and Cdc42	Ruffling, stress fiber formation	–
Intracellular trafficking	Tetanus toxin	Cleavage of VAMP/synaptobrevin	Spastic paralysis	+ (B domain)
	Botulinum neurotoxins B, D, G, and F	Cleavage of VAMP/synaptobrevin	Flaccid paralysis	–
	Botulinum neurotoxins A and E	Cleavage of SNAP-25	Flaccid paralysis	+
	Botulinum neurotoxin C	Cleavage of syntaxin	Flaccid paralysis	–
	H. pylori vacuolating cytotoxin	Alteration in the endocytic pathway	Vacuole formation	–
Cytoskeleton	C. botulinum C2	ADP-ribosylation of monomeric G-actin	Failure of actin polymerization	–
Toxins directly injected into eukaryotic cells				
Signal transduction	YpkA	Protein serine/threonine kinase	Inhibition of phagocytosis	–
	YopH	Tyrosine phosphatase	Inhibition of phagocytosis	–
	Tir	Receptor for intimin	Actin nucleation and pedestal formation	–
	SopE	Rac and Cdc42 activation	Membrane ruffling, cytoskeletal reorganization, proinflammatory cytokine production	–
Cytoskeleton	ExoS	ADP-ribosylation of Ras	Collapse of cytoskeleton	–
	C. botulinum C3	ADP-ribosylation of Rho	Breakdown of cellular actin stress fibers	–
	SopB	Inositol phosphate phosphatase	Increased chloride secretion (diarrhea)	–
	YopE	Unknown	Cytotoxicity, actin depolymerization	–
Mediators of apoptosis	YopP/YopJ	Unknown	Apoptosis	–
	AvrRxv	Unknown	Necrotic lesions	–
	IpaB	Binding to ICE	Apoptosis	–
Unknown	ExoU	Unknown	Cytotoxicity (cell death)	–

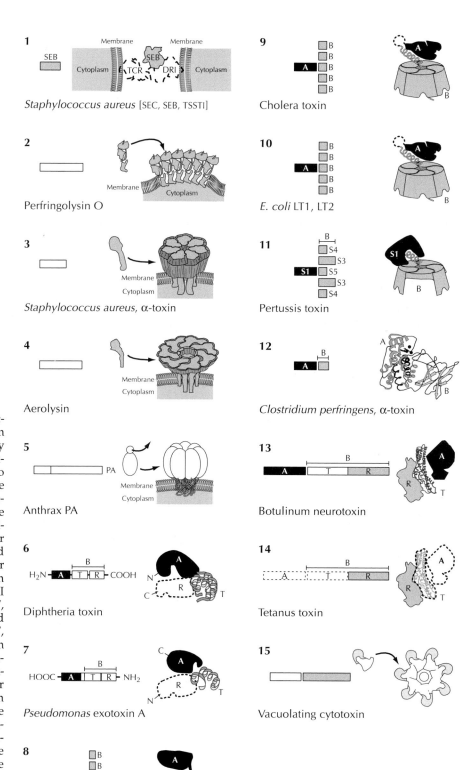

Figure 11.1 Structural features of bacterial toxins whose structures have been solved. (Left) Scheme of the primary structure of each toxin. For the A/B toxins, the domain composition is also shown. The A (or S1 in PT) represents the catalytic domain, whereas the B represents the receptor-binding domain. The A subunit is divided into the enzymatically active A1 domain and the A2 linker domain in Shiga toxin, CT, *E. coli* LTI and LTII, and PT. The B domain has either five subunits, which are identical in Shiga toxin, CT, and *E. coli* LTI and LTII and different in size and sequences in PT, or two subunits, the translocation (T) and the receptor-binding (R) subunits, in DT, *Pseudomonas* exotoxin A, botulinum toxin, and tetanus toxin. (Right) Schematic representation of the three-dimensional organization of each toxin. For *Staphylococcus* enterotoxin B, the protein is shown in the ternary complex with the human class II histocompatibility complex molecule (DR1) and the T-cell antigen receptor (TCR). For all toxins the schematic representation is based on the X-ray structure, except that for VacA, whose structure has been solved by quick-freeze, deep-etch electron microscopy.

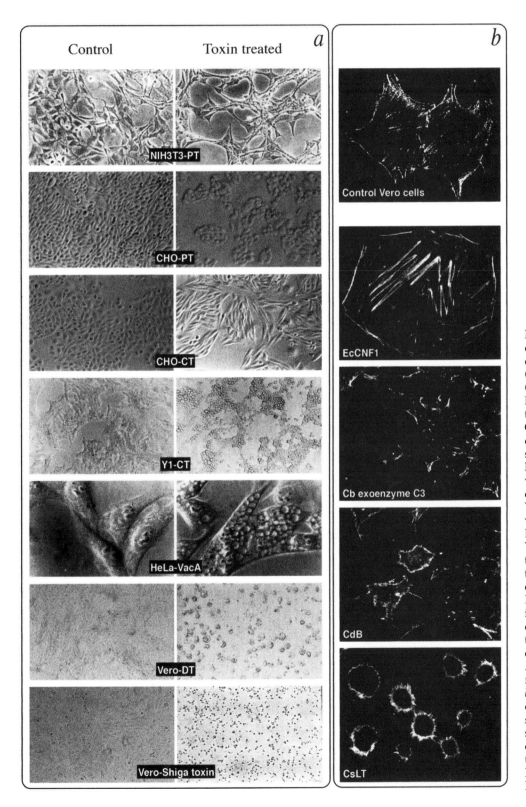

Figure 11.2 Morphology changes induced by toxins in cell lines. **(a)** Untreated cells (left) and toxin-treated cells (right). Abbreviations: NIH3T3-PT, NIH 3T3 cells treated with pertussis toxin; CHO-PT, Chinese hamster ovary cells treated with pertussis toxin; CHO-CT, Chinese hamster ovary cells treated with cholera toxin; Y1-CT, Y1 cells treated with CT; HeLa-VacA, HeLa cells treated with vacuolating cytotoxin A; Vero-DT, Vero cells treated with DT; Vero-Shiga toxin, Vero cells treated with Shiga toxin. Photographs from our laboratory or kindly provided by Ida Luzzi, Istituto Superiore di Sanità, Rome, Italy. **(b)** Immunofluorescence micrograph of untreated or toxin-treated Vero cells. Abbreviations: EcCNF1, *E. coli* cytotoxin-necrotizing factor; Cb exoenzyme C3, *C. botulinum* C3; CdB, *C. difficile* cytotoxin B; CsLT, *Clostridium sordellii* lethal toxin. Cells were stained with palloidin-fluorescein to visualize F-actin. Photographs kindly provided by Patrice Boquet, INSERM, Nice, France.

mediated signal transduction (Box 11.1), which was discovered in the early 1980s; it could be understood only because pertussis toxin specifically blocked the inhibitory G proteins. The second pathway is the molecular mechanism of neurotransmitter release, which was discovered when tetanus toxin was found to specifically cleave vesicle-associated membrane protein, a key molecule in intracellular vesicular trafficking.

Toxins have a target in most compartments of eukaryotic cells. For simplicity, we divide the toxins into three main categories (Figure 11.3): (i) those that exert their powerful toxicity by acting on the surface of eukaryotic cells simply by touching important receptors, by cleaving surface-exposed molecules, or by punching holes in the cell membrane, thus breaking the cell permeability barrier (Figure 11.3, groups 1 and 2); (ii) those that have an intracellular target and hence need to cross the cell membrane (these toxins need at least two active domains, one to cross the eukaryotic cell membrane and the other to modify the toxin target) (Figure 11.3, group 3); and (iii) those that are directly delivered by the bacteria into eukaryotic cells (Figure 11.3, group 4).

Toxins Acting on the Cell Surface

Some toxins deliver their toxic message just by binding to receptors on the surface of eukaryotic cells. The most common of these toxins are a family of related molecules known as superantigens. The binding domains (B domains) of the AB toxins sometimes also belong to this group. Some toxins act by cleaving important molecules exposed on the surface of eukaryotic cells. Another group of molecules that bind to the cell surface and in many cases affect cell function contains the pore-forming toxins, which act by punching holes in the cell membrane and breaking the cell permeability barrier.

BOX 11.1

Toxins act like the hero of "Independence Day"

More than 50% of bacterial toxins act on signal transduction. This suggests that perhaps the best way to poison a living organism is to interfere with cell communication. The vital importance of communication is obvious to our generation, which is dominated by multimedia and global internet connections. The importance of communication is even stressed by popular movies like "Independence Day," where the only way to defeat the extraterrestrial invaders is to interfere with their communication system, and "Tomorrow Never Dies," where the most dangerous criminal combated by

James Bond can get global power only by taking over communications. Like the hero of "Independence Day" and the criminal in "Tomorrow Never Dies," bacterial toxins have found it very convenient to attack cell communication systems. In eukaryotic organisms, cell communication is mediated by (i) tyrosine phosphorylation of the cytoplasmic carboxy-terminal part of a receptor that recruits SH2 signal transducers and initiates a cascade of intracellular signaling events or (ii) modification of a receptor-coupled GTP-binding protein that transduces the signal to enzymes releasing secondary messengers such as cAMP, inositol-triphosphate, and diacylglycerol, which also initiate a cascade of

intracellular signaling events. Interestingly, bacterial toxins almost exclusively target signal transduction mediated by GTP-binding proteins and only in rare cases attack receptors that transmit signals by direct tyrosine phosphorylation of the cytoplasmic portion of the receptor. Why is this? The answer is likely to be associated with the fact that GTP-binding proteins function as servers that transmit signals from many different receptors while tyrosine phosphorylation transmits the signal from one receptor only. Therefore, a toxin targeting tyrosine phosphorylation would be like a criminal taking over the local radio station of a village—nothing compared to the bad guys of the Bond movies.

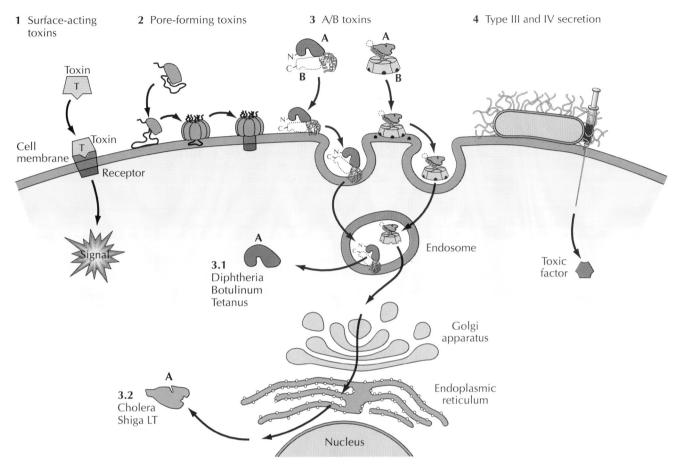

1 Surface-acting toxins   **2** Pore-forming toxins   **3** A/B toxins   **4** Type III and IV secretion

Toxin
T

Cell membrane   Toxin
T
Receptor

Signal

3.1
Diphtheria
Botulinum
Tetanus

Endosome

Toxic factor

Golgi apparatus

Endoplasmic reticulum

3.2
Cholera
Shiga LT

Nucleus

Figure 11.3   Schematic representation of the four groups of bacterial toxins. Group 1 toxins act by binding receptors on the cell membrane and sending a signal to the cell. Group 2 toxins act by forming pores in the cell membrane, perturbing the cell permeability barrier. Group 3 toxins are A/B toxins, composed of a binding domain (B subunit) and an enzymatically active effector domain (A subunit). Following receptor binding, the toxins are internalized and located in endosomes, from which the A subunit can be transferred directly to the cytoplasm by using a pH-dependent conformational change (3.1) or can be transported to the Golgi and the ER (sometimes driven by the KDEL ER retention sequence), from which the A subunit is finally transferred to the cytoplasm (3.2). Group 4 toxins are injected directly from the bacterium into the cell by a contact-dependent secretion system (type III or type IV secretion system).

Figure 11.4   Schematic representation of the interaction of superantigens with the MHC class II molecule and T-cell receptor.

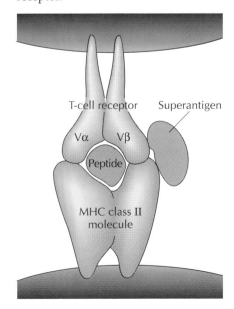

Superantigens

Superantigens are produced mostly by *Staphylococcus aureus* and *Streptococcus pyogenes*. They are bivalent molecules that bind two distinct molecules (Figure 11.4), the major histocompatibility complex (MHC) class II molecule and the variable part of the T-cell receptor (Vβ or Vγ). The cross-linking between the MHC and the T-cell receptor is able to activate T cells even in the absence of a specific peptide. Therefore, superantigens are potent polyclonal activators of T cells both in vitro and in vivo. In vivo, the potent polyclonal activation results in a massive release of cytokines such as interleukin-1 and tumor necrosis factor, which are believed to play an important role in diseases such as the toxic shock syndrome induced by toxic shock syndrome toxin 1 (TSST-1), vomiting and diarrhea caused by staphylococcal enterotoxins, and the exanthemas caused by the pyrogenic streptococcal exotoxins.

The staphylococcal enterotoxins (SE) comprise a family of homologous proteins of approximately 230 amino acids, with one central disulfide bond, named staphylococcal enterotoxins A through H (SEA, SEB, SEC1, SEC2, SEC3, SED, SEF, SEG, and SEH). TSST-1 also belongs to the same family but is shorter (194 amino acids) and has no cysteines. The streptococcal erythrogenic toxins include streptococcal pyrogenic enterotoxin A and C (SPEA and SPEC) and streptococcal superantigen A (SSA).

Other bacterial toxins have been proposed to be superantigens. One of them is the streptococcal pyrogenic exotoxin B (SpeB), a virulence factor with cysteine protease activity produced by group A streptococci. SpeB is expressed as a 40-kDa protein and is converted to a 28-kDa active protease by proteolytic cleavage. It causes a cytopathic effect on human endothelial cells and represents a critical virulence factor in human infection and in mouse models of invasive disease.

A role proposed for the cysteine protease activity of SpeB is a posttranslational modification of the *Streptococcus* surface-expressed M protein. Removal of 24 amino acids from the N terminus of mature M protein alters the bacterial immunoglobulin-binding properties. Sequence analysis of the SpeB gene in 200 group A *Streptococcus* isolates has identified three main variants. One of them contains an Arg-Gly-Asp (RGD) motif and preferentially binds human integrins $\alpha_v\beta_3$ and $\alpha_{IIb}\beta_3$.

Binding Domains of A/B Toxins

A/B toxins (discussed in detail in "Soluble toxins with an intracellular target," below) are composed of a toxic A domain and a carrier binding domain (B domain), which binds the toxin receptor on the cell surface and aids in translocation of the A domain across the cell membrane. Since the primary toxicity of A/B toxins is mediated by their A subunits, the role of the B domain is often neglected. However, there are several instances where receptor binding is sufficient to affect cell function. The amount of toxin necessary to achieve these effects is usually orders of magnitude greater than that required for the intoxication mediated by the A subunit, and therefore these effects are unlikely to play any major role in vivo. Nevertheless, they may be very useful for in vitro studies. The B subunits for which a cell function is known are those of pertussis toxin (PT), cholera toxin (CT), and the related *Escherichia coli* heat-labile enterotoxin (LT).

The B subunit of PT is a polyclonal mitogen for T cells. In vitro the effect requires at least 0.3 µg of PT or its B subunit. The effect has not been observed in vivo, possibly because the dose required for activity is never achieved. It also has hemagglutination activity. In vitro, the effect requires 0.3 µg of wild-type PT or its B subunit per ml. In addition, the B oligomer is able to induce signal transduction through inositol phosphate pathway. CT and LT B subunits are systemic and mucosal immunogens, and this property is dependent on the binding to the GM_1 receptor. Moreover, they are strong inhibitors of T-cell activation and are able to induce apoptosis of $CD8^+$ T cells and, to a lesser extent, of $CD4^+$ T cells. Also, this effect has been described only in vitro, and there is no evidence so far of a role in vivo.

Toxins Cleaving Cell Surface Molecules

Bacteroides fragilis enterotoxin (BFT) is a protein of 186 residues that is secreted into the culture medium. The toxin has a zinc-binding consensus motif (HEXXH), characteristic of metalloproteases and other toxins such as tetanus and botulinum toxins. In vitro, the purified enterotoxin undergoes autodigestion and can cleave a number of substrates including gelatin, actin, tropomyosin, and fibrinogen. When added to cells in tissue culture, the toxin cleaves the 33-kDa extracellular portion of E-cadherin, a 120-kDa transmembrane glycoprotein responsible for calcium-dependent cell-cell adhesion in epithelial cells, that also serves as a receptor for *Listeria monocytogenes* (for details see chapters 5 and 1, respectively). In vitro, BFT does not cleave E-cadherin, suggesting that the membrane-embedded form of E-cadherin is necessary for cleavage.

BFT causes diarrhea and fluid accumulation in ligated ileal loops. In vitro, it is nonlethal but causes morphological changes such as cell rounding and dissolution of tight clusters of cells. The morphological changes are associated with F-actin redistribution. In polarized cells, BFT is more active from the basolateral side than from the apical side, decreases the monolayer resistance, and causes dissolution of some tight junctions and rounding of some of the epithelial cells, which can separate from the epithelium. In monolayers of enterocytes, BFT increases the internalization of many enteric bacteria such as *Salmonella, Proteus, E. coli,* and *Enterococcus* but decreases the internalization of *L. monocytogenes.*

BFT belongs to a large family of bacterial metalloproteases that usually cleave proteins of the extracellular matrix. *Pseudomonas aeruginosa* and *Staphylococcus epidermidis* elastases and *Clostridium histolyticum* collagenase are the best-known examples.

Pore-Forming Toxins

Pore-forming toxins work by punching holes in the plasma membrane of eukaryotic cells, thus breaking the permeability barrier that keeps macromolecules and small solutes selectively within the cell. These toxins are often identified as lytic factors (lysins). Since erythrocytes have often been used to test the activity of these toxins, many of them are known as hemolysins. However, it should be kept in mind that the hemolytic activity was only a convenient way to measure their ability to punch holes in the cell membrane and release the intracellular hemoglobin, whose red color is easy to detect even with the naked eye. Therefore, while erythrocytes are a convenient target of hemolysins in vitro, they are never the main physiological target of this class of toxins in vivo, since the toxins exert their virulence effect mostly by permeabilizing other cell types. To generate channels and holes in the cell membrane, this class of proteins must be amphipathic, with one part interacting with the hydrophilic cavity filled with water and the other interacting with the lipid chains or the nonpolar segments of integral membrane proteins. The ultimate consequence of cell permeabilization is usually death, whereas the early consequences of losing the permeability barrier are often the release of cytokines, activation of intracellular proteases, and sometimes induction of apoptosis.

Large-Pore-Forming Toxins

Streptolysin O (SLO) is the best-known member of a toxin family containing 19 different secreted proteins of 50 to 60 kDa that have the common property of binding specifically to cholesterol on the cell membrane. They contain a common motif (ECTGLAWEWWR) located approximately 40 amino acids from the carboxy terminus. Oxidation of the cysteine residue, included in this motif, abolishes the lytic activity; this activity can be restored by adding reducing agents such as thiols. These proteins, listed in Table 11.2, are produced by *Streptococcus pyogenes, S. pneumoniae, Bacillus,* a variety of clostridia including *Clostridium tetani,* and *Listeria.* They all bind cholesterol-containing membranes in a nonsaturable fashion and then polymerize to form very large pores, which can be up to 35 nm in diameter and which make the cell membrane permeable to small solutes and large macromolecules (Figure 11.5A), thus leading to rapid cell death. Resolution of the three-dimensional structure of perfringolysin O, a member of this family produced by *C. perfringens,* has revealed a novel mechanism of membrane insertion and pore formation. The membrane-bound receptor, cholesterol, plays a significant role in monomer targeting, oligomerization, membrane insertion, and stabilization of the membrane pore. SLO is widely used as a tool in cell biology to produce in vitro large pores in cell membranes, thus allowing the introduction of large molecules into the cell cytoplasm.

Small-Pore-Forming Toxins

The small-pore-forming family of toxins form very small pores (1 to 1.5 nm in diameter) in the membrane, allowing the selective permeabilization of cells to solutes that have a molecular mass lower than 2,000 Da.

Alpha-toxin and leukotoxins belong to a family of related 33-kDa toxins secreted by most pathogenic strains of *S. aureus*. This family contains leukotoxins such as leukocidin F, leukocidin S, and α-lysin A and C, which bind to the membranes, where they assemble into heptameric structures that form very small pores (approximately 1 nm in diameter). The X-ray

Table 11.2 Large-pore-forming toxins

Bacterial genus	Species	Toxin name
Streptococcus	*S. pyogenes*	Streptolysin O
	S. pneumoniae	Pneumolysin
	S. suis	Suilysin
Bacillus	*B. cereus*	Cereolysin O
	B. alvei	Alveolysin
	B. thuringiensis	Thuringiolysin O
	B. laterosporus	Laterosporolysin
Clostridium	*C. tetani*	Tetanolysin
	C. botulinum	Botulinolysin
	C. perfringens	Perfringolysin O
	C. septicum	Septicolysin O
	C. histolyticum	Histolyticolysin O
	C. novyi A (*oedematiens*)	Novyilysin
	C. chauvoei	Chauveolysin
	C. bifermentans	Bifermentolysin
	C. sordellii	Sordellilysin
Listeria	*L. monocytogenes*	Listeriolysin O
	L. ivanovii	Ivanolysin
	L. seeligeri	Seeligerolysin

A

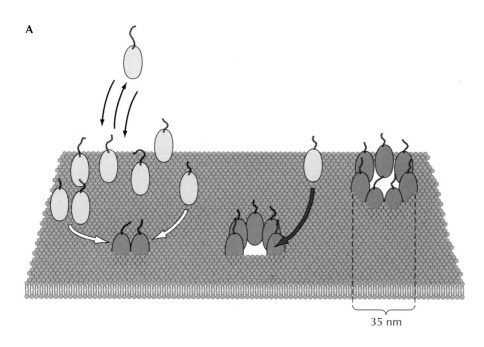

35 nm

B

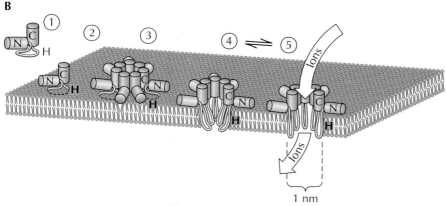

1 nm

Figure 11.5 Schematic representation of membrane interaction and oligomerization of large-pore-forming toxins **(A)** and small-pore-forming toxins **(B).**

structure of the transmembrane pore of alpha-toxin has been solved and has confirmed the heptameric structure of the oligomer and the self-assembly of the glycine-rich region (H domain in Figure 11.5B) to form the pore. Cells become permeable only to molecules smaller than the pores (up to 2,000 Da). Ions like Ca^{2+} are able to enter the cells, so that electrical permeability is established, while large molecules such as all cytoplasmic enzymes are retained within the cell. At this stage, the toxin induces a number of changes in the cell such as production of eicosanoids, activation of endonucleases, release of cytokines, and early apoptotic events. At high concentrations, the toxin can cause membrane rupture and cell lysis, thus killing the cells. The toxin causes membrane damage in a variety of cells including erythrocytes, platelets, and leukocytes. In erythrocytes, for instance, the intoxication proceeds in distinct steps: binding to cell membrane, ion leakage, and eventually rupture of the cell membrane with release of larger molecules.

Aerolysin is secreted by *Aeromonas hydrophila* as a 52-kDa protoxin that is activated by proteolytic cleavage of the carboxy-terminal peptide to yield a 48-kDa active toxin. The toxin binds to a family of GPI-anchored specific receptor on the surface of target cells, including the T-cell receptor RT6. Following binding, the protein oligomerizes and forms heptameric structures that insert into the cell membrane, forming pores of approximately 1.5 nm in diameter.

Membrane-Perturbing Toxins

Delta-toxin or delta-hemolysin is secreted into the medium by *S. aureus* strains at the end of the exponential phase of growth. It is a 26-amino-acid peptide (MAQDIISTIGDLVKWIIDTVNKFTKK) that has the general structure of soap with a nonpolar segment followed by a strongly basic carboxy-terminal peptide. The peptide has no structure in aqueous buffers but acquires an α-helical structure in low-dialectric-constant organic solvents and membranes. The α-helix has a typical amphipathic structure, which is necessary for the toxin to interact with membranes. The toxin binds nonspecifically parallel to the surface of any membrane without forming transmembrane channels. At high concentration, the peptide self-associates and increases the perturbation of the lipid bilayer that eventually breaks into discoidal or micellar structures. It is very interesting that mellitin, which is also a 26-amino-acid lytic peptide produced by *S. aureus*, has no sequence homology with delta-toxin but has identical distribution of charged and nonpolar amino acids. These toxins are active in most eukaryotic cells. Cells first become permeable to small solutes and eventually swell and lyse, releasing cell intracellular content.

E. coli Hemolysin

E. coli hemolysin is a 110-kDa protein secreted into the culture supernatant by some pathogenic *E. coli* strains during exponential growth. The protein belongs to a large family of Ca^{2+}-dependent hemolysins known as RTX toxins because they contain a conserved repeated glycine- and aspartate-rich motif of 9 amino acids [XLXGGXG(N/D)D (Repeat in Toxins)]. They are synthesized as inactive protoxins; activation requires a specific bacterial gene (*hlyC*) that mediates fatty acid acylation of lysines 594 and 690. This family contains hemolysins from *Proteus vulgaris* and *Morganella morganii*, cytotoxins from *Actinobacillus pleuropneumoniae* and *A. suis*, leukotoxins from *Pasteurella haemolytica* and *A. actinomycetemcomitans*, and the bifunctional adenylate cyclase-hemolysin from *Bordetella pertussis*.

The toxin is encoded by four genes, one of which, *hlyA*, encodes the 110-kDa hemolysin, while the others are required for its posttranslational fatty acid acylation (*hlyC*) and secretion (*hlyB* and *hlyD*). HlyA contains a 50- to 60-amino-acid amino-terminal domain that is recognized by the products of *hlyB* and *hlyD* and is required for secretion. Binding to the target cells requires binding of Ca^{2+} to the repeat domain. Following binding, the N-terminal domain formed by hydrophobic and amphipathic transmembrane sequences is inserted into the membrane and forms transmembrane pores that are small (1 nm in diameter), unstable, hydrophilic, and cation specific. The toxin lyses erythrocytes, leukocytes, endothelial cells, renal epithelial cells, granulocytes, monocytes, and human T lymphocytes. Like for other small-pore-forming toxins, the initial permealization causes the loss of small solutes, uptake of water, and osmotic cell lysis.

Anthrax Protective Antigen
Anthrax protective antigen (PA) is one of the three proteins of the anthrax toxin complex secreted by *Bacillus anthracis;* this complex comprises the receptor-binding protective antigen (PA), the edema factor (EF), and lethal factor (LF) (Figure 11.6). PA can be considered as a B domain for two distinct A subunits such as EF and LF. The three subunits, encoded by a large plasmid, are synthesized and secreted independently. PA is a protein of 735 amino acids that binds the receptors on the cell surface. After binding, PA is activated by proteolytic removal of the 20-kDa amino terminus, generating the PA C-terminal 63-kDa peptide (PA_{63}), which remains associated with the membrane. PA_{63} binds the amino terminus of EF or LF. Following receptor-mediated endocytosis, the low pH causes a conformational change in PA that allows the translocation of EF or LF across the cell membrane. During the normal receptor-mediated endocytosis and endosome acidification, PA undergoes conformational changes, i.e., oligomerization into rings with sevenfold symmetry and formation of pores in artificial lipid bilayers and membranes. The crystallographic structure has revealed how PA can be assembled into heptamers and has suggested how some of the domains can undergo pH-driven conformational changes.

Soluble Toxins with an Intracellular Target

The group of toxins with an intracellular target (A/B toxins) contains many toxins with different structures that have only one general feature in common: they are composed of two domains generally identified as A and B. A is the active portion of the toxin; it usually has enzymatic activity and can recognize and modify a target molecule within the cytosol of eukaryotic cells. B is usually the carrier for the A subunit; it binds the receptor on the cell surface and facilitates the translocation of A across the cytoplasmic membrane. Depending on their target, these toxins can be divided into different groups that act on protein synthesis, signal transduction, actin polymerization, and vesicle trafficking within eukaryotic cells.

Toxins Acting on Protein Synthesis
The toxins that inhibit protein synthesis, causing rapid cell death, at extremely low concentrations are diphtheria toxin, *Pseudomonas aeruginosa* exotoxin A, and Shiga toxin.

Diphtheria Toxin
Diphtheria toxin (DT) is a 535-amino-acid polypeptide encoded by a bacteriophage that lysogenizes *Corynebacterium diphtheriae.* The gene is regulated by an iron-binding protein, and therefore the toxin is expressed only in the absence of iron. Following cleavage at a protease-sensitive site, the toxin is divided into two fragments (A and B) that are held together by a disulfide bridge. The A fragment (DTA) is an enzyme with ADP-ribosylating activity that binds NAD and transfers the ADP-ribose group to elongation factor 2 (EF2) according to the reaction shown below:

$$NAD + EF2 \xrightarrow{\text{DTA}} ADPR\text{-}EF2 + nicotinamide + H^+$$

The target site in EF2 is a unique amino acid resulting from the posttranslational modification of histidine at position 715; it is named diphthamide. The region containing diphthamide 715 is very close to the anticodon rec-

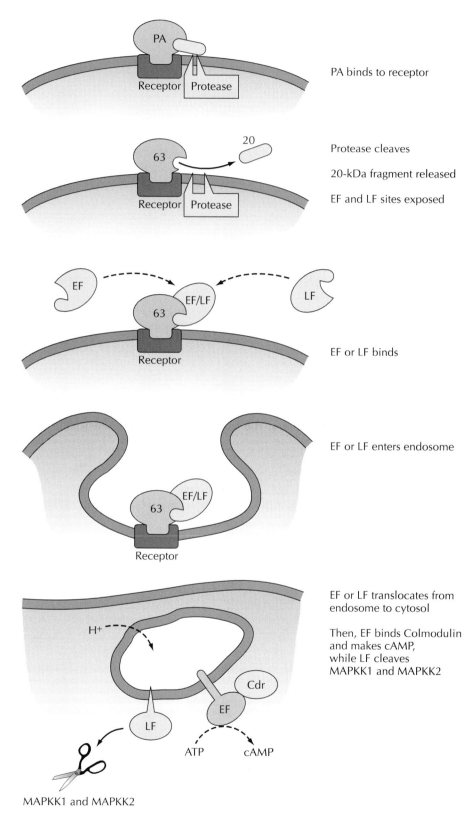

PA binds to receptor

Protease cleaves

20-kDa fragment released

EF and LF sites exposed

EF or LF binds

EF or LF enters endosome

EF or LF translocates from endosome to cytosol

Then, EF binds Colmodulin and makes cAMP, while LF cleaves MAPKK1 and MAPKK2

MAPKK1 and MAPKK2

Figure 11.6 Mechanism of PA-mediated entry and intoxication of anthrax LF and EF toxins.

ognition domain of EF2. This suggests that ADP-ribosylation interferes with EF2 binding to the tRNA. The ADP-ribosylated EF2 is no longer able to support protein synthesis, and the cells die. The lethal dose of DT is extremely low: 100 ng/kg of body weight. The B fragment can be further divided into two domains: the R (receptor binding) and T (transmembrane) domains. R binds to the heparin-binding epidermal growth factor-like precursor and is internalized by receptor-mediated endocytosis. The toxin receptor is present in most mammalian cells; however, rodents are not susceptible to DT because the receptor has a few amino acid changes that abolish binding. Following acidification of the endosomes, the T domain, which is composed mostly of hydrophobic α-helices, changes conformation and penetrates the membrane and somehow facilitates the translocation of the A subunit to the cytoplasm. DT is used to make the diphtheria vaccine, where it is present after chemical detoxification by formaldehyde treatment.

Another activity described for DT is apoptosis of target cells. This activity is apparently mediated by the A fragment but is not linked to the enzymatic activity. In fact, apoptosis has also been described for cross-reacting material (CRM197), an enzymatically inactive, nontoxic mutant of DT. Whether apoptosis plays a role in toxicity in vivo is unclear; however, it cannot be a major role because the mutants active in apoptosis but enzymatically inactive are nontoxic in vitro and in vivo.

Pseudomonas Exotoxin A

Pseudomonas exotoxin A (ExoA) is a 66-kDa single-chain protein with a mechanism of action (ADP-ribosylation of EF2) identical to that of DT. However, the two proteins are totally unrelated and have no primary sequence homology. Nevertheless, the folding and three-dimensional structure of the catalytic site are conserved and superimposable on those of all mono-ADP-ribosylating enzymes known, including the A fragments of DT, CT, LT, and PT (Figure 11.7), suggesting that these enzymes either evolved from a common ancestor or had a convergent evolution.

The toxin is secreted into the supernatant by *P. aeruginosa* and can be divided into three functionally different domains: the amino-terminal part, R, binds the α$_2$-microglobulin receptor on the surface of target cells; the central portion, T, is composed mostly of hydrophobic α-helix and mediates translocation of the catalytic domain to the cytosol; and the carboxy-terminal part has ADP-ribosylating activity. Following receptor-mediated endocytosis, the toxin undergoes retrograde transport to the endoplasmic reticulum (possibly thanks to the C-terminal REDLK sequence, which resembles the endoplasmic reticulum retention signal KDEL), where it crosses the membrane and is translocated to the cytoplasm to reach the EF2 target.

Due to the potent lethal activity, the catalytic domains of DT and *Pseudomonas* toxin have been widely used to construct fusion proteins that are able to specifically bind and kill tumor cells or other types of dangerous cells. The first of its kind, fusion between interleukin-2 and the A subunit of DT, was able to specifically kill activated T cells and has been tested for the treatment of cutaneous T-cell lymphoma.

Shiga Toxin

Shiga toxin (also known as verotoxin) is a prototype of a number of related toxins produced by the causative agents of dysentery (*Shigella dysenteriae*) and hemorrhagic colitis (*E. coli* producing Shiga toxins types 1 and 2). The toxins are encoded by an operon which carries the genes for the A and B

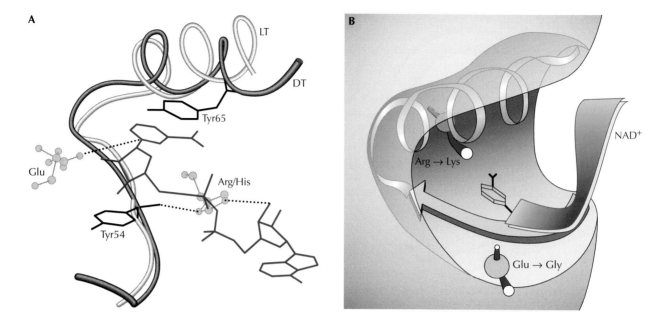

Figure 11.7 Catalytic and NAD-binding site of ADP-ribosylating toxins. **(A)** Diagram of the three-dimensional structure of the catalytic site of DT and LT showing the NAD molecule (red) inside the cavity and the two amino acids, Glu and Arg/His, important for catalysis. **(B)** Schematic representation of the catalytic site of PT showing the NAD and the two catalytic amino acids which have been mutagenized to generate the genetically detoxified PT9K/129G mutant.

subunits and can be located in the chromosome or on bacteriophages. Shiga toxin is a typical A/B toxin, with an enzymatically active A peptide of 35 kDa that is responsible for the toxicity. This has an *N*-glycosidase activity, which cleaves an adenine residue from the 28S RNA, altering the function of the ribosomes, which are no longer able to interact with elongation factors EF1 and EF2. This results in an inability to carry out protein synthesis and cell death. Interestingly, the plant toxin ricin has an identical mode of action. The B subunit is a pentamer composed of five identical monomers of 7,700 Da that bind to the globotriaosylceramide (Gb) eukaryotic cell receptor. The organization of the B pentamer is remarkably similar to that found in CT and LT, suggesting that this type of pentameric structure is favorable and has been independently and recurrently adopted during evolution. After binding to Gb$_3$, the toxin is internalized by receptor-mediated endocytosis and is transported to the Golgi and to the endoplasmic reticulum, from which the A subunit is translocated to the cytoplasm, where it can gain access to the ribosomal target.

Toxins Acting on Signal Transduction

Communication between and within cells is essential for any living organism. In eukaryotic cells, signals from outside stimulate receptors on the surface of the cells that transmit the signal across the cell membrane mainly by two mechanisms: (i) tyrosine phosphorylation of the cytoplasmic carboxy-terminal part of the receptor that recruits SH$_2$-signal transducers and initiates a cascade of intracellular signaling events; and (ii) modification of a receptor-coupled GTP-binding protein that transduces the signal to enzymes, releasing secondary messengers such as cyclic AMP (cAMP), inositol triphosphate, and diacylglycerol, which also initiate a cascade of intracellular signaling events. Bacterial toxins act at all levels of signal transduction: they modify GTP-binding proteins (PT, CT, and LT), produce secondary messengers such as cAMP (CyaA and EF), and act on the intracellular signal cascade either by modifying small GTP-binding proteins such as Rho, Rac, Cdc42 (*Clostridium difficile* toxins A and B) or by modifying kinases such as mitogen-activated MAPKK1 (LF). Other toxins (described in the next section) that are directly injected by the bacteria into cells also work on signal transduction: these can act on the intracellular signal cascade, modifying small GTP-binding proteins such as Ras and Rho (ExoS and C3) or modifying receptors by phosphorylating or dephosphorylating them (YopH and YopO).

Toxins Acting on G Proteins

CT and *E. coli* LT-I and LT-II

CT and *E. coli* LT-I and LT-II have an identical mechanism of action and homologous primary and three-dimensional structures. CT is produced by *Vibrio cholerae,* the causative agent of cholera, and LT-I and LT-II are produced by enterotoxigenic *E. coli* (ETEC) isolated from humans with traveler's diarrhea, from pigs (LT-I), from animals with no evident disease, or from food (LT-II). The toxins are generally described as having an AB_5 composition, meaning that they contain one enzymatically active A subunit and a pentameric B subunit composed of five identical monomers. The toxins are encoded by a bicistronic operon located on a filamentous bacteriophage (CT) or a plasmid (LT). CT is secreted into the culture medium by *V. cholerae,* while LT is exported to and accumulated in the periplasm of *E. coli.*

The A subunit is a 27-kDa enzyme with ADP-ribosyltransferase activity, an enzymatic activity shared with DT, ExoA, PT, and *P. aeruginosa* ExoS. The enzyme binds NAD and transfers the ADP-ribose group to an arginine residue present in an -LRX**R**VXT- motif located in the central part of all G proteins, thus modifying their ability to transduce the signals from the coupled receptors. To be fully active, the A subunit has to bind to an intracellular small GTP-binding protein that is involved in vesicular trafficking (ADP-ribosylation factor [ARF]). The main targets of the enzyme are G_s, G_t, and G_{olf}. The effects on G_s are the best known: the ADP-ribosylation causes permanent activation of the G_s-regulated adenylate cyclase, inducing an increase in the intracellular content of the secondary messenger cAMP. This induces an alteration of ion transport, with an increase in chloride secretion and inhibition of sodium absorption. When the toxins are released in the intestine during infection, the ultimate consequence is intestinal fluid accumulation and the watery diarrhea typical of the diseases. If the toxins are experimentally introduced into other tissues, they usually cause massive release of fluids and edema. The A subunits of CT, LT-I, and LT-II have quite a high sequence and three-dimensional homology, with LT-II being the least closely related. They also have the common fold and structure of the active site of ADP-ribosyltransferases described for DT and ExoA (Figure 11.7A).

The B subunit has a ring-like structure composed of five monomers of 11,500 Da each, with a central hole that houses the C-terminal portion of the A subunit. Each monomer binds a receptor molecule that for CT is mostly ganglioside GM_1, while LT also binds other galactose-containing structures such as other glycosphingolipids, glycoprotein receptors, poly-glycosylceramides, and paragloboside.

Following receptor binding, the toxins are internalized by receptor-mediated endocytosis and undergo retrograde transport through the Golgi to the ER (possibly guided by the KDEL ER retention sequence). At this stage, the A subunits cross the membrane and end up in the cytoplasm, where they are activated by binding ARF and can finally reach and modify their targets. The B subunits of CT and LT-I have a high degree of homology, although it is lower than that of the A subunit. Interestingly, the B subunit of LT-II, although having an identical structure, has no homology to the B subunit of CT and LT-I, indicating that the pentamer is a good evolutionary solution that can be achieved with different primary structures. In addition to carrying the A subunits, the pentameric B subunits have other biological activities on their own, such as induction of apoptosis of $CD8^+$ and $CD4^+$ T cells (as described above). CT and LT are perhaps the best mucosal immunogens and adjuvants known to date. This property can be exploited to develop mucosal vaccines against cholera and ETEC infection and to induce a mucosal response against the antigens that are coadministered. To develop safe vaccines that can be used in humans without carrying the toxic features of CT and LT, enzymatically inactive mutants have been developed by site-directed mutagenesis. These mutants are very promising candidates as mucosal adjuvants and vaccines.

PT

PT is a chromosomally encoded, 105-kDa virulence factor secreted into the culture supernatants by *B. pertussis*, the bacterium that causes whooping cough. The toxin is composed of five noncovalently linked subunits named S1 through S5, which are organized into two functionally different domains called A and B. The five subunits are individually secreted to the periplasmic space, where the toxin is assembled and then released to the culture medium by a specialized type IV secretion apparatus (see chapter 14). The A domain, which is composed of the S1 subunit, is an enzyme that shares the active-site enzymatic activity and structure with DT, ExoA, CT, and LT (Figure 11.7A) and intoxicates eukaryotic cells by ADP-ribosylating their GTP-binding proteins. The protein transfers the ADP-ribose group to a cysteine located in the carboxy-terminal-X**C**GLX motif of the α-subunit of many G proteins such as G_i, G_o, G_t, and G_{gust}. G_s and G_{olf} have a tyrosine instead of the cysteine and are not substrates for PT. The consequence of ADP-ribosylation is the uncoupling of G proteins from their receptors, with alteration of all signals that are transduced by them. This may have different consequences in different tissues, the most common of which are lymphocytosis, increased insulin secretion, and sensitization to histamine.

The B domain is a nontoxic oligomer formed by subunits S2, S3, S4, and S5 in a 1:1:2:1 ratio. This domain binds to glycoproteins having a branched mannose core with attached *N*-acetylglucosamine. The toxin is then internalized by receptor-mediated endocytosis and is likely to follow the retrograde transport through the Golgi as in the case of CT; however, supportive data are not available. In addition to the intoxication mediated by the enzymatic activity, the toxin has other biological activities, such as mitogenic activity on T cells, which are mediated only by the binding of the B domain to the receptors. These are described above. PT is one of the main components of acellular vaccines against whooping cough. In these vaccines the toxin is chemically detoxified (by formaldehyde or hydrogen peroxide) or genetically detoxified (by changing the two catalytic amino acids [Figure 11.7B] by site-directed mutagenesis). The nontoxic mutant obtained by site-directed mutagenesis (PT-9K/129G) was found in clinical trials to be more immunogenic than any other form of detoxified PT and to induce protection from disease in infants.

Toxins Generating cAMP

Adenylate Cyclase

The adenylate cyclase produced by *B. pertussis*, *B. parapertussis*, and *B. bronchiseptica* is a 177.7-kDa bifunctional protein composed of a cell-invasive and calmodulin-dependent adenylate cyclase domain (400 amino-terminal residues) fused to a pore-forming hemolysin consisting of 1,306 residues. The carboxy-terminal hemolytic domain (described as a hemolysin above) binds the receptors and delivers the amino-terminal catalytic domain into the cell, where it binds calmodulin and catalyzes high-level synthesis of the second messenger cAMP, thereby disrupting cellular functions. The target cells are believed to be mainly alveolar macrophages and leukocytes.

Anthrax Edema Factor

Anthrax edema factor (EF) is one of the three proteins of the anthrax toxin complex which contains the receptor-binding PA, which has a B domain for two distinct A subunits such as EF and LF. The three subunits, encoded by a large plasmid, are synthesized and secreted independently. EF is secreted as a precursor of 800 amino acids, and cleavage of the 33-amino-acid signal peptide produces the 767-amino-acid mature protein. The N-terminal domain of EF is responsible for the binding to PA, whereas the C-terminal region contains the catalytic domain. EF binds to proteolytically activated and receptor-bound PA (Figure 11.6). Following receptor-mediated endocytosis, the low pH causes a conformational change in PA, allowing the translocation of EF across the cell membrane. Once inside the cells, it binds calmodulin and catalyzes the synthesis of the second messenger cAMP, thereby perturbing the cell regulatory mechanisms.

Toxins Inactivating Cellular Kinases

Anthrax LF causes the shock-like symptoms observed in systemic anthrax infection, by inducing macrophages to overexpress proinflammatory cytokines. In animals, it induces sudden death. LF binds to proteolytically activated and receptor-bound PA. Following receptor-mediated endocytosis, the low pH causes a conformational change in PA that allows the translocation of LF across the cell membrane. Once in the cytoplasm, LF (a metalloprotease that, like clostridial toxins, contains the consensus sequence -HEXXH-) cleaves the amino terminus of MAPKK1 and MAPKK2 (Figure 11.6). The cleavage inactivates MAPKK1 and inhibits the MAPK signal transduction pathway, a conserved pathway that controls cell proliferation and signal transduction. It is not clear whether the cleavage of MAPKK1 and MAPKK2 is also responsible for the observed cell death or whether this is due to the cleavage of a substrate that has not yet been identified.

Toxins Inducing Arrest in the G_2 Phase of the Cell Cycle

The cytolethal distending toxin produced by *Haemophilus ducreyi* (HdCDT) induces cell enlargement followed by cell death. This effect is similar to that induced by CDT produced by *E. coli, Shigella, Campylobacter,* and *Actinobacillus actinomycetemcomitans.* HdCDT is a complex of three proteins of 25, 30, and 20 kDa, encoded by a cluster of three linked genes. The proteins show a high homology to the products of *E. coli* and *Actinobacillus* CDT genes. HdCDT intoxicates eukaryotic cells by causing a three- to fivefold gradual distension and induces cell cycle arrest in the G_2 phase. Transition of cells from G_2 into mitosis requires activation of the cyclin-dependent kinase p34^{cdc2} by dephosphorylation. Treatment of cells with HdCDT induces an increased level of the tyrosine-phosphorylated form (inactive) of p34^{cdc2} and hence leads to cell cycle arrest. It is hypothesized that the toxin acts directly on some kinase or phosphatase in the signaling network controlling the p34^{cdc2} activity.

Toxins with Phospholipase C Activity

Alpha-toxin is the key virulence factor produced by *Clostridium perfringens,* which is responsible for gas gangrene and is involved in the pathogenesis of sudden death syndrome in young animals. The toxin is a zinc metalloenzyme that has phospholipase C (PLC) activity. *C. perfringens* knockout mutants are unable to cause disease, and vaccination with a genetically detoxified toxoid induces protection against disease. Originally, the toxin was identified as an enzyme with PLC activity, but subsequently it was discovered that not all toxins with PLC activity were toxic (*Bacillus cereus* phosphatidylcholine-specific PLC [PC-PLC] is not toxic), leading to the conclusion that enzymatic activity alone was not sufficient for toxicity.

The three-dimensional structure of *C. perfringens* alpha-toxin has recently been solved, and it has been shown that the 370 residues are organized in two domains: an α-helical N-terminal domain that contains the active site and is highly homologous to *B. cereus* PC-PLC toxin, and a β-stranded C-terminal domain that is involved in membrane binding and shows high structural homology to eukaryotic calcium-binding C2 domains. This C-terminal domain is present in hemolytic bacterial PLCs (*C. bifermentans* PLC, *C. novyi* gamma-toxin) but is absent in phospholipases which have no hemolytic activity (PC-PLC). It has been proposed that the loss in the hemolytic activity is due to the absence of the membrane-interacting regions located in the C-terminal domain of *C. perfringens* alpha-toxin.

Recently, other bacterial PLCs, like those from *L. monocytogenes* and *Mycobacterium tuberculosis,* have been implicated in the pathogenesis of a number of diseases.

Toxins Modifying Small GTP-Binding Proteins

Clostridium difficile Toxins A and B
Enterotoxin A (308 kDa) and cytotoxin B (270 kDa) are secreted into the culture supernatant by *C. difficile,* bind to eukaryotic cells, and are taken up by receptor-mediated endocytosis. Intracellularly, they monoglucosylate small GTP-binding proteins such as Rho, Rac, and Cdc42. The site of modification is threonine 37 in Rho and threonine 35 in Rac and Cdc42. Other small GTPases such as Ras, Rab, Arf, and Ran are not glucosylated. The monoglucosylation results in breakdown of the cellular actin stress fibers. These toxins are described in more detail in chapter 12.

E. coli CNF
Cytotoxin necrotizing factors (CNF) are single-chain proteins of 1,014 amino acids that are produced by a number of pathogenic *E. coli* strains.

They induce ruffling, stress fiber formation, and cell spreading in cultured cells by activating the small GTP-binding proteins Rho, Rac, and Cdc42, which control assembly of actin stress fibers. Activation is achieved by deamidation of glutamine 63 of Rho and glutamine 61 of Cdc42 (see chapter 12 for details). Two forms are known: CNF1 is chromosomally encoded, while CNF2 is located on a large, transmissible F-like plasmid called Vir. The two factors have 85% identical amino acid sequences and show similarities to the dermonecrotic toxins of *Pasteurella multocida* and *B. pertussis*.

Toxins Acting on Vesicular Trafficking

Vital cellular processes such as receptor-mediated endocytosis and exocytosis use specialized vesicles either to internalize portions of the plasma membrane and address them to different destinations by using specialized sorting compartments or to transport molecules synthesized in the ER and modified in the Golgi apparatus to the cell surface. In some tissues, dedicated vesicle transport can be used for the local release of special cargo. A typical example of this is the local release of neurotransmitters at the presynaptic membrane. Recently, great progress in the understanding of this field has been made by the discovery that key molecules involved in vesicle docking and membrane fusion, such as VAMP/synaptobrevin, SNAP-25, and syntaxin, are specific targets of neurotoxins produced by *Clostridium tetani* and *Clostridium botulinum*, which are specific zinc-dependent proteases. The vacuolating cytotoxin of *Helicobacter pylori* has recently been discovered to act on vesicle traffic at the level of late endosomes, a compartment whose molecular mechanisms are still totally unknown. Hopefully, the study of this toxin will help to elucidate this pathway.

Proteases of VAMP/Synaptobrevin

Tetanus toxin.   Tetanus toxin is a 150-kDa protease, produced by toxigenic strains of *C. tetani*, that is responsible for the spastic paralysis typical of clinical tetanus. The toxin is extremely potent, with a 50% lethal dose (LD_{50}) of 0.2 ng/kg. Following protease cleavage, it is divided into two fragments (H [heavy] and L [light]), held together by a disulfide bridge, a structure typical of A/B toxins. The heavy chain is composed of two fragments, H_C (or fragment C), which binds to a still unknown receptor, and H_N, which is involved in the transmembrane translocation of the L chain to the cytosol. The L chain is a 447-amino-acid fragment containing the -HEXXH- motif typical of metalloproteases. It binds zinc and specifically cleaves VAMP/synaptobrevin, a molecule essential for the docking of the neurotransmitter containing vesicles to the cell membrane (Figure 11.8). The specific metal-dependent proteolytic activity is shared with the neurotoxins of *C. botulinum* and *B. anthracis* LF. During intoxication, tetanus toxin is internalized at the presynaptic terminal of the neuromuscular junction and migrates retroaxonally within the motoneuron to the spinal cord, where it is released into the intersynaptic space located between the motoneuron and the inhibitory interneuron cell, penetrates the latter, is translocated to the cytosol following acid-induced conformational change, and blocks neuroexocytosis by cleaving VAMP. Following inactivation of the toxicity by formaldehyde treatment, tetanus toxin is used for vaccination against tetanus, a practice applied for more than 60 years. Today, very promising results have been obtained by using the nontoxic recombinant fragment C for vaccination.

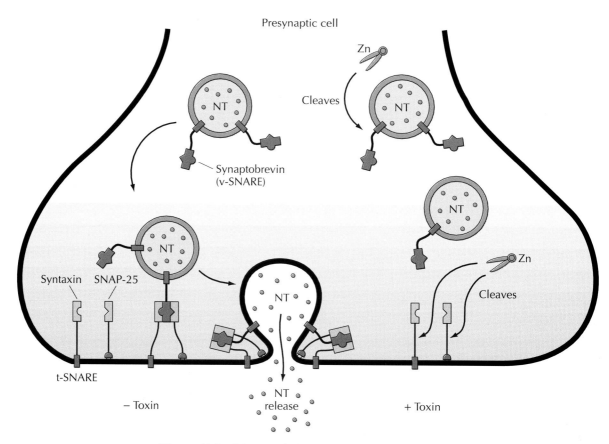

Figure 11.8 Scheme of a postsynaptic membrane showing the normal process of neurotransmitter (NT) release (left) and the mechanism of action of the neurotoxins (right). Vesicles containing the neurotransmitter have a transmembrane protein (synaptobrevin or v-SNARE), that binds specifically two proteins enclosed on the cell membrane (syntaxin, SNAP-25, or t-SNARE). The initial interaction becomes increasingly stronger, forcing the vesicle and cellular membranes to become in close contact and finally to fuse, thus releasing the neurotransmitters into the intercellular space. The mechanism of action of tetanus and botulinum neurotoxins is shown on the right. They cleave the v-SNARE and/or the t-SNARE, thus preventing the docking and fusion of neurotransmitter-containing vesicles.

***C. botulinum* neurotoxin.** *C. botulinum* neurotoxin serotypes A to G are usually ingested with food and are responsible for the flaccid paralysis typical of clinical botulinum intoxication. Like tetanus toxin, they are synthesized as 150-kDa precursors that are activated by proteolytic cleavage and divided into two fragments, H and L, of 100 and 50 kDa, respectively, held together by a disulfide bridge. H binds to receptors present in the motor neurons at the neuromuscular junction and translocates the active L form into the cytosol. L chains of neurotoxins B, D, F, and G are zinc-dependent proteases that specifically cleave VAMP/synaptobrevin; they are composed of approximately 450 amino acids, with an overall homology of 30 to 60%, and contain the -HEXXH- motif typical of metalloproteases.

Proteases of SNAP-25 and Synaptobrevin
Like tetanus toxin, botulinum neurotoxins A, E, and C are synthesized as 150-kDa precursors that are activated by proteolytic cleavage and divided

into two fragments, H and L, of 100 and 50 kDa, respectively, held together by a disulfide bridge. H binds to receptors present in the motor neurons at the neuromuscular junction and translocates the active L form into the cytosol. L chains are zinc-dependent proteases that specifically cleave SNAP-25 (synaptosome-associated protein of 25 kDa), at positions 197 (type A) or 180 (type E). SNAP-25 is located in the cytoplasmic face of the plasma membrane and, during exocytosis, binds VAMP, a protein located in the external face of the secretory vesicle membrane. The L chains are composed of approximately 450 amino acids, with an overall homology to the remaining botulinum neurotoxins of 30 to 60%, and they contain the -HEXXH- motif typical of metalloproteases. The gene coding for neurotoxin A is located on a plasmid, while the gene coding for neurotoxin E is on a bacteriophage. Type C toxin has the general H and L structure of the other botulinum neurotoxins; the only difference is that it cleaves syntaxin, another protein involved in the exocytotic machinery. Type C is also unique because it can also cleave SNAP-25, thus being the only one able to cleave two substrates.

Toxin Acting on Late Endosomes/Vacuoles

Helicobacter pylori, a pathogen living in the stomachs of humans and responsible for peptic ulcers, produces vacuolating cytotoxin (VacA), a toxin that causes massive growth of vacuoles within epithelial cells. The toxin, which is synthesized as a 140-kDa precursor, is secreted from the bacterium by the 45-kDa carboxy-terminal region, using a secretory mechanism similar to that of immunoglobulin A (IgA) proteases of *Neisseria* (see chapter 13). The toxin purified from the supernatant has a flowerlike structure consisting of seven monomers of 95 kDa, each of which can be cleaved at a protease-sensitive site into two fragments of 37 and 58 kDa. The gene coding for the amino-terminal 40-kDa portion is approximately 90% homologous in the different isolates and induces vacuoles when the gene is placed under the control of a strong eukaryotic promoter and transfected into epithelial cells, suggesting that the active site must be located in this region (the mechanism of action is unknown). The carboxy-terminal part binds the receptors on eukaryotic cells. The toxin has amino-terminal and carboxy-terminal portions with 90% identity across strains and a central portion present mainly in two different genotypes (M1 and M2), which have an overall identity of only 55%. M1 and M2 target the toxin to different cells. The toxin induces an alteration of the endocytic pathway that results in selective swelling of the vesicles having the typical markers of late endosomes. The small GTP-binding protein Rab7 is necessary for vacuole formation. It is believed that VacA must modify an unknown but fundamental effector of membrane traffic. The identification of the VacA target may help to dissect the molecular mechanisms of this still obscure part of intracellular trafficking. It cannot be excluded that the VacA toxicity is mediated only by the channel-forming activity of the toxin or cell membranes.

Toxins Acting on the Cytoskeleton

The cytoskeleton, a structure composed of microfilaments, microtubules, and intermediate filaments, controls the shape and spatial organization of the cells and is involved in cell movement, endo- and exocytosis, vesicle transport, cell contact, and mitosis (see chapter 10). Rapid structural changes of these cytoskeletal proteins are based on their ability to polymerize and depolymerize. Bacterial toxins act mostly on microfilaments that are 7 to 9 nm in diameter and are built by polymerized actin. Many of the toxins somehow affect the structure of the cytoskeleton. Most of them do it by modifying the regulatory, small G proteins, such as Ras, Rho, Rac, and Cdc42, which control cell shape. These toxins, which have a dramatic but indirect effect on the cytoskeleton and are described in the section discussing the toxins affecting signal transduction, are *E. coli* CNF and *C. difficile* enterotoxins A and B (also see chapter 12). Other toxins acting on regulatory G proteins are exoenzyme S, C3, and YopE, which are described below as toxins that are directly injected into the eukaryotic cells. Other bacterial molecules that cannot be strictly considered toxins but that have a powerful ability to polymerize actin are ActA and IcsA of *Listeria* and *Shigella*, respectively. These are described elsewhere in this volume (see chapter 10). Another toxin acting indirectly on the cytoskeleton is the zonula occludens toxin (Zot) produced by *V. cholerae*, a toxin with an unknown mechanism of action that modifies the permeability of tight junctions. It has recently been shown that the Zot protein has homologies to a phage structural protein. This evidence leads to the question whether Zot is a phage protein that evolved into a toxin or whether it is just a phage protein that happens to have some biological activities. Other bacterial factors, including those that induce tyrosine phosphorylation and actin polymerization at the bacterium-cell contact site, ruffling of the cell surface, and bacterial internalization and formation of pseudopodia, are discussed earlier in this volume (see chapters 9 and 10). Here we consider only toxins that have the cytoskeleton as a direct target. The only toxin shown to affect directly the cytoskeleton is the C2 toxin of *C. botulinum*, which ADP-ribosylates monomeric actin, making it unable to polymerize. A second protein that has recently been described as being able to bind actin and stabilize the fibers supporting the ruffles induced by the *Salmonella* type III secretion system is SipA (see chapters 10 and 13).

C. *botulinum* C2 toxin is a member of a family of "binary" cytotoxins that ADP-ribosylate monomeric G-actin at arginine 177. Since the arginine is a contact site between actin monomers, the binding of the ADP-ribose makes actin unable to polymerize. C2 is composed of two separate molecules, the 50-kDa enzymatically active toxin and the binding component, which is synthesized as a 100-kDa precursor and is proteolytically cleaved to generate a 75-kDa fragment that binds the cell receptor on the cell surface. This organization closely resembles that of the EF, LF, and PA of *B. anthracis*. Toxins related to C2 are *C. perfringens* iota toxin, *C. spiriforme* toxin, and a similar molecule produced by *C. difficile*.

Toxins Directly Delivered by Bacteria into the Cytoplasm of Eukaryotic Cells

Most of the molecules described in this section are usually not found in discussions of bacterial toxins. Here we describe molecules that, generally, have been discovered recently and are part of a fascinating, rapidly expanding field. In the classical view, toxins were believed to be molecules that cause intoxication when released by bacteria into the body fluids of multicellular organisms (these are described earlier in this chapter). This definition of toxins provided a rationale for the pathogenicity of the so-called "toxinogenic bacteria" but failed to explain the pathogenicity of many other virulent bacteria such as *Salmonella, Shigella,* and *Yersinia,* which did not release toxic proteins into the culture supernatant. Today we know that these bacteria also intoxicate their hosts by using proteinaceous weapons. The difference is that the confrontation does not take place between the whole bacterial population and the whole eukaryotic organism. Instead, each bacterium engages in an individual fight with one eukaryotic cell. The final result is not different from what happens after the bite of a poisonous insect or snake or after the release of a potent bacterial toxin into the infected host. However, in this case the battle involves one eukaryotic cell and one prokaryotic cell. These bacteria intoxicate individual eukaryotic cells by using a contact-dependent secretion system to inject or deliver toxic proteins into the cytoplasm of eukaryotic cells. This is done by using specialized secretion systems that in gram-negative bacteria are called type III or type IV, depending on whether they use a transmembrane structure similar to flagella or conjugative pili, respectively (see chapters 13 and 14).

Proteins Acting on Phosphorylation

Yersinia YpkA and YopH

Phosphorylation is central to many regulatory functions associated with the growth and proliferation of eukaryotic cells. Bacteria have learned to interfere with these key functions in several ways. The best-known system is that of *Yersinia*, where a protein kinase (YpkA) and a protein tyrosine phosphatase (YopH) are injected into the cytoplasm of eukaryotic cells by a type III secretion system to paralyze the macrophages before they can kill the bacterium. While no information is available on the target eukaryotic proteins, we know that these toxic proteins are essential for virulence. In fact, challenge with a YpkA knockout mutant causes a nonlethal infection, whereas all mice challenged with wild-type *Y. pseudotuberculosis* die.

EPEC Tir

A unique protein acting on signal transduction in eukaryotic cells is Tir (translocated intimin receptor) of enteropathogenic *E. coli* (EPEC). This is a protein containing two predicted transmembrane domains and six tyrosines. The 78-kDa protein is transferred (by a type III secretion system-dependent mechanism) to eukaryotic cells, where it becomes an integral part of the eukaryotic cell membrane and functions as receptor for the bacterial adhesin, intimin. At this stage, the protein mediates attachment of bacteria to the eukaryotic cells and is tyrosine phosphorylated, resulting in an apparent molecular mass of 90 kDa. Following tyrosine phosphorylation, the protein mediates actin nucleation, resulting in pedestal formation and triggering tyrosine phosphorylation of additional host proteins, including phospholipase C-γ. Tir is essential for EPEC virulence and is the first bacterial protein described that is tyrosine phosphorylated by host cells.

Proteins Acting on Small G Proteins

S. typhimurium SopE

SopE is injected into eukaryotic cells by a type III secretion system of *Salmonella typhimurium,* and it binds and activates the small G proteins Rac and Cdc42. Transfection of eukaryotic cells with the *sopE* gene, under the control of a eukaryotic cell promoter, induces profuse membrane ruffling and actin cytoskeleton reorganization, similarly to *E. coli* CNF (see above). In vivo, the activation of Rac and Cdc42 causes membrane ruffling and cytoskeletal rearrangements that mediate the uptake and internalization of the bacterium within the host eukaryotic cells. In addition, SopE stimulates nuclear responses that induce the synthesis of proinflammatory cytokines that contribute to the induction of diarrhea.

P. aeruginosa Exoenzyme S

P. aeruginosa exoenzyme S is a 49-kDa protein with ADP-ribosylating activity that ADP-ribosylates the small G protein Ras at arginine 41. The protein contains the motifs typical of the catalytic site of ADP-ribosylating enzymes, as described above for DT. To become enzymatically active, ExoS requires an interaction with a cytoplasmic factor named factor-activating exoenzyme S, or 14-3-3 protein. The toxin is injected into eukaryotic cells by a type III secretion system. When cells are transfected with the *exoS* gene under the control of a eukaryotic cell promoter, the cytoskeleton collapses and the cell morphology changes, resulting in rounding of the cells.

C. botulinum Exoenzyme 3

Exoenzyme C3 is not a true toxin and does not necessarily belong in this discussion. It is a protein of 211 amino acids that is produced by *C. botulinum* and that in vitro ADP-ribosylates the small regulatory protein Rho at asparagine 41, inactivating its function. The enzymatic activity is identical to that of all ADP-ribosylating enzymes (described above for DT). If the protein is microinjected into cells or if the cells are transfected with the C3 gene under a eukaryotic promoter, actin stress fibers are disrupted, the cells are rounded, and arborescent protrusions form. However, we do not know whether C3 alone is able to enter cells and intoxicate them, because no mechanism of cell entry has been found. It is described here in the hope that a new mechanism of this type will be discovered for it in the near future. Toxins with activity similar to C3 have been identified in *S. aureus* (EDIN), *Clostridium limosum*, *B. cereus*, and *Legionella pneumophila*.

Proteins Altering Inositol Phosphate Metabolism

SopB, a protein secreted by a type III secretion system (see chapter 13) of *Salmonella dublin*, has sequence homology to mammalian inositol phosphate 4-phosphatase and has inositol phosphate phosphatase activity in vitro. The enzyme hydrolyzes phosphatidylinositol triphosphate (PIP_3), which is a messenger molecule that inhibits chloride secretion, thus favoring fluid accumulation and diarrhea. The protein is essential for enteropathogenicity. Mutants in which SopB is either deleted or inactive are still invasive but are unable to cause fluid secretion and neutrophil accumulation into calf ileal loops.

SopB is homologous to the *Shigella* virulence factor IpgD, suggesting that a similar mechanism of virulence is also present in *Shigella*.

Apoptosis-Inducing Proteins

YopP is a 288-amino-acid (30-kDa) protein that is encoded by a large plasmid of *Y. enterocolitica* and is injected into eukaryotic cells by a type III secretion system. The protein is called YopJ in *Y. pseudotuberculosis* and is homologous to the SipB protein encoded by the chromosome of *Salmonella*, to IpaB of *Shigella*, to AvrA of *Salmonella enterica*, and to the AvrRxv protein of *Xanthomonas campestris*. It has been shown that *Y. enterocolitica* induces apoptosis in infected macrophages and that the gene coding for YopP is the only one necessary for this activity, suggesting that YopP is indeed the effector protein. It has been recently demonstrated that YopJ is sufficient to cause down-regulation of multiple mitogen-activated protein kinases in host cells, but the mechanism of action is still unknown. The homologous AvrRxv protein of *X. campestris* is also translocated into host plant cells, where it activates a plant defense mechanism leading to the formation of local necrotic lesions due to programmed cell death. This is the first example where animal- and plant-pathogenic bacteria share a type III secretion-dependent effector that elicits apoptosis in their hosts.

IpaB, a protein secreted by *Shigella*, also induces macrophage apoptosis by binding to the interleukin-1-converting enzyme precursor (ICE or caspase I), that induces the processing to the active caspase, a well-known mediator of apoptosis. The SipB protein of *Salmonella* also induces apoptosis by binding ICE.

Proteins Acting on the Cytoskeleton

YopE, a protein encoded by the *Yersinia* pathogenicity island and secreted by the type III secretion system, is known to paralyze macrophage phagocytosis by causing actin depolymerization and disrupting the cell stress fibers, with consequent rounding of the cells and loss of cell shape. The mechanism of action of YopE is still unknown.

Cytotoxins with Unknown Mechanisms of Action

Several extracellular products secreted by the *P. aeruginosa* type III secretion system (see chapter 13) are responsible for virulence. Among these, the 70-kDa protein ExoU is responsible for causing acute cytotoxicity in vitro and in epithelial lung injury. The mechanism of action is not known.

Selected Readings

Alouf, J. E., and J. R. Freer (ed.). 1999. *The Comprehensive Sourcebook of Bacterial Protein Toxins.* Academic Press, Ltd., London, United Kingdom.

Collier, R. J. 1967. Effect of diphtheria toxin on protein synthesis: inactivation of one of the transfer factors. *J. Mol. Biol.* **25:**83–98.

Duesbery, N. S., C. P. Webb, S. H. Leppla, V. M. Gordon, K. R. Klimpel, T. D. Copeland, N. G. Ahn, M. K. Oskarsson, K. Fukasawa, K. D. Paull, and G. F. Vande Woude. 1998. Proteolytic inactivation of MAP-kinase-kinase by anthrax lethal factor. *Science* **280:**734–737.

Flatau, G., E. Lemichez, M. Gauthier, P. Chardin, S. Paris, C. Fiorentini, and P. Boquet. 1997. Toxin-induced activation of the G protein p21 Rho by deamidation of glutamine. *Nature* **387:**729–733.

Just, I., J. Selzer, M. Wilm, C. von Eichel-Streiber, M. Mann, and K. Aktories. 1995. Glucosylation of Rho proteins by *Clostridium difficile* toxin B. *Nature* **375:**500–503.

Li, H., A. Liera, and R. A. Mariuzza. 1998. Structure-function studies of T-cell receptor-superantigen interactions. *Immunol. Rev.* **163:**177–186.

Pizza, M., A. Covacci, A. Bartoloni, M. Perugini, L. Nencioni, M. T. De Magistris, L. Villa, D. Nucci, R. Manetti, M. Bugnoli, F. Giovannoni, R. Olivieri, J. T. Barbieri, H. Sato, and R. Rappuoli. 1989. Mutants of pertussis toxin suitable for vaccine development. *Science* **246:**497–500.

Rappuoli, R., and C. Montecucco (ed.). 1997. *Guidebook to Protein Toxins and Their Use in Cell Biology.* Sambrook and Tooze Publication, Oxford University Press, Oxford, United Kingdom.

Schiavo, G., B. Poulain, O. Rossetto, F. Benfenati, L. Tauc, and C. Montecucco. 1999. Tetanus toxin is a zinc protein and its inhibition of neurotransmitter release and protease activity depend on zinc. *EMBO J.* **11:**3577–3583.

Schmidt, G., P. Sehr, M. Wilm, J. Selzer, M. Mann, and K. Aktories. 1997. Gln63 of Rho is deamidated by *Escherichia coli* cytotoxic necrotizing factor-1. *Nature* **387:**725–729.

Silhavy, T. J. 1997. Death by lethal injection. *Science* **278:**1085–1086.

12

Bacterial Protein Toxins as Tools in Cell Biology and Pharmacology

Klaus Aktories

At least four properties of bacterial protein toxins make them suitable as cell biological and pharmacological tools (Tables 12.1 and 12.2). First, the toxins enter cells without damaging the cell integrity. This ability is typical of intracellularly acting protein toxins and depends on very complex uptake mechanisms, which are described in detail elsewhere. In general, uptake of toxins depends on at least three steps: receptor binding, endocytosis of the receptor-bound toxin, and translocation of the toxin through the vesicle/compartment membrane into the cytosol. Of course, this does not apply to toxins that explicitly act on the cell surface and form membrane pores.

Second, the toxins possess high specificity. A high cell specificity is most often based on a toxin-specific membrane-binding domain and on specific receptors present on the surface of eukaryotic target cells. This is, for example, the reason for the cell specificity of neurotoxins. The cell surface receptors for botulinum neurotoxins, which are still unknown (and this holds true also for tetanus toxin), are found exclusively on neuronal cells. Other cell types that may contain the protein target are not affected, because the toxins are not able to enter these cells. Most importantly, many bacterial protein toxins are characterized by a very specific target recognition. Two examples illustrate this fact. First, diphtheria toxin and *Pseudomonas* exotoxin A selectively ADP-ribosylate diphthamide, a posttranslationally modified histidine residue found exclusively in eukaryotic elongation factor 2, but not in other proteins. Second, *Clostridium botulinum* C2 toxin ADP-ribosylates nonmuscle actin but not skeletal muscle actin, although these two actin isoforms are more than 95% identical in their amino acid sequence and share the acceptor amino acid arginine 177.

Third, many bacterial protein toxins are remarkably potent and extremely efficient. It has been estimated that one molecule of diphtheria toxin per cell is sufficient to kill the cell by ADP-ribosylation of elongation factor 2 and subsequent inhibition of protein synthesis. Moreover, botulinum neurotoxins are by far the most potent agents known. A high affinity of the toxins for their specific cell receptor may contribute to this high potency. However, in most cases, the high potency is believed to be due to the en-

Table 12.1 Bacterial toxins frequently used as tools

Study of signal transduction involving heterotrimeric G proteins
  G_s protein activation by cholera toxin and related toxins
  $G_{i,o}$-protein inhibition by pertussis toxin
Study of signal transduction involving low-molecular-mass GTPases
  Rho subfamily protein activation by CNF and DNT
  RhoA, RhoB, and RhoC inactivation by C3-like transferases
  Rho/Ras subfamily protein inhibition by large clostridial cytotoxins
Study of the involvement of the actin cytoskeleton in cellular processes
  Depolymerization of actin by *C. botulinum* C2 toxin and *C. perfringens* iota toxin
  Modulation of actin regulation via Rho proteins (see above)
Study of the involvement of synaptic peptides in exocytosis
  Cleavage of synaptic peptides by botulinum neurotoxins and tetanus toxin
Construction of cellular protein delivery systems
  Single- or dichain transport systems consisting of the binding-translocation
    subunits of diphteria toxin and *Pseudomonas* exotoxin A
  Binary transport systems consisting of the separate components of anthrax
    toxin, *C. botulinum* C2 toxin, or *C. perfringens* iota toxin
Selectively killing of target cells
  Inactivation of elongation factor 2 by fusion toxins or immunotoxins consisting
    of diphtheria toxin or *Pseudomonas* exotoxin A
Permeabilization of eukaryotic cells
  Streptolysin O
  S. aureus alpha-toxin
Tracing of neurons
  By retrograde transport of CT
  By retrograde transport of neurotoxins

zyme activities of the toxins. Therefore, toxins are endoproteases, *N*-glycosidases, ADP-ribosyltransferases, glucosyltransferases, or deamidases (see also Table 12.2), which covalently modify their protein substrates with high efficiency. Although most toxin-induced enzymatic reactions are reversible in vitro and can be driven back under specific experimental conditions, the toxin-induced modifications are essentially irreversible and complete in intact cells and under physiological conditions. Moreover, the covalent modification by toxins occurs even in protein mixtures, in which the target protein is present only in small amounts. Therefore, the reaction is exploited to selectively radiolabel target proteins by ^{32}P ADP-ribosylation or ^{14}C-glycosylation to facilitate the purification of these targets or to allow analysis by sodium dodecyl sulfate-polyacrylamide gel electrophoresis.

Fourth, the remarkable cell biological efficiency of bacterial toxins is not based simply on the specific kinetics of the enzyme reactions catalyzed by the toxins. A pivotal physiological role of the eukaryotic target is also important. Thus, very often the toxins encounter cellular "master" regulators, thereby inhibiting or stimulating crucial signal pathways absolutely necessary for the normal physiological function of the cells targeted. This aspect is of particular importance for the use of toxins as pharmacological and cell biological tools and has been extensively exploited during recent years. In fact, research on the action of novel bacterial toxins or on the molecular mechanisms of toxins which were hitherto obscure has been motivated greatly by the prospect of unraveling crucial biological pathways or novel pivotal regulators of the target cells.

Table 12.2 Enzyme activities and targets of the bacterial protein toxins most frequently used as tools

Toxin	Activity or target
ADP-ribosyltransferases	
Diphtheria toxin	Elongation factor 2
Pseudomonas exotoxin A	Elongation factor 2
CT	G_s proteins
PT	$G_{i,o}$ proteins
C. botulinum C2 toxin	Nonmuscle β/γ-actin, smooth muscle γ-actin
C. perfringens iota toxin	All actin isoforms
C3-like transferases (e.g., *C. botulinum* C3 exoenzyme)	RhoA, RhoB, and RhoC
Glycosyltransferases	
C. difficile toxins A and B	Rho, Rac, and Cdc42
C. sordellii lethal toxin	Rac, Cdc42, Ral, Ras, and Rap
Deamidases	
CNF1 and CNF2	Rho, Rac, and Cdc42
Metalloendoproteases	
Botulinum neurotoxins A, C, and E	SNAP25
Botulinum neurotoxins B, D, F, and G	Synaptobrevin (VAMP)
Botulinum neurotoxin C	Syntaxin
Tetanus toxin	Synaptobrevin (VAMP)
Anthrax toxin	MAPKK

Toxins as Tools To Study Nucleotide-Binding Proteins

For unknown reasons, GTP-binding proteins are the preferred substrates for bacterial toxins. For example, heterotrimeric G proteins, which couple to heptahelical membrane receptors and are involved in various signaling processes, are ADP-ribosylated by cholera toxin (CT), the related heat-labile *Escherichia coli* enterotoxins, and pertussis toxin (PT). Various evidently unrelated bacterial toxins modify low-molecular-mass GTPases (e.g., Rho proteins) by either ADP-ribosylation, glycosylation, or deamidation. Finally, elongation factor 2, the eukaryotic protein substrate of diphtheria toxin and *Pseudomonas* exotoxin A, is a GTP-binding protein. Actin, another important eukaryotic substrate for ADP-ribosylation by bacterial toxins, is not a GTP-binding protein but an ATP-binding protein. Because all these nucleotide-binding proteins are functionally important cellular proteins, the toxins which allow their selective covalent modification are widely used as tools.

Cholera Toxin and Pertussis Toxin as Tools To Study G-Protein-Mediated Signaling

The G-protein ADP-ribosylating toxins CT and PT are the "classical" protein toxins which are used as pharmacological tools. Both toxins are used to study signal pathways involving heterotrimeric G proteins (Figure 12.1 and Box 12.1).

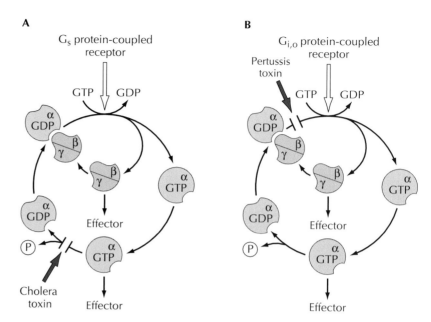

Figure 12.1   ADP-ribosylation of G proteins by CT and PT. **(A)** CT ADP-ribosylates the α subunits of G_s proteins. ADP-ribosylation blocks the GTPase activity of $G_{s\alpha}$ and activates the G protein persistently. CT is used as a tool to radiolabel G_s proteins and to manipulate the signaling via G_s. **(B)** PT ADP-ribosylates the α subunits of $G_{i,o}$ proteins, thereby blocking the receptor-mediated activation of the G protein. Thus, inhibition of a specific signal transduction process by PT indicates the involvement of "PT-sensitive" G proteins in this signaling pathway.

Cholera Toxin

CT, which is produced by *Vibrio cholerae,* is the principal cause of the watery diarrhea of cholera. The toxin (85 kDa) is composed of an enzyme component (A subunit, ~28 kDa) and a binding component which is formed of five identical B subunits (each ~11.6 kDa). The enzyme component consists of an A1 subunit (~21 kDa) and an A2 subunit (~6 kDa), which are linked by a disulfide bridge. A1 possesses ADP-ribosyltransferase activity, whereas A2 mediates the interaction with the B component. The physio-

---

BOX 12.1

G proteins, targets of CT and PT

Whereas CT activates G_s protein, PT inhibits the activation of $G_{i,o}$ proteins (Figure 12.1). The functional consequences of the modification of G proteins depend on the signal transduction systems of the specific cell targeted by the toxins.

G_s proteins are, for example, coupled to adrenergic (β_{1-3}), adenosine (A2), dopamine (D1 and D5), hista-

mine (H2), serotonin (5-HT4), and prostaglandin (DP, EP2, EP4, and IP) receptors. $G_{i,o}$ proteins are coupled to adrenergic (α_2), muscarinic (M2 and M4), dopamine (D2, D3, and D4), serotonin (5-HT1), opioid (OP1), and somatostatin (SST1) receptors.

The α subunit of G_s activates adenylyl cyclase (all isoforms) and causes an increase in Ca^{2+} currents. The α subunit of $G_{i,o}$ induces inhibition of adenylyl cyclase, an increase in K^+ currents, and a decrease in Ca^{2+} cur-

rents. Moreover, the free β/γ subunits of heterotrimeric G proteins interact with effectors. They are released by CT-induced activation of G_s and remain associated with $G_{i,o}$ after inactivation of the G protein by PT. The β/γ subunits activate phospholipase C-β and adenylyl cyclase isoforms II, IV, and VII. They inhibit adenylyl cyclase isoform I and increase K^+, Ca^{2+}, and Na^+ currents. Thus, all these signal transduction pathways are affected by CT and PT.

logical target of CT (the same holds true for the related *E. coli* heat-labile enterotoxins) is the $G_{s\alpha}$ subunit of heterotrimeric G proteins. The toxin ADP-ribosylates arginine 201 (or equivalent arginine residues depending on the splice variants) of $G_{s\alpha}$. ADP-ribosylation inhibits the intrinsic GTPase of $G_{s\alpha}$, induces dissociation from $\beta\gamma$ subunits, and renders the α subunit persistently active. In enterocytes, stimulation of adenylyl cyclase and increase in intracellular cyclic AMP levels caused by the persistently active $G_{s\alpha}$ subunits contribute to the fluid and electrolyte loss that occurs in the course of cholera. At least in vitro, CT also ADP-ribosylates other G proteins such as transducin and even $G_{o/i}$ (under specific conditions). Substrates also include cell proteins with a suitable arginine residue, as well as small arginine derivatives like agmatine, and, finally, the toxin itself is modified by an auto-ADP-ribosylation reaction. Moreover, like all other ADP-ribosyltransferases, CT possesses NAD glycohydrolase activity and splits NAD into ADP-ribose and nicotinamide in the absence of a suitable substrate. However, this activity is very low compared with the ADP-ribosyltransferase activity. In vitro, ADP-ribosylation by CT is stimulated largely by the presence of cofactors called ADP-ribosylation factors, which have been identified as members of a subfamily of small GTPases (ARF subfamily of GTPases). Although the specificity of CT in in vitro experiments is not very impressive, in intact cells $G_{s\alpha}$ appears to be the preferred substrate of the toxin.

In addition to its usage as a tool to manipulate signal transduction pathways involving G_s protein or to label $G_{s\alpha}$ protein, CT has been successfully used as an adjuvant applied with an unrelated antigen to stimulate mucosal immune responses. Moreover, CT is used as a tracer for retrograde transport in neuronal cells.

Pertussis Toxin

PT (formerly called islet-activating protein) is one of the exotoxins of *Bordetella pertussis,* the causative agent of whooping cough. It is a hexameric toxin of 105 kDa and consists of the enzyme subunit S1 (~26 kDa, A subunit) and the binding pentamer (B subunit) formed of S2 (~22 kDa), S3 (~22 kDa), two S4 subunits (~12 kDa), and S5 (~12 kDa). The enzyme subunit (S1) of PT ADP-ribosylates G_α isoforms of the G_i subfamily of heterotrimeric G proteins in the presence of $\beta\gamma$ subunits (Figure 12.1). The toxin catalyzes the ADP-ribosylation of a cysteine residue located 4 amino acid residues from the carboxy-terminal end of the α subunit of sensitive G proteins (e.g., G_{i1-3}, $G_{o1,2}$, and G_t). About 20 years ago, PT was most important for the identification of the G_i protein as an additional G protein besides G_s. The consequence of toxin-catalyzed ADP-ribosylation is a functional uncoupling of G proteins from their membrane receptors with subsequent blockade of the G-protein-transduced signal pathways. Importantly, G proteins of the G_s, G_q, G_{12}, and G_z subfamilies and a splice variant of G_{i2} [$G_{i2(L)}$] are not modified by pertussis toxin. Thus, the toxin is widely used to test whether a signal transduction pathway involves so-called "PT-sensitive" G proteins.

Toxins as Tools To Study the Regulation of the Actin Cytoskeleton

Actin, which is one of the most abundant proteins in eukaryotic cells, is the major component of the microfilament system. It regulates the architecture of cells and is involved in various motile processes. Besides its function in skeletal muscle contraction, it plays important roles in migration, phagocytosis, endocytosis, secretion, and intracellular transport. The actin cytoskeleton is the target of various bacterial toxins that affect the microfilament protein either directly by ADP-ribosylation or indirectly by modifying the regulatory mechanisms involved in the organization of the actin cytoskeleton. In the latter case, Rho proteins, which are regulators of the cytoskeleton, are especially important targets of bacterial protein toxins.

Actin consists of a single polypeptide chain of 375 amino acids (~43 kDa) and binds divalent cations and adenine nucleotides (ATP and ADP) with high affinity. Almost all cellular functions of actin depend on the ability of the microfilament protein to polymerize. Actin filaments are polar structures with two nonequivalent ends. The polymerization is faster at the plus ends (barbed ends) and slower at the minus ends (pointed) of filaments. In nonmuscle cells, actin occurs in the monomeric and polymeric states. The distribution between G-actin and F-actin depends largely on the presence of actin-binding proteins such as gelsolin, profilin, vinculin, and insertin.

Actin ADP-Ribosylating Toxins

Actin is the specific substrate for the family of actin ADP-ribosylating toxins. This family has four members: *Clostridium botulinum* C2 toxin, *C. perfringens* iota toxin, *C. spiroforme* toxin, and the ADP-ribosylating toxin from *C. difficile*. All these toxins are binary in structure and consist of separate enzyme and cell-binding/translocation subunits which are not linked by either covalent or noncovalent bonds. After attachment of the binding component to the cell surface and induction of a binding site for the enzyme component take place, the toxin subunits will interact with each other (Figure 12.2).

C. botulinum C2 toxin and *C. perfringens* iota toxin are used as cell biological tools. The binding component (C2II) of *C. botulinum* C2 toxin is a ca. 80-kDa protein that has to be activated by trypsin to release a ca. 65-kDa active fragment. The molecular mass of the enzyme component (C2I) is about 50 kDa. C2I ADP-ribosylates nonmuscle β/γ actin and γ-smooth muscle actin but not α-actin isoforms at arginine 177. In contrast, *C. perfringens* iota toxin catalyzes the ADP-ribosylation of all mammalian actin isoforms known. Interestingly, the difference between the various mammalian isoforms is less than 5%. The modification by the toxins is highly specific. Neither G proteins, which are substrates for the arginine-modifying CT, nor other cytoskeletal elements such as tubulin are ADP-ribosylated. ADP-ribosylation of actin has several functional consequences. Most important is the inhibition of the actin polymerization (Figure 12.2). The acceptor amino acid of actin is located at an actin-actin contact site, and it is believed that the attachment of the ADP-ribose moiety inhibits the interaction of the actin monomers. This is also the reason why G-actin but not polymerized actin is a substrate for ADP-ribosylation. Second, ADP-ribosylated actin behaves like a capping protein that binds to the barbed ends of actin filaments, thereby inhibiting the polymerization of unmodified G-actin at the plus ends of filaments. In contrast, ADP-ribosylated actin does not interact with the minus ends of filaments. Therefore, the critical actin concentration for polymerization increases to values that correspond to the critical actin concentration at the minus end of actin filaments. This

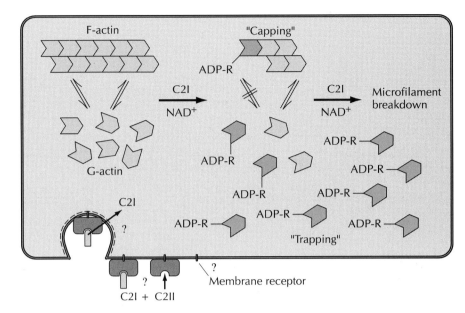

Figure 12.2 Model of the action of the actin-ADP-ribosylating *C. botulinum* C2 toxin. The activated binding component (C2II) binds to the cell surface receptor of the target cell. Thereby, a binding site for the enzyme component C2I is initiated. C2I is internalized and translocates into the cytosol, where monomeric actin is ADP-ribosylated. ADP-ribosylation of actin blocks its polymerization ("trapping" of the monomeric form). Moreover, ADP-ribosylated actin monomers bind like capping proteins to the plus ends of actin filaments, thereby inhibiting polymerization of unmodified actin monomers ("capping"). Because the minus ends of filaments are free, actin can be released at this site. The released actin is immediately ADP-ribosylated and trapped. Thus, the major consequence of ADP-ribosylation is breakdown of the microfilaments.

means that actin filaments depolymerize. However, the released monomeric actin is immediately ADP-ribosylated by the toxin and trapped in the nonpolymerizable form. Third, ADP-ribosylation of actin completely blocks actin ATPase activity and increases the ATP exchange. Finally, ADP-ribosylation affects the interaction of actin with actin-binding proteins, which are essential for regulation of actin polymerization. For example, the nucleation activity of the gelsolin-actin complex is blocked after ADP-ribosylation of actin. All these functional consequences of the ADP-ribosylation of actin finally result in morphological changes of targeted cells (rounding up) (Figure 12.3), redistribution of the microfilaments, depolymerization of F-actin, and an increase in the amount of G-actin. Thus, the actin ADP-ribosylating toxins are the most powerful tools to induce depolymerization of microfilaments in intact cells. Actually, the complete actin cytoskeleton of cells can be depolymerized. Therefore, the toxins have been used to study the role of actin in migration, secretion, endocytosis, superoxide anion production, and intracellular vesicle transport.

   C. botulinum C2 toxin was used to investigate the activation of neutrophils by chemotactic agents. In neutrophils, complement C5a, *N*-formyl peptides, or leukotriene B_4 induce cellular responses such as cell shape change, adhesion, migration, degranulation, and phagocytosis. It was shown by using the toxin that all these events depend on the redistribution of the cytoskeleton and on changes in actin polymerization. For many years,

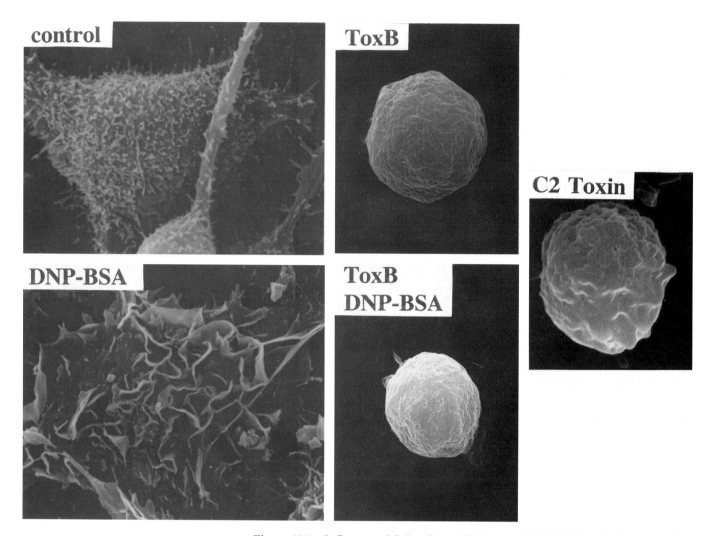

Figure 12.3 Influence of *C. botulinum* C2 toxin and *C. difficile* toxin B on morphology of RBL cells. Stimulation of immunoglobulin E-primed RBL cells by antigen (2,4-dinitrophenyl-bovine serum albumin [DNP-BSA]) via the high-affinity antigen receptor (FcεRI) induces membrane ruffling. Pretreatment of cells with *C. botulinum* C2 toxin, which depolymerizes actin by ADP-ribosylation, causes dramatic morphological changes but increases regulated serotonin release. *C. difficile* toxin B, which inactivates Rho family proteins by glucosylation, induces a similar morphology of RBL cells but completely blocks antigen-induced serotonin release. The experiment shows that inhibition of serotonin release by toxin B is not simply caused by an action on the actin cytoskeleton and indicates that Rho proteins are essentially involved in the signal transduction of the FcεRI receptor. Scanning electron micrographs courtesy of J. Wilting, Freiburg, Germany.

the mycotoxins phalloidin, which induces polymerization, and cytochalasin, which causes depolymerization of F-actin, were used as tools to elucidate the physiological functions of actin. Whereas phalloidin is hampered by its poor cell accessibility, the shortcomings of cytochalasins are the incomplete depolymerization of microfilaments and additional nonspecific effects. Therefore, it seems more appropriate to use actin-ADP-ribosylating toxins as tools for selective and complete depolymerization of actin. By using C2 toxin, it was shown that redistribution of the actin cytoskeleton largely affects the activity of superoxide anion-producing NADPH oxidase of neutrophils, thereby confirming previous results obtained with cytochalasins.

C2 toxin has been used in studies on exocytosis. Actin is suggested to function as a subcortical barrier, preventing vesicle fusion with the cell membrane. In fact, in various cell types C2 toxin increases the regulated exocytosis. This was shown for the N-formyl peptide-stimulated release of N-acetylglucosaminidase and of vitamin B_{12}-binding protein from neutrophils, for steroids in Y-1 adrenal cells, for serotonin in RBL cells, and for insulin in isolated rat islets. In general, similar results were obtained with cytochalasin. In all these cases, C2 toxin did not affect the basal release of mediators. However, biphasic effects of C2 toxin were observed on the release of noradrenaline from PC12 cells. Treatment of the cells with the toxin for up to 1 h increased noradrenaline release, whereas longer incubation reduced the mediator release induced by depolarization or by carbachol. In hamster insulinoma HIT-T15 cells, C2 toxin treatment inhibited insulin release by about 50%. Inhibition of insulin release was more pronounced in the second phase of the biphasic insulin release. These findings can be interpreted to indicate that actin filaments are involved in the recruitment of vesicles to the releasable pool and that C2 toxin blocks this action. Also, in suspended mast cells, C2 toxin inhibited regulated mediator release. In these cells, however, the inhibitory effect of C2 toxin turned into a stimulatory effect after adhesion of the mast cells, suggesting a complex role of actin which cannot be explained simply by a barrier function or vesicle recruitment.

C2 toxin was used to investigate the regulation of ileal smooth muscle contraction. It was shown that C2 toxin inhibited the contraction in ileal longitudinal smooth muscle preparations induced by electrical stimulation, by agonists like bradykinin or carbachol, and even by Ba^{2+} ions, which directly stimulate smooth muscle contraction. Since the toxin modifies G- but not F-actin, these data were taken as an indication that G/F-actin transition is important in smooth muscle contraction. Moreover, C2 toxin was used to investigate the autoregulation of the actin synthesis that appears to depend on the levels of G-actin and on the G/F-actin ratio. In all these studies, proper control of the specific action of C2 toxin was necessary. For example, the single toxin components (C2I or C2II) applied should not induce any effect in the system studied. The toxin effect should occur with some delay of at least 15 to 30 min. This time is necessary for the translocation of the toxin. Last but not least, it should be tested whether actin is in fact ADP-ribosylated by the toxin.

Toxins That Affect Small GTPases

During recent years, various toxins that modify low-molecular-mass GTP-binding proteins have been identified (Figure 12.4). Among these are the ADP-ribosylating C3-like transferases, the glycosylating large clostridial toxins, and a rather novel group of deamidating toxins. Because these toxins (especially C3-like transferases) were most important for the functional characterization of small GTPases, these proteins are briefly reviewed.

At least five families of small GTPases can be described; these are the Ras, Rab, Arf, Ran, and Rho subfamilies, which have about 30% sequence identity. Members within a specific GTPase subfamily exhibit at least 50%

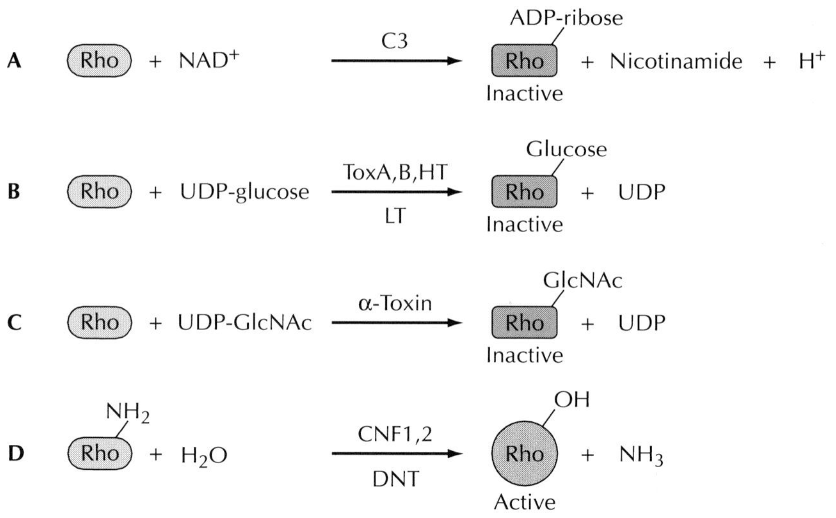

Figure 12.4 Covalent modification of Rho proteins by bacterial protein toxins. **(A)** C3-like transferases ADP-ribosylate RhoA, RhoB, and RhoC by using NAD as a cosubstrate. **(B)** Rho family proteins are glucosylated by *C. difficile* toxin A and B and by the hemorrhagic (HT) and lethal (LT) toxins of *C. sordellii*. The cosubstrate is UDP-glucose. While *C. difficile* toxins A and B and HT glucosylate all Rho family members, LT modifies Rac and Cdc42 (depending on the producer strain) but not Rho. Additionally, Ras family proteins (e.g., Ras, Rap, and Ral) are substrates for glucosylation by LT. **(C)** The alpha-toxin from *C. novyi* catalyzes an N-acetylglucosaminylation. **(D)** Rho proteins (Rho, Rac, and Cdc42) are activated by cytotoxic necrotizing factors (CNF1 and CNF2) from *E. coli* and by the dermonecrotic toxin (DNT) from *Bordetella* species.

identity in the amino acid sequence. Like heterotrimeric G proteins, these low-molecular-mass or "small" GTP-binding proteins are regulated by a GTPase cycle. They are inactive with GDP bound and activated after GDP-GTP exchange. Hydrolysis of bound GTP terminates the active state of the GTPases.

It has been shown that especially Rho subfamily GTPases are targets for bacterial protein toxins. This family comprises more than 10 GTPases (including RhoA, RhoB, RhoC, RhoD, RhoE, RhoG, Rac1, Rac2, Rac3, Cdc42Hs, and G25K). Rho proteins are controlled by three groups of regulatory proteins. Guanine nucleotide exchange factors induce the activation of Rho GTPases by facilitating the GDP-GTP exchange. GTPase-activating proteins stimulate and catalyze the GTP hydrolysis, thereby inactivating the GTPases, and, finally, guanine nucleotide dissociation inhibitors (GDIs) block nucleotide exchange and keep the inactive form of Rho proteins in the cytosol (Figure 12.5).

The small GTPases of the Rho family (Rho, Rac, and Cdc42) are involved in the regulation of the actin cytoskeleton. Rho induces the formation of stress fibers and adhesion complexes. Rac is involved in lamellipodium formation and induces adhesion complexes, which appear to be different from those induced by Rho. Finally, Cdc42 induces the formation of microspikes. Beside their roles in the organization of the actin cytoskeleton, Rho proteins act as molecular switches in various signal transduction processes (see below) and may play essential roles in invasion of eukaryotic cells by various bacteria (e.g., *Salmonella* and *Shigella*).

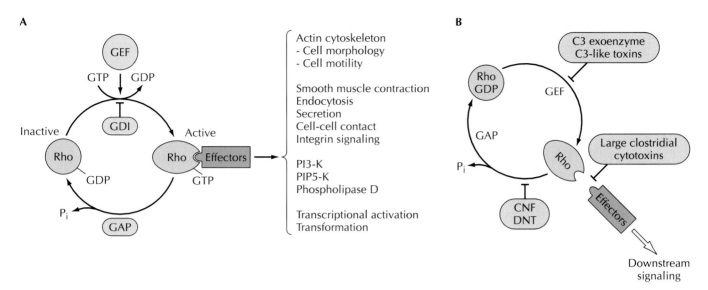

Figure 12.5 **(A)** Regulatory GTPase cycle of Rho proteins. Rho GTPases are inactive in the GDP-bound form and active after GTP-GDP exchange. The nucleotide exchange is stimulated by guanine nucleotide exchange factors (GEF) and inhibited by guanine nucleotide dissociation inhibitors (GDI). The active state of the GTPase is terminated by hydrolysis of bound GTP. GTP hydrolysis is stimulated by GTPase-activating proteins (GAP). Some of the various processes regulated by Rho GTPases are indicated. **(B)** Effects of Rho-modifying toxins. C3 exoenzyme and C3-like transferases ADP-ribosylate Rho at Asn 41, inhibiting the activation of Rho. Glucosylation of Rho GTPases by large clostridial cytotoxins blocks the interaction of active Rho with effectors and causes inactivation of Rho GTPase-dependent processes. The cytotoxic necrotizing factor CNF and the dermonecrotic toxin DNT inhibit the switch-off mechanism of activated Rho GTPases by deamidation of a glutamine residue (Gln 63 of Rho), which is essential for GTP hydrolysis resulting in constitutively activated Rho.

Rho-Inactivating Toxins

C3-Like ADP-Ribosyltransferases

Rho GTPases are the specific substrates of the C3-like ADP-ribosyltransferases. The C3-like transferases are all basic proteins (IP > 9) with molecular masses of about 25 kDa, having a sequence identity of 30 to 70%. ADP-ribosylation of Rho by C3 transferases needs no additional factors than NAD. Rho ADP-ribosylation is affected by guanine nucleotides, divalent cations, detergents, and temperature. Purified endogenous Rho, recombinant Rho proteins, and the membranous Rho protein are better substrates for ADP-ribosylation when bound to GDP than when bound to GTP. By contrast, the ADP-ribosylation of cytosolic Rho by C3 is increased by GTP or stable GTP analogs because the GTP-Rho has a lower affinity for GDI and the GDI-Rho complex is hardly modified by C3. Similarly, detergents and phospholipids increase ADP-ribosylation most probably by dissociating the GDI-Rho complex.

C3 transferases ADP-ribosylate RhoA, RhoB, and RhoC at asparagine 41, thereby blocking the biological activity of Rho (Figure 12.5). Although asparagine 41 is located in the so-called effector region of the Rho protein, the precise mechanism of Rho inhibition by ADP-ribosylation is not entirely clear. For example, ADP-ribosylated Rho is still able to bind to protein kinase N, one putative effector of RhoA. Whether other Rho effectors are also able to interact with ADP-ribosylated Rho remains to be clarified. Although C3 modifies Rac by up to 5% in vitro, all experimental evidence indicates that in intact cells C3-like transferases ADP-ribosylate RhoA, RhoB, and RhoC but not Rac and Cdc42. This is most important because C3 has been widely used to study the cell biological role of Rho.

C3 transferases appear to lack any specific binding and translocation subunit. Therefore, the cell accessibility of these ADP-ribosyltransferases is rather poor. Most probably this is also the reason why C3 shows low toxicity compared with other toxins. For example, intraperitoneal injection of 100 μg of C3 into mice has no obvious consequences. Thus, it appears questionable to describe C3 as a "real" toxin. Notably, various isoforms of C3 may exist which may differ in their ability to enter cells. Nevertheless, the structural data available suggests that C3 has only a catalytic subunit. To circumvent the problem of poor cell entry, several approaches have been used. First, C3 was introduced into cells by osmotic shock. In fact, all methods used to permeabilize cells can be applied to allow C3 to enter cells. Mostly, pore-forming toxins like streptolysin O were used. Many excellent studies, especially from the laboratory of Alan Hall, used microinjection techniques. Even more sophisticated approaches involve the use of chimeric toxins. To this end, a fusion toxin that consists of C3 and the receptor-binding and translocation subunits of diphtheria toxin was constructed. Although this fusion toxin is able to enter all diphtheria toxin-sensitive cells, the uptake pathway of the fusion toxin differs from that of diphtheria toxin. Another chimeric toxin was constructed of *C. botulinum* C2 toxin and C3 transferase. C3 was fused to the N-terminal part of the enzyme component of C2I, which is apparently involved in the interaction of C2I with C2II. This C3-C2IN fusion toxin could be introduced into cells via the binding and translocation component C2II. Similarly, as observed with the diphtheria toxin-C3 fusion protein, in intact cells the Rho-ADP-ribosylating activity of C3-C2IN was increased several hundredfold compared with that of the native C3 transferase. However, most studies took advantage of the fact that C3 enters most cells when used at high concentrations for a long time (e.g., 24 to 48 h). For comparison, the effect of the C3-C2IN fusion protein was observed 1 to 2 h after addition to the culture medium.

As mentioned above, the use of C3 was most important in elucidating the functions of Rho proteins. Because the C3-induced effects on Rho appear to be very specific, the inhibition of a specified signaling pathway by using C3 can be taken as an indication that Rho is involved in the signaling process. Thereby, it was shown that RhoA specifically control stress fiber formation and induction of cell adhesions. C3 was used in studies to show that Rho participates in the control of cell aggregation, integrin signaling, control of phosphatidylinositol 3-kinase, phosphatidylinositol-4-phosphate 5-kinase, and phospholipase D. Moreover, C3 transferases were used to characterize the role of Rho proteins in endocytosis, secretion, and control of transcription, cell cycle progression, and cell transformation. Finally, studies on the involvement of Rho proteins in neurite outgrowth and nerve growth were largely supported by the usage of C3 (Figure 12.5).

Large Clostridial Cytotoxins

Recently, it has been shown that Rho proteins are the targets of various clostridial cytotoxins which modify the GTPases by glycosylation. These toxins have molecular masses between 250 and 308 kDa and are therefore called "large" clostridial cytotoxins. From a medical point of view, *C. difficile* toxins A and B, which are the major virulence factors in antibiotic-associated diarrhea, are the most important. Toxin B is several hundredfold more potent in inducing cytotoxic effects in cell culture than is toxin A and is therefore designated a cytotoxin. However, toxin A also induces cytotoxic effects. The main difference between the two toxins may be due to the use of different cell membrane receptors. After parenteral administration, both toxins are lethal at an identical dose.

In most cells, *C. difficile* toxins induce depolymerization of the actin cytoskeleton, leading to a morphology similar to that induced by C3-like transferases. It has been shown that Rho proteins are also substrates for *C. difficile* toxins (Figures 12.2 and 12.5). However, these large clostridial toxins catalyze the glucosylation of Rho GTPases by using UDP-glucose as a co-substrate. Notably, not only RhoA, RhoB, and RhoC but also other members of the Rho subfamily like Cdc42 and Rac are substrates. Other small GTPases like Ras, Rab, Arf, or Ran are not glucosylated. Glucosylation occurs at Thr37 of Rho and Thr35 of Rac and Cdc42. Threonine 35/37 is conserved in all small GTPases and is involved in the coordination of the divalent cations and of nucleotides. Moreover, this residue is located in the switch-I region of the proteins. This region undergoes major conformational changes upon nucleotide binding and participates in effector coupling. It has been demonstrated that glucosylation inhibits the RhoA-effector interaction, explaining the functional inactivation of the GTPase after toxin treatment. Moreover, the acceptor amino acid residue for glucosylation is located very close to the site of ADP-ribosylation (Asn41 in RhoA). This explains why glucosylated Rho is no longer ADP-ribosylated by C3 and vice versa. Thus, failure to label Rho in lysates of toxin B-treated cells by C3-induced [^{32}P]ADP-ribosylation indicates that all Rho was already modified by toxin B in intact cells.

Because *C. difficile* toxins A and B glucosylate all Rho subfamily GTPases, they are used to screen whether Rho GTPases are involved in certain regulatory pathways. For example, toxin B was used to study the involvement of Rho proteins in signal transduction of the FcεRI receptor and in histamine or serotonin secretion from mast cells or RBL cells (Figure 12.5). Toxin B inhibited regulated secretion induced by antigen, compound 48/80, and even the calcium ionophore A23187. Because Rho proteins are involved in various diverse regulatory pathways including regulation of the actin cytoskeleton, it is important to clarify whether the inhibitory effect of the toxin on secretion is secondary and induced by depolymerization of actin. In this case, another toxin that selectively inhibits actin polymerization but has no effect on Rho proteins, such as *C. botulinum* C2 toxin, is useful. In the above-mentioned case of histamine or serotonin secretion from RBL cells, it was shown that C2 toxin largely increased regulated secretion, excluding the possibility that the effect by toxin B is caused merely by disturbing the actin cytoskeleton.

Other members of the family of large clostridial cytotoxins are also glycosyltransferases but differ in cosubstrate or protein substrate specificity. *C. novyi* alpha-toxin modifies Rho subfamily proteins like *C. difficile* toxin but catalyzes N-acetylglucosaminylation by using UDP-GlcNAc as a co-substrate (Figure 12.2). *C. sordellii* lethal toxin, which is about 90% similar to *C. difficile* toxin B, also uses UDP-glucose as a cosubstrate but differs in its protein targets. Whereas Rac is a very good substrate for this toxin, Rho is poorly modified (most probably not at all in intact cells). In addition, lethal toxin glucosylates and inactivates Ras subfamily proteins like Ras,

Rap, and Ral. Thus, the lethal toxin interferes with the Ras-signaling pathway and blocks activation of the mitogen-activated protein kinase (MAPK) cascade by growth factors.

Rho-Activating Toxins

Constitutively active Rho proteins can be constructed by exchange of amino acid residues which are essential for GTPase activity. The exchanges of glycine 14 and glutamine 63 of Rho for valine and leucine, respectively, to obtain Rho with very low GTP-hydrolyzing activity are well known. These mutant proteins were often used in microinjection studies to elucidate the role of Rho in organization of the actin cytoskeleton. Rho proteins are also activated by bacterial protein toxins. Recently, it has been shown that Rho GTPases are activated by the cytotoxic necrotizing factors CNF1 and CNF2 from *E. coli* and by the dermonecrotic toxin produced by various *Bordetella* species (Figure 12.4). CNF1 and CNF2 are 115-kDa proteins which are 99% similar. The toxins cause formation of multinucleated cells and induce the polymerization of actin and the formation of stress fibers, microspikes, and lamellipodia. CNF catalyzes the deamidation of glutamine 63 of RhoA to glutamic acid. Because glutamic acid is not able to fulfill the function of glutamine, the intrinsic and GAP-stimulated GTPase activity of Rho is blocked. Most probably, CNF not only deamidates Rho but also deamidates Cdc42 and Rac, explaining the formation of microspikes, lamellipodia, and membrane ruffles after toxin treatment. Similarly to CNFs, dermonecrotic toxin (154 kDa) causes deamidation of Rho proteins. Thus, both toxins can be used to activate Rho proteins, for example in cell culture.

Toxins as Part of a Transmembrane Carrier System

Many toxins affecting eukaryotic cells must be translocated into the cytosol to find their target. Some of these translocated toxins are among the biggest toxins known (e.g., large clostridial cytotoxins with molecular masses of 250 to 308 kDa). The ability of these toxins to cross the cell membrane and, in whole or in part, to enter the cytosol has been exploited for use as a carrier system for proteins not related to the toxins themselves. As mentioned above, these toxins are A/B toxins and consist of the biologically active component (A, enzyme component) and the binding and translocation component (B). Actually, the binding and translocation domains are functionally and structurally separate entities, and therefore the toxins are composed of three major domains. In most toxins, these functional domains are located on a single toxin chain (e.g., *Pseudomonas* exotoxin A), are positioned on different chains linked by disulfide bonds (e.g., diphtheria toxin, botulinum neurotoxins), or are located on specific components which are noncovalently associated (e.g., CT and PT). Another group of toxins are characterized by nonlinkage of the enzyme and binding-translocation components. These toxins are also called binary toxins. To construct a protein delivery system, all three toxin components can be used. If it is desired to target a particular cell type, the toxin receptor-binding part must be changed (see below). On the other hand, it might be desirable to transfer into the cytosol an enzyme which otherwise is not able to cross the cell membrane. In this case, the receptor and translocation domains of the toxins are fused to the enzyme. This type of fusion toxin is exemplified by the above-mentioned constructs of C3 with diphtheria toxin or with C2 toxin.

Also, the light chain of botulinum neurotoxin has been introduced into cells by means of an anthrax toxin fusion protein.

Clostridial Neurotoxins as Tools To Study Exocytosis

Tetanus toxin and botulinum toxins consist of a heavy chain (~100 kDa), which is responsible for binding and membrane translocation, and a light chain (~50 kDa), which is biologically active and harbors the enzyme activity. The chains are linked by a disulfide bond. The clostridial neurotoxins are metalloendoproteases. They selectively cleave proteins (synaptobrevin [VAMP] syntaxin, and SNAP-25) which are involved in exocytosis of synaptic vesicles, thereby blocking neurotransmitter release from presynaptic nerve endings. Whereas tetanus toxin occurs in a single serotype, at least seven different serotypes of botulinum neurotoxins (types A, B, C1, D, E, F, and G) have been identified. The toxins differ by their protein substrate specificity; e.g., synaptobrevin is cleaved by tetanus toxin and botulinum toxins B, D, F, and G; SNAP25 is cleaved by botulinum toxins A, C, and E; and syntaxin is cleaved by botulinum toxin C. Moreover, most toxins cleave the synaptic proteins at different sites. Although tetanus toxin and botulinum neurotoxins act on the molecular level in a very similar manner, their actions differ on the anatomic level, a fact that causes completely different symptoms of intoxication. Whereas botulinum neurotoxins act on peripheral nerves and cause flaccid paralysis, tetanus toxin acts on the central nervous system and induces spastic paralysis (tetanus).

The delineation of the molecular mechanisms and the identification of the targets of the clostridial neurotoxins have promoted the use of these

---

BOX 12.2

Toxins as therapeutic agents

Bacterial protein toxins are used to produce toxoids for vaccination. A most effective vaccination is that by tetanus toxoid, which might have prevented the deaths of hundreds of thousands of people. For other diseases, complex toxoid preparations were used, often containing several components (e.g., anti-pertussis vaccination), some of which might be responsible for the side effects of vaccination. Therefore, recombinant mutant toxins (e.g., PT) which were inactivated by site-directed mutagenesis were clinically tested, with promising results.

For several years, botulinum neurotoxins, which are the most potent toxins known (in mice, the 50% lethal dose is 100 pg/kg), have been approved as therapeutic agents in a variety of ophthalmological and neurologic disorders. The first published report concerned the use of these toxins in strabismus, but their usage has been extended to many diseases including blepharospasm, hemifacial spasm, and several types of dystonias like spasmodic dysphonia and cervical dystonia. Other potential applications are certain types of tremor, urinary retention, and anismus.

Toxin fusion proteins exploit the potent cytotoxic effects of bacterial protein toxins in efforts to kill cancer cells. Therefore, the enzymatic portion of a toxin and the toxin domain that causes membrane translocation are fused to a receptor-binding protein which specifically binds to the target cell. When antibodies directed against a specific cell surface receptor are used as the receptor-binding domain of the fusion toxin, these toxins are called immunotoxins. The enzyme and translocation domain of diphtheria toxin and *Pseudomonas* exotoxin A are most often used in these fusion toxins. Problems with these immunotoxins are large size, instability, inadequate biological specificity, insufficient penetration into solid tumors, and antigenicity of the constructs. Another promising but still experimental approach is the use of these fusion toxins to purge stem cell preparations of cancer cells ex vivo before autologous transplantation.

extremely potent agents as cell biological tools to study the involvement of their target proteins in exocytosis from neuronal cells or synaptosomes. Moreover, in combination with permeabilization methods (e.g., usage of pore-forming toxins), clostridial neurotoxins are now also used in studies on signal secretion coupling of nonneuronal cells (e.g., insulin-secreting cells). These cells are otherwise insensitive toward neurotoxins because they lack the specific membrane receptor. Notably, botulinum neurotoxins, which are not only the most potent toxins but also the most potent biologically active substances ever identified, are now used as therapeutic agents (Box 12.2).

Toxins as Tools for Permeabilization of Eukaryotic Cells

Pore-forming toxins are widely used as biological tools to permeabilize eukaryotic cells. The aim of this approach is to manipulate the intracellular ionic milieu, to introduce small molecules which are otherwise membrane impermeable (e.g., nucleotides) into the cytosol, or even to transfer peptides and proteins such as antibodies or toxin fragments (e.g., the light chain of tetanus toxin) into the cell. Most often used are *S. aureus* alpha-toxin and streptolysin O from *Streptococcus pyogenes*. However, several other pore-forming toxins have been described (e.g., the RTX family, with *E. coli* hemolysin being one of the best-studied pore-forming toxins), which are of potential importance as cell biological tools.

S. aureus alpha-toxin is a hydrophilic polypeptide of 293 amino acids, whose crystal structure was analyzed recently. The model derived from the crystallographic data shows a mushroom-shaped homo-oligomeric heptamer with a transmembrane domain composed of a 14-strand antiparallel β-barrel with a pore diameter of 2.6 nm. The pore formation caused by alpha-toxin is limited to the cell membrane because the toxin monomers (~33 kDa) are not able to pass through the pore.

Streptolysin O (~60 kDa) interacts with membrane cholesterol, oligomerizes and forms much larger pores (up to 30 nm), which are permeable for large molecules (>150 kDa). Therefore, the toxin can also interact with internal membranes. To avoid damage to intracellular membranes, the binding of and pore formation by streptolysin O can be dissociated by a temperature shift. Thus, binding is carried out at low temperature (0°C). Thereafter, surplus streptolysin O is removed by washing and the permeabilization is initiated by increasing the incubation temperature to 37°C. Pore-forming toxins have been widely used in studies on exocytosis from various secretory cells, on calcium regulation, on signal-contraction coupling in smooth muscle cells, and on intracellular membrane trafficking. Researchers using pore-forming toxins must consider that these toxins induce a wide spectrum of biological effects, many of which are explained by changes in cellular ion fluxes and appear to be triggered by monovalent ion fluxes and by Ca^{2+} influx (e.g., secretion, generation of lipid mediators, and cytoskeletal redistribution). However, some of the effects of pore-forming toxins cannot be explained in this way. Examples of the latter effects are activation of G-protein signaling and proteolytic shedding of membrane proteins. All these effects must be considered when interpreting data obtained with pore-forming toxins as tools.

Finally, it should be mentioned that pore-forming toxins play an immense commercial role as pesticides against insects. These members of a large family of pore-forming toxins (Cry families) are produced by *Bacillus thuringiensis*. The various related toxins specifically kill insect larvae but are apparently not harmful to mammals. Therefore, spraying plants with spores of *B. thuringiensis* seems to be environmentally safe. Importantly, the various toxins are specific for certain insects only. Recently, the genes for toxins were introduced into some crop plants in an effort to protect them from insect attack.

Selected Readings

Cassel, D., and T. Pfeuffer. 1978. Mechanism of cholera toxin action: covalent modification of the guanyl nucleotide-binding protein of the adenylate cyclase system. *Proc. Natl. Acad. Sci. USA* **75:**2669–2673.

Chardin, P., P. Boquet, P. Madaule, M. R. Popoff, E. J. Rubin, and D. M. Gill. 1989. The mammalian G protein rho C is ADP-ribosylated by *Clostridium botulinum* exoenzyme C3 and affects actin microfilament in Vero cells. *EMBO J.* **8:**1087–1092.

Just, I., J. Selzer, M. Wilm, C. Von Eichel-Streiber, M. Mann, and K. Aktories. 1995. Glucosylation of Rho proteins by *Clostridium difficile* toxin B. *Nature* **375:**500–503.

Ridley, A. J., and A. Hall. 1992. The small GTP-binding protein rho regulates the assembly of focal adhesions and actin stress fibers in response to growth factors. *Cell* **70:**389–399.

Schiavo, G., F. Benfenati, B. Poulain, O. Rossetto, P. Polverino de Laureto, B. R. DasGupta, and C. Montecucco. 1992. Tetanus and botulinum-B neurotoxins block neurotransmitter release by proteolytic cleavage of synaptobrevin. *Nature* **359:**832–835.

13

Type III Secretion Systems in Animal- and Plant-Interacting Bacteria

KURT SCHESSER, MATTHEW S. FRANCIS, ÅKE FORSBERG,
AND HANS WOLF-WATZ

A multitude of activities performed by bacteria, as well as eukaryotic cells, are dependent upon the transport of proteins through membranes. In bacteria, these activities include the synthesis of extracellular organelles such as flagella, the degradation of nutrients too large to ingest directly, and the delivery of toxins directed against eukaryotic cells to their site of action. Transporting a protein through a lipid bilayer is not a trivial task, considering the relative physical dimensions of proteins compared to small solutes such as water and sodium ions. This task is further complicated by the fact that most proteins destined to be transported across a membrane must be prevented from assuming their three-dimensional structure prior to transport. Although many of the details of the protein transport process are now known, several key fundamental questions remain. This chapter discusses the type III secretion system possessed by several gram-negative bacteria that live, for at least part of their life cycle, in close association with eukaryotic cells. What makes type III secretion unique is that it apparently functions to translocate bacterially encoded proteins not just across the bacterial membranes but also across the eukaryotic membrane.

Gram-negative bacteria possess several protein secretion systems (Figure 13.1). The best-studied systems at the mechanistic level are the *sec*-dependent type II and V secretion systems, which most probably represent evolutionarily ancient systems since homologues of the components of these secretion machines also exist in gram-positive bacteria, the archaebacteria, and the eukaryotes. Proteins secreted by these systems possess N-terminal signal sequences of 16 to 26 residues (the signal sequences in gram-positive bacteria can be a little longer) that consist of a basic N-terminal domain, a hydrophobic central core segment, and a distal domain that contains a cleavage site in which the signal sequence is removed during the transport across the inner membrane. The basic and hydrophobic domains are essential for the recognition event between the protein to be transported and components of the secretion complex located on the cytoplasmic face of the inner membrane. Following this initial recognition event, the protein destined to be secreted is transported across the inner membrane by a

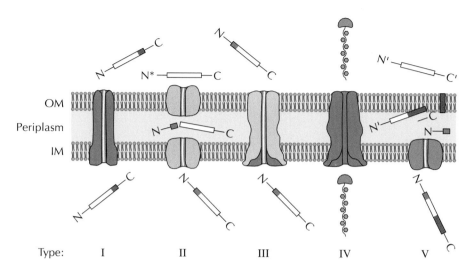

OM

Periplasm

IM

Type: I   II   III   IV   V

Figure 13.1 Schematic diagram of bacterial type I to V secretion systems. Recognition sequences that target the proteins to the respective secretion complexes are indicated either in red (for N-terminal signal sequences) or grey (for C-terminal signal sequences). Proteins secreted by the type I and III pathways traverse the inner membrane (IM) and outer membrane (OM) in one step, whereas proteins secreted by the type II and V pathways cross the inner membrane and outer membrane in separate steps. The N-terminal signal sequences of proteins secreted by the type II and V systems are enzymatically removed upon crossing the inner membrane, in contrast to proteins secreted by the type I and III systems, which are exported intact. Although proteins secreted by the type II and V systems are similar in the mechanism by which they cross the inner membrane, differences exist in how they traverse the outer membrane. Proteins secreted by the type II system are transported across the outer membrane by a multiprotein complex, whereas those secreted by the type V system autotransport across the outer membrane by virtue of a C-terminal sequence which is enzymatically removed upon release of the protein from the outer membrane. Type IV pathways secrete either polypeptide toxins (directed against eukaryotic cells) or protein-DNA complexes between either two bacterial cells or between a bacterial and eukaryotic cell. Shown in the figure is the protein-DNA complex delivered by *Agrobacterium tumefaciens* into a plant cell that consists of a 20-kb single-stranded DNA molecule and the VirD2 (grey) and VirE2 (light red) proteins.

mechanism that has been extensively studied and is fairly well understood. After transport across the inner membrane, a protein may either remain in the periplasm, integrate in the outer membrane, or continue its journey across the outer membrane. Proteins secreted by the type II and V pathways differ from each other in how they traverse the outer membrane. Proteins secreted by the type V secretion system, which include the immunoglobulin A proteases of *Neisseria gonorrhoeae* and the *Helicobacter pylori* vacuolating cytotoxin, mediate their own secretion ("autotransport") across the outer membrane. In contrast, proteins secreted by the type II secretion system, such as the extracellular degradative enzyme pullulanase of *Klebsiella oxytoca*, require additional proteins for their secretion across the outer membrane.

Proteins secreted by the type I secretion system are mainly toxins such as hemolysin of *Escherichia coli* and cyclolysin produced by *Bordetella pertussis*. Unlike proteins secreted by the type II secretion system, these proteins contain no signal sequence at their N termini but instead contain domains at their C termini that are necessary for recognition by the type I secretion complex. This complex is relatively simple, consisting of only three proteins: an ATPase that provides the energy for transport as well as forms a pore through the inner membrane, a protein that spans the periplasmic space, and a protein located in the outer membrane. Both the nature of the type I secretion signal and the transport mechanism are currently unclear. Furthermore, it is unknown why gram-negative bacteria evolved a dedicated system for the secretion of these toxins.

Type IV secretion systems in gram-negative bacteria are specialized for the cell-to-cell transfer of a variety of biomolecules (see chapter 14 for details). This secretion system is used for the transfer of protein-DNA complexes between either two bacterial cells (e.g., during bacterial conjugation) or between a bacterial and eukaryotic cell (e.g., *Agrobacterium tumefaciens*-mediated DNA transfer into plant cells). Additionally, *B. pertussis* and *H. pylori* possess type IV secretion systems which mediate the secretion of pertussis toxin and interleukin-8-inducing factor, respectively. Approximately 10 *A. tumefaciens* proteins are required to transfer a single-stranded 20-kb DNA molecule (referred to as the T-complex) into plant cells. Based on several different types of experiments, these proteins are thought to form a transmembrane pore spanning the bacterial inner and outer membranes (and perhaps the plant membrane as well). The structural basis of the type IV secretion machine that allows it to secrete polynucleotide as well as polypeptide molecules is not known.

Several gram-negative bacteria that interact with either plant or animal cells possess a secretion system, referred to as the type III secretion system, that functions to deliver proteins directly from the bacterial cell into the cytosol of the eukaryotic cell. This secretion system was first described and characterized in *Yersinia* spp. that are pathogenic for humans. In the mid-1950s it was noted that the ability of *Y. pestis*, the causative agent of bubonic plague, to grow at 37°C in media lacking calcium correlated with a loss of virulence. Subsequently it was found that virulence was dependent on the presence of a 70.5-kb extrachromosomal plasmid (pCD1). This plasmid in turn was shown to direct the massive secretion of around 10 proteins into the culture supernatant by bacteria incubated under the "nongrowing" conditions (i.e., 37°C in media lacking calcium). Genetic analysis of the 70.5-kb virulence plasmid revealed that it encoded both the proteins that were secreted and approximately 20 proteins that comprised the secretion complex (Figure 13.2). For a long time the function of these secreted proteins was unknown (although almost all of them were shown to be required for virulence in the mouse model system), since in soluble form they had no apparent effects on eukaryotic cells. Recently, it has been demonstrated that these virulence proteins are in fact delivered by bacteria directly into the eukaryotic cytoplasm, from where they exert their effects. Since these proteins pass directly from the bacterium into the eukaryotic cell, the "secretion" of these proteins by bacteria grown in culture is most probably artifactual.

It also turned out that type III secretion systems were involved in another phenomenon that was first described in the middle part of this century: the gene-for-gene hypothesis underlying the plant hypersensitivity response toward bacterial pathogens. A major breakthrough in the study of host-parasite interactions came with the discovery that several gram-negative bacteria that infect plants possess secretion systems that are clearly evolutionarily related to the type III secretion systems possessed by bacteria that infect animals. This key finding immediately suggested that plant- and animal-infecting bacteria have common strategies at the cellular and molecular levels in how they deal with eukaryotic cells. The wealth of knowl-

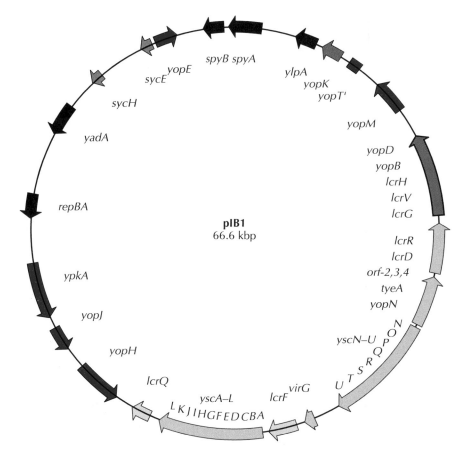

Figure 13.2 The type III secretion system of *Y. pseudotuberculosis* is encoded on the virulence plasmid pIB1. The genes encoding components of the type III system and those involved in plasmid replication (*repAB*) and plasmid partitioning (*spyAB*) are depicted in the map as arrows showing the direction of transcription. A nonfunctional, truncated gene has no arrowhead and is indicated by an apostrophe to denote the point of truncation. Genes in light grey (*yscA* to *yscU*, *virG*, *lcrDR*, *yopN*, and *tyeA*) either encode components of the secretion apparatus or those involved in controlling protein secretion. The genes shown in dark grey (*lcrGVH yopBD* and *yopK*) encode proteins involved in controlling translocation of effectors into target cells. The genes encoding the secreted effector proteins (*yopE*, *yopH*, *yopJ*, *yopM*, *yopT'*, and *ypkA*) are shown in red. The genes encoding the chaperones required for efficient secretion of effector Yops (*sycE* and *sycH*) are shown in pink, and important regulatory genes are depicted in light red. Note that some proteins may have dual functions. For example, several proteins encoded by the *lcrGVH yopBD* operon are involved in both controlling translocation and regulating the system. Moreover, TyeA is involved in control of Yop secretion but is also required for translocation of the YopE and YopH effector molecules. The DNA sequence of pIB1 on which this map is based is unpublished data provided by Peter Cherepanov and Thomas Svensson, Department of Microbiology, FOA NBC-Defence, Umeå, Sweden.

edge derived from research on plant-bacterium systems has served to, and undoubtedly will continue to, guide certain aspects of animal-bacterium research.

Type III secretion systems have been found in several genera of bacterial animal and plant pathogens (Table 13.1). Interestingly, it has recently become apparent that nonpathogenic plant-infecting bacteria utilize type III secretion systems as well. The nitrogen-fixing plant symbiont *Rhizobium*

Table 13.1 Type III secretion systems of animal- and plant-interacting bacteria

Organism	Location	Type III system	Typical substrates	Function in virulence
Animal-interacting bacteria				
Yersinia spp.	Virulence plasmid	Ysc (*Yersinia* secretion)	Yop (*Yersinia* outer proteins)	Blocks phagocytosis and induction of cytokine expression, induces apoptosis
Pseudomonas aeruginosa	Chromosome	Psc (*Pseudomonas* secretion)	Exo (exoenzyme toxins) Pop (*Pseudomonas* outer proteins)	Blocks phagocytosis, induces cytotoxicity
Shigella flexneri	Virulence plasmid	spa/mxi (surface presentation of Ipa-antigens/membrane expression of Ipa)	Ipa (invasion protein antigens)	Induces uptake into eukaryotic cells, apoptosis, and vacuole membrane lysis
Salmonella typhimurium, Salmonella dublin	Chromosome (SPI-1)	inv/spa (invasion/see *Shigella*)	Sip (secreted invasion proteins), Sop (*Salmonella* outer proteins)	Induces uptake into epithelial cells, apoptosis, cytokine expression, and fluid secretion in the intestine
Salmonella typhimurium	Chromosome (SPI-2)	ssa (secretion system apparatus)	?	Essential for virulence
EPEC	Chromosome (LEE)	sep (secretion of *E coli* proteins)	Esp (*E. coli* secreted protein) Tir (transmembrane intimin receptor)	Induces tight adherence between the bacterium and its target cell
Plant-interacting bacteria				
Pathogens				
Erwinia amylovora, Ralstonia solanacearum, Pseudomonas syringae, Xanthomonas campestris	Chromosome or plasmid[a]	Hrc (Hrp conserved)[b]	Harpin, Avr (avirulence proteins)	Virulence-associated proteins; elicit hypersensitive response in resistant plants
Symbionts				
Rhizobium spp.	Plasmid	Hrc (?)	?	Induces nodule formation in legumes

[a]*Ralstonia solanacearum.*
[b]Hrp, hypersensitive reaction and pathogenicity.

uses a type III secretion system to secrete proteins involved in the formation of root nodules (where nitrogen fixation occurs). Remarkably, it appears that in a few cases there are similarities between the proteins secreted by the type III secretion systems of plant- and animal-infecting bacteria. Perhaps it should not come as a complete surprise that a bacterial plant symbiont would possess a type III secretion system, since in many respects the way in which *Rhizobium* induces legumes to form nodules resembles the infection process observed in host-pathogen systems. Future research may even reveal that members of the normal animal intestinal microflora may contain a type III-like secretion system since these bacteria, much like the pathogenic species discussed above, live in close association with eukaryotic cells.

Among the bacterial species possessing type III secretion systems, there is a high degree of similarity between the various type III secretion systems both in terms of genetic organization and at the level of the individual genes. The genes encoding the type III secretion apparatus are clustered in blocks that largely display a conserved genetic order, suggesting that these genes have spread in toto by horizontal gene transfer. This hypothesis is further strengthened by the fact that in many cases these gene clusters either are contained on potentially mobilizable plasmids (for example, see Figure 13.2) or, for chromosomally located gene clusters, are flanked by insertion elements (the latter are usually referred to as pathogenicity islands). Among the individual genes themselves, which encode the type III secretion apparatus, some are found in virtually all species while others are more restricted in their distribution. Thus, type III secretion systems from different species can be grouped into subfamilies based on sequence similarities and genetic organization. Although caution should be exercised in drawing conclusions strictly from DNA sequence comparisons, it does appear that if type III-encoding genes have indeed been acquired as intact genetic clusters, certain individual genes within these clusters either have been lost completely or have diverged to a point where they are no longer clearly related to genes found in other species.

Divergence of type III gene sequences between species could be due, at least in part, to different selection pressures that are imposed on the bacteria during their coevolution with their respective hosts. In this scenario, a particular bacterial species would first acquire a type III-like secretion system and then modify that system, through natural selection, to fit its specific needs. These needs would be shaped by numerous factors such as the host defense system and the strategies used by the bacterium to grow at the expense of the host organism. The intestinal pathogen *Salmonella* provides an illustrative example of the adaptation of type III gene clusters. This pathogen possesses two entirely separate type III secretion systems, encoded by different gene clusters, which, based on genetic evidence, apparently were acquired at different times during evolution. These two systems are required for different stages of the *Salmonella* infection process. One system secretes proteins involved in the invasion of eukaryotic cells, while the other system is required for the bacterium to survive once it is inside the eukaryotic cell. It is currently unclear why *Salmonella* requires two independent type III secretion pathways.

Unlike the clustered blocks of genes which encode the type III secretion machine, the proteins translocated into the eukaryotic cell by this type III secretion machine (commonly referred to as effector proteins) are genetically unlinked. This may indicate that at least some of the genes encoding the effector proteins could have been acquired independently of the genes encoding the secretion apparatus. In fact, some of the genes encoding these effector proteins may have been acquired (or "captured") from eukaryotes (Box 13.1). It is now becoming clear that several of the proteins secreted by this system inactivate or at least retard the innate immune response of animal cells and trigger hypersensitivity responses in plant cells following their delivery into eukaryotic cells. Apparently, the injected proteins target key processes within eukaryotic cells that are normally activated upon exposure to gram-negative bacteria. The working hypothesis is that the effector proteins have evolved to exploit chinks in the eukaryotic host defense armor. A substantial amount of research is currently focused on identifying the eukaryotic processes that are targeted by the type III effector proteins. By determining how these bacterial proteins function within eukaryotic cells, much will undoubtedly be learned about the animal and plant defense systems.

Type III Secretion Systems

Type III secretion systems have a number of features in common with the flagellum-specific secretion pathway. Since flagellar motility is an ancient system that probably existed before the divergence of archaebacteria and prokaryotes, it is likely that type III secretion systems evolved from the flagellum-specific secretion pathway as a means of secreting and translocating antihost factors into host cells. Since flagellar motility was probably present in most free-living bacteria, this could explain why the type III secretion systems are so widespread among gram-negative animal and plant pathogens. The secretion and assembly of flagellar components have been reasonably well characterized. The flagellum-specific secretion pathway mediates the secretion of all proteins that assemble into a continuous axial structure, from proximal rod to hook to filament and, finally, to the filamentous cap. A feature common to both systems is that protein secretion appears to occur by a continuous process without any detectable peri-

---

BOX 13.1

Eukaryotic motifs found in type III secreted proteins

Several bacterial proteins secreted by type III systems contain domains that are common in eukaryotic proteins. Examples include Ser/Thr protein kinase domains, leucine-rich repeats, and tyrosine phosphorylation sites. It is believed that animal- and plant-infecting bacteria have either acquired ("captured") or independently evolved proteins harboring these eukaryotic-like structural or enzymatic motifs for the purpose of using them against eukaryotic cells. Although several type III proteins have been physically demonstrated to be delivered into eukaryotic cells (see the text), the idea that these proteins are active within eukaryotic cells is strengthened by the fact that in a few cases the proteins possess eukaryotic-like motifs that have no clear function within the bacterial cell. For example, the type III protein AvrBs3 of the plant pathogen *Xanthomonas campestris* contains a nuclear localization sequence that is required for the recognition of this protein by the plant, which presumably occurs within the plant nucleus. The *Yersinia* YopH and the *Salmonella* SptP proteins are broad-spectrum protein tyrosine phosphotases (PTPases) whose catalytic domains contain invariant residues present in all eukaryotic PTPases. *Yersinia* mutant strains containing a catalytically inactive YopH-encoding gene are no longer virulent in the mouse model, indicating that tyrosine dephosphorylation of host proteins plays an important role in the virulence of *Yersinia*. Like the nuclear localization sequence of AvrBs3, the activity of YopH (and presumably SptP) is most probably restricted to eukaryotic cells since tyrosine phosphorylation is not thought to occur in bacteria.

plasmic protein intermediates. Flagellar components are assembled at the distal end of the nascent structure, with evidence suggesting that assembly is mediated by protein traffic through a channel within the growing structure. A central channel is observed in the filament and most probably also in the rod. In addition, neither the flagellar proteins secreted during biosynthesis nor the virulence proteins secreted by a type III-dependent mechanism are processed during secretion. In both systems, the recognition sequence targeting proteins for secretion resides in the N terminus of the individual proteins, but no motif common to them has been identified (for details see below).

The proteins required for flagellar secretion and assembly localize mainly to the cytoplasmic membrane or remain within the cytoplasm. This suggests a mechanism where the secretion apparatus identifies and targets the flagellar proteins to the secretion channel in the proximal rod and provides the required energy to transport the proteins out to the bacterial surface. Since homologues to several proteins which constitute the flagellum-specific secretion pathway are present in all type III secretion systems so far identified, it is considered that secretion of virulence effectors by type III pathways occurs by an analogous mechanism. Additional proteins unique to subfamilies of type III secretion systems are also known. These proteins may collectively participate to form a rod-like structure extending through the periplasm to the pore in the outer membrane. This would allow continuous secretion of the virulence proteins to the exterior, avoiding the formation of periplasmic protein intermediates. Thus, the role of type III secretion systems, or at least the part shared with the flagellum-specific secretion apparatus, is complex. Proteins destined for secretion are identified and then targeted to the secretion channel to permit secretion. The energy required to drive the channeling of proteins to the exterior is also provided by the secretion apparatus. In addition, the channel is gated to prevent accidental leakage through this structure.

Secretion Apparatus

Flagella are extracellular appendages that mediate bacterial motility. A flagellum is composed of three structural components: a thin helical filament of variable length, a hook, and a basal body. The basal body is embedded in the bacterial envelope and serves to anchor the flagellum filament to the bacterial surface via the hook protein linkage (a probable universal joint). The basal body consists of a central continuous rod surrounded by several ring structures (Figure 13.3). The L ring (for lipopolysaccharide) resides in the outer membrane, and the P ring (for peptidoglycan) is embedded in the peptidoglycan layer. The MS ring (for membrane-supramembrane) is situated in the inner membrane, and the bell-shaped C ring (for cytoplasm), harboring the motor/switch proteins (FliG, FliM, and FliN), is associated with the MS ring in the inner membrane but projects into the cytoplasm. Another component of the flagellar biosynthetic machinery consists of a family of proteins located in either the inner membrane or the cytoplasm, which collectively function as the flagellum-specific secretion apparatus responsible for the sequential secretion of proteins forming the rod, hook, and filaments.

Interestingly, structures that contained proteins known to be part of the SPI-1 (for *Salmonella* pathogenicity island 1)-encoded type III secretion complex of *Salmonella* were isolated and revealed features resembling the flagellar basal body (Figure 13.3). These structures were observed only on

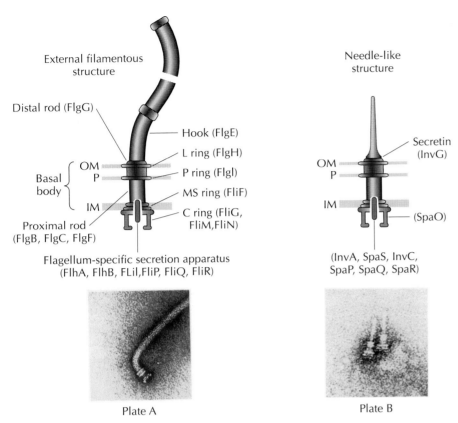

External filamentous
structure

Needle-like
structure

Distal rod (FlgG)

Hook (FlgE)

L ring (FlgH)

OM
P

Basal
body

P ring (FlgI)

MS ring (FliF)

IM

C ring (FliG,
FliM,FliN)

Proximal rod
(FlgB, FlgC, FlgF)

Flagellum-specific secretion apparatus
(FlhA, FlhB, FLiI,FliP, FliQ, FliR)

Secretin
(InvG)

OM
P

IM

(SpaO)

(InvA, SpaS, InvC,
SpaP, SpaQ, SpaR)

Plate A

Plate B

Figure 13.3 Structural comparison between the enteric bacterial flagellum and the *Salmonella* type III secretion system. Three major components of the flagellum structure are shown: external filamentous structure, basal body, and flagellum-specific secretion apparatus. The structure of the type III secretion system spanning the bacterial envelope is represented like the basal body, as deduced from electron microscopic and scanning electron microscopic studies (see the micrograph inset of a purified flagellum [plate A] and the *Salmonella* SPI-1 type III apparatus [plate B]). Nevertheless, these structures are shown in grey to indicate limited sequence similarity among the corresponding protein components. Conversely, proteins of the type III apparatus that have extensive similarities to components of the flagellum-specific secretion apparatus, located within the bell-shaped C ring (proteins indicated in parentheses; see Table 13.2), are shown in red. OM, outer membrane; P, peptidoglycan; IM, inner membrane.

bacteria deficient in flagellum production but expressing a functional type III secretion system. In particular, the part embedded in the bacterial envelope appeared physically similar to the ring structures in the inner and outer membranes of the basal body. The proteins constituting the MS and C rings in the basal body have been identified, but these display only limited similarity to particular proteins of the type III secretion systems of pathogenic bacteria. Nevertheless, in view of the basal body-like structures of *Salmonella*, it is likely that proteins specific to type III secretion systems form similar ring structures in the bacterial inner membrane.

A significant feature of the *Salmonella* type III secretion complex is that instead of the extracellular flagellar appendage, a hollow needle-like structure, 80 nm long and 13 nm wide, extends from the bacterial surface. This needle-like structure may function to bridge the gap between the bacterium and the host cell to establish translocation of virulence proteins across the host cell plasma membrane. In support of this, a 7- to 9-nm-diameter, EspA-containing filamentous surface organelle of enteropathogenic *Escherichia coli* (EPEC) physically connects bacteria to infected eukaryotic cells. This structure, encoded by the EPEC type III secretion system located on the pathogenicity island designated LEE (for "locus of enterocyte effacement"), is required for translocation of effector proteins into the target cell. By analogy, there is a substantial amount of indirect evidence suggesting that proteins secreted by the type III secretion machinery of plant-interacting bacteria, like bacterium-animal cell systems, are injected directly into the plant cell cytoplasm. Thus, it is not surprising that the Hrp (for "hypersensitive reaction and pathogenicity") type III secretion system of the plant pathogen *Pseudomonas syringae* also produces a novel pilus-like surface appendage, termed Hrp pilus. The Hrp pilus has a diameter of 8 nm and is at least 2 μm long. Therefore, this structure may have the capacity to penetrate plant cells through pores in the usually impermeable polysaccharide cell wall that surrounds individual cells. This would enable the direct injection of Avr (for "avirulence") (Box 13.2) proteins into the plant cytosol. However, injection of Avr proteins into plant cells through such structures remains to be shown.

BOX 13.2

Type III secretion system and plant defense responses

Like animals, plants are under constant attack by microbial pathogens and have an elaborate defense system to prevent viral, bacterial, and fungal pathogens from establishing a systemic infection. Two possible outcomes can occur when a bacterial pathogen enters a plant through a stomata or wound. In a so-called "susceptible" plant host, the bacterial pathogen is able to colonize the plant and cause disease. If, on the other hand, the bacterium is unfortunate enough to find itself within a "resistant" plant host, it is rapidly recognized and killed by any one of a number of plant defense responses. These responses include the hypersensitivity response, which entails a programmed cell death localized to the area of the plant surrounding the initial site of bacterial penetration, that serves to prevent the bacterium from spreading and establishing a systemic infection. (Hypersensitivity responses are commonly

observed as brown spots on the leaves of houseplants.)

What is the basis for the recognition of the bacterium by the plant, and why is the recognition often restricted to specific plant species or lines? For 50 years it has been known that the recognition event requires the expression of complementary plant and bacterial genes designated the resistance (*R*) and avirulence (*avr*) genes, respectively. This phenomenon is commonly referred to as the "gene-for-gene" hypothesis. If either the plant possesses a mutated *R* gene (i.e., an *r* allele) or the bacterium lacks the corresponding *avr* gene, the plant does not recognize the bacterium, and consequently the bacterium is able to spread and cause disease. Thus, the term "avirulence" refers to the fact that if the *avr* gene (or a specific allele) expressed by the bacterium is matched with its plant *R* gene counterpart, the bacterium is rendered avirulent due to activation of the plant defense responses.

What type of proteins do the bacterial *avr* genes encode? These genes en-

code either components of the type III secretion system or proteins that are believed to be secreted by the type III system. Several indirect lines of evidence suggest that bacterial plant pathogens, analogous to bacterial animal pathogens, inject Avr proteins directly into plant cells, where they physically interact with plant R proteins, triggering (in a resistant plant) the plant defense responses. The roles that Avr proteins play in the pathogenicity of susceptible plants are not known, although some type of important function is envisioned, since otherwise there would apparently be a strong selection pressure for bacterial pathogens to lose these genes. Since plant- and animal-infecting bacteria have similar type III secretion systems and provoke similar type III-mediated responses of eukaryotic cells (programmed cell death), it is possible that animal cells possess a mechanism that is similar to the plant recognition system triggered (in a resistant plant) by bacterial Avr proteins.

While there is no significant similarity among proteins comprising the flagellar basal body and type III secretion systems, seven proteins that constitute the flagellum-specific secretion apparatus show extensive similarity to particular proteins of type III secretion systems. The properties of these proteins are summarized in Table 13.2. These flagellum-specific secretion proteins localize in, or are associated with, the inner membrane and appear to be linked to the basal-body ring structures, possibly localizing to the central pore inside the MS ring. As suggested in Figure 13.3, it is very likely that a similar structure is formed by protein components of the type III apparatus specific for secretion of virulence proteins. Additionally, based on the observation of continuous secretion (no periplasmic protein intermediates) by type III complexes, it has been postulated that a flagellar rod-like structure is associated with the ring structures in the inner membrane and extends through the periplasm to the outer membrane. However, no proteins with significant sequence similarity to the flagellar rod proteins FlgB, FlgC, FlgF, and FlgG have been observed for any component of type III secretion systems. Nevertheless, as discussed above, it is possible that some of the proteins unique to type III secretion systems are involved in forming a structure that connects the postulated rings in the inner and outer membranes.

Although most proteins of type III secretion systems appear to be located in the bacterial inner membrane or cytoplasm, a subfamily of proteins including YscC and MxiD, from *Yersinia* and *Shigella*, respectively, are clearly located in the outer membrane and are required for type III secretion. These proteins display extensive similarity to a family of outer membrane proteins denoted secretins. Secretins mediate the transport across the outer membrane of various large molecules, including extracellular enzymes via the type II secretion system (PulD of *Klebsiella* spp.), filamentous bacteriophages (the pIV family), and type IV pili (PilQ of *Pseudomonas aeruginosa*). Interestingly, a common feature among this diverse protein family is that they contain an N-terminal signal sequence allowing their transport to the outer membrane via a *sec*-dependent mechanism. In addition to the *sec* system, the PulS lipoprotein is required to prevent periplasmic degra-

Table 13.2 Proteins common to the flagellum-specific secretion apparatus and the *Salmonella* type III secretion system[a]

SPI-1	Flagellum	Location	Function and structure
InvC	FliI	Cytoplasm	F_0F_1 proton-translocating ATPase; energizer of flagellar secretion and assembly
SpaO	FliN/Y[b]	Cytoplasmic face of the inner membrane	Probable component of the C ring
SpaP	FliP	Inner membrane	Secretion apparatus
SpaQ	FliQ	Inner membrane	Secretion apparatus
SpaR	FliR	Inner membrane	Secretion apparatus
SpaS	FlhB	Inner membrane	Secretion apparatus; suppression of hook multimer formation
InvA	FlhA	Inner membrane	Secretion apparatus; channel formation?

[a] While components of the SPI-1 type III secretion system of *Salmonella* are illustrated, all type III secretion systems encoded by animal- and plant-infecting bacteria contain proteins with extensive similarity to these components of the flagellum-specific secretion apparatus.
[b] FliN (*Escherichia coli, Salmonella typhimurium*); FliY (*Bacillus subtilis*).

dation of PulD and to ensure correct outer membrane positioning of this enzyme. While no proteins specific to type III secretion systems appear similar to the PulS chaperone, the VirG lipoprotein from *Yersinia* performs an analogous function to PulS, being responsible for localizing YscC to the outer membrane. Significantly, the YscC protein family is not related to the proteins that form the L and P rings in the outer membrane of the flagellar basal body. This is not surprising considering that different proteins are assembled on the L ring of the flagellar basal body compared to the secretin-like proteins in the outer membrane of type III secretion systems. Secretin multimers from several systems form pore-like structures with external diameters of around 200 Å and internal diameters of between 50 and 80 Å, sufficient for passage of large proteins and bacteriophages through the outer membrane. A similar-sized pore in the outer membrane of *Yersinia* is formed by the multimerization of the YscC protein. It is believed that the outer-membrane pore structure formed by type III secretion systems permits the channeling of virulence effectors through the outer membrane to the exterior. Since this pore contains a large aqueous channel, it would be expected that a protein(s) would be required to gate this channel, preventing nonspecific passage of components in or out of the cell. Like the YscC protein family, another constituent of the type III secretion complex, including the YscJ protein of *Yersinia*, is processed and transported across the inner membrane by a *sec*-dependent mechanism. While the YscJ family contains N- and C-terminal features that suggest that these proteins could span the entire bacterial envelope, no specific function has been elucidated for this family of proteins.

The emerging picture for the mechanism of type III secretion is that proteins destined for secretion are identified and targeted to the secretion apparatus at the inner membrane via a mechanism similar to that for flagellar proteins (Figure 13.3). This recognition signal could well be a stem-loop structure in the mRNA encoding the secreted proteins (see below for details). The additional proteins of the secretion apparatus may be required to gate pores present in both the inner and outer membranes connected by a channel extending through the periplasm, as well as to provide the energy required to channel the proteins to the exterior.

Secreted Proteins

Type III secretion is a specific mechanism used by numerous bacteria to establish an infection and/or symbiotic relationship with eukaryotic host cells by mediating the directed translocation of effector proteins into the host cell cytosol. The secretion signal of proteins targeted either by the type III mechanism or by the flagellum-specific secretion apparatus resides in the N-terminal coding region of the respective genes. For example, the first 11 amino acids of the YopE cytotoxin from pathogenic *Yersinia* spp. were sufficient to ensure secretion. Interestingly, no common sequence motif or structure has been identified for the secreted proteins, and when the N-terminal coding region of the secreted YopE and YopN proteins of *Yersinia* was systematically mutagenized, no mutants with abolished secretion were observed. However, the predicted structures of the mRNA of the secretion signal for YopE and YopN both contained stem-loop structures incorporating the start codon and the ribosome-binding site within a predicted base-paired duplex (Figure 13.4). Therefore, it is possible that type III secretion systems recognize the mRNA structure and that this interaction can relieve the presumed blockage in translation of the protein destined for secretion.

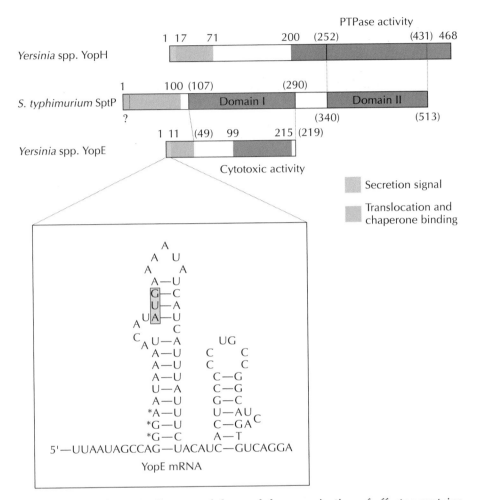

Figure 13.4 Schematic diagram of the modular organization of effector proteins injected into eukaryotic cells by type III secretion systems. The *Yersinia* YopE and YopH anti-phagocytic proteins and SptP from *S. typhimurium* are given as examples. The minimal N-terminal domain required for secretion of YopE (amino acids 1 to 11) and YopH (amino acids 1 to 17) are shown in light grey. The domain immediately downstream of the secretion signal (shown in pink) represents the minimal region of YopE (amino acids 1 to 49) and YopH (amino acids 1 to 71) required for the proteins to be injected into eukaryotic cells. These translocation domains are also overlapped by high affinity chaperone binding sites. The C-terminal region of YopE (amino acids 99 to 215) and YopH (amino acids 200 to 468) are essential for their cytotoxic (grey) and phosphatase (PTPase; red) effector functions, respectively. The SptP protein also displays a modular organization, having regions with extensive similarity to the cytotoxic domain of YopE (indicated as domain I) and the PTPase domain of YopH (domain II). Amino acid residues encompassing these domains within all three proteins are shown in parentheses. The N-terminal domain of SptP (amino acids 1 to 100) is essential for secretion, translocation, and high-affinity binding of its cognate chaperone. Like YopE and YopH, it is likely that SptP contains a short N-terminal secretion signal indicated by a question mark. Also shown is the predicted mRNA structure for the YopE secretion signal. The putative ribosome-binding site (asterisks) and the AUG start codon (pink box) are incorporated into the base-paired duplex structure. Presumably, an interaction between the mRNA and a component of the secretion apparatus denatures the duplex RNA structure and leads to coupling between effector protein translation and secretion.

This indicates that translation and secretion may be intimately coupled at the very site of the secretion apparatus.

Some of the secreted proteins are not only secreted through the two bacterial membranes but also translocated across the plasma membrane of the eukaryotic target cell (for details, see below). This appears to be a universal feature of type III secretion systems. The translocation process requires that all proteins contain a distinct translocation motif independent of the secretion signal and additional accessory proteins also secreted by type III secretion systems. The recognition sequence for translocation of Yop effectors of *Yersinia* resides in the region immediately downstream of the N-terminal secretion signal. It is important to stress that although translocation requires protein secretion, these are mutually exclusive events and it follows that different regions of the translocated protein encode the information necessary for each event.

Thus, virulence effector proteins injected into the interior of the host cell by type III secretion systems of pathogenic bacteria are modular proteins usually segregated into three functionally distinct domains (Figure 13.4). For example, the extreme N-terminal coding region of Yops produced by *Yersinia* contains the specific signal that determines a protein for type III secretion, presumably specified by the mRNA structure rather than the amino acid sequence. Immediately downstream of the secretion domain resides a larger signature domain (usually between 50 and 80 amino acids) that identifies a protein targeted for translocation across the eukaryotic plasma membrane. For several effector proteins, this region is protected or capped by specific chaperones in the cytoplasm of the bacterium (see below). Finally, the C-terminal regions of the proteins contain effector domains, which in some cases display enzymatic activities normally associated with eukaryotic cell function (Box 13.1).

Chaperone Proteins

Small cytoplasmic accessory proteins (chaperones) specifically interact with their cognate type III secreted protein to ensure presecretory stabilization and efficient secretion. However, chaperones may not be needed for presentation of the protein to the secretion apparatus. Unlike *sec*-dependent secretion, type III secretion systems encode several dedicated chaperones that specifically interact with one or at most two partner proteins destined for secretion. In *Yersinia*, the region in which chaperones interact with their partner Yop also overlaps with the motif necessary for translocation. Thus, chaperone-Yop interactions in the cytoplasm of bacteria presumably "caps" the region required for translocation to prevent premature interaction with other proteins of the type III secretion apparatus and/or self-aggregation prior to secretion. In several type III secretion systems, the lack of a specific chaperone reduces secretion of the partner protein due to degradation of the aggregated protein in the bacterial cytoplasm. This is best exemplified by the IpgC chaperone from *Shigella* that independently binds to IpaB and IpaC in the cytoplasm. These specific interactions serve to inhibit the premature association and subsequent degradation of an IpaBC protein complex before secretion. Hence, the term "bodyguards" has also been used to describe the function of type III specific chaperones while the cognate proteins still reside in the cytoplasm.

Delivery and Effectors

Injection System

The proteins secreted by the type III secretion system attack eukaryotic cells in several ways. In most systems so far studied, the effector proteins are translocated across the eukaryotic plasma membrane into the cytoplasm by extracellular bacteria (Figure 13.5). For *Yersinia* and EPEC, it is thought that one function of the effector proteins is to either prevent phagocytosis (*Yersinia*) or form a tight adherence between the bacterium and the target cell that induces the formation of so-called pedestal structures (EPEC). In contrast, *Shigella* and *Salmonella* deliver effector proteins into the eukaryotic cytoplasm that actually induce the eukaryotic cell to internalize the bacterium. In addition, these two pathogens utilize their type III secretion machines to deliver effector proteins following their internalization within the eukaryotic cell. These examples illustrate that gram-negative bacteria have evolved effector proteins that induce and/or modulate eukaryotic host responses that occur during various stages of the infection process.

▶ *For Figure 13.5, see color insert.*

In all cases so far studied, the delivery of effector proteins is dependent upon physical contact between the bacterium and the eukaryotic cell. The bacterial adhesion molecule mediating this tight interaction may not necessarily be an integrated component of the type III secretion system. In *Yersinia*, following the intimate binding between the bacterium and the surface of the eukaryotic cell, Yop effector proteins traverse the eukaryotic cell membrane only at the interface between the two cells. Consequently, no effector proteins are found in the surrounding culture fluid. Thus, the translocation process appears to be polarized. Additionally, eukaryotic cell contact somehow also induces *Yersinia* to increase the expression of Yop effector proteins (Box 13.3). Similarly, *Salmonella* and *Shigella* also have a cell contact-induced delivery system of the Sip/Sop and Ipa effector proteins, respectively. In contrast to *Yersinia,* however, eukaryotic cell contact does not result in increased *sop/sip* and *ipa* gene expression, nor does it appear that the type III secretion systems of *Salmonella* and *Shigella* deliver proteins into eukaryotic cells in a polarized fashion (i.e., in infected-cell cultures, Sip/Sop/Ipa proteins are found in the tissue culture media as well as in the cytoplasm of the infected cells).

Most of what is known about how the type III secretion systems translocate proteins across the eukaryotic membrane (alternatively referred to as "injection") has been derived from studies on *Yersinia.* The injection process in *Yersinia* is dependent on at least two accessory Yops, YopB and YopD, that are themselves secreted to the exterior of the bacterial cell by the type III secretion machine. Bacterial mutants deficient in either of these proteins are still able to secrete the other Yops from the bacterial cell but are unable to inject these proteins into eukaryotic cells; instead, these proteins accumulate at the zone of contact between the bacterial and eukaryotic cells (Figure 13.5). The YopB protein has two potential membrane-spanning domains and is involved in the lysis of erythrocytes. It has been proposed that a complex of YopB and YopD forms a pore in eukaryotic cell membranes through which the other Yops are injected into the eukaryotic cytosol. However, unlike YopB, YopD itself is injected and can be detected in cultured eukaryotic cells infected with *Yersinia.* This may indicate that in addition to facilitating the injection of the other Yops, YopD has an activity within eukaryotic cells.

BOX 13.3

How does *Yersinia* utilize target cell contact to induce Yop synthesis and promote their polarized translocation? A model

Contact of *Yersinia* with a HeLa cell results in the induction of *yop* transcription and the subsequent polarized injection of Yops across the target cell plasma membrane at the zone of contact between the interacting cells. Presumably, to facilitate the induction of Yop synthesis, a signal must be sensed by the bacterium and then transmitted from the surface into the bacterial cytoplasm. With this in mind, how does the bacterium sense the interaction with a eukaryotic cell surface ligand? The type III secretion system per se is essential not only to secrete Yops but also to restrict Yop synthesis to concise periods during bacterial infection. It is proposed that the Ysc secretion machine transports a protein to the bacterial surface that specifically functions to sense physical contact between the pathogen and the host cell. This protein may also block the secretion channel such that only during physical contact would the blockage be released. This would limit Yop translocation to the zone of contact between the interacting cells—a process termed polarized translocation.

Since Yop synthesis is dependent upon a functional Ysc secretion apparatus, how is the signal transmitted through the secretion machine to facilitate this tight regulatory control? Significantly, a mutation in the *lcrQ* gene is always derepressed for Yop synthesis, although when overexpressed in *trans*, LcrQ down-regulates *yop* expression. Furthermore, LcrQ is also secreted via the Ysc secretion machine. Secretion of LcrQ into the extracellular medium correlates with induction of Yop synthesis and secretion. However, in *ysc* secretion mutants normally repressed for Yop synthesis, LcrQ is found only in the bacterial cytoplasm. It follows that upon introduction of an *lcrQ* mutation into these *ysc* secretion mutants, Yop synthesis is derepressed. The following model for contact-dependent Yop regulation by *Yersinia* has been proposed based on the phenotypes of the various mutants. Contact between the pathogen and its target cell results in the opening of the gated secretion apparatus. This allows the rapid secretion of LcrQ, significantly reducing its intra-cellular concentration. This releases the LcrQ-mediated repression of *yop* expression and results in increased Yop synthesis and secretion. This model explains how the bacterium coordinates up-regulation of *yop* expression and polarized secretion and also suggests that these two regulatory events are controlled by the same sensing and gating mechanism.

Interestingly, the parallel between the flagellum-specific secretion pathway and type III secretion system can be extended to how each system regulates gene expression. Just as Yop synthesis and secretion are induced by the rapid and specific secretion of the *yop* regulatory protein LcrQ through the Ysc secretion machine, expression of flagellum-specific genes is induced upon the secretion of the anti-σ factor, FlgM, through the flagellum itself. The release of the antagonist FlgM permits the flagellum-specific σ factor (FliA) to interact with DNA to induce gene transcription. While LcrQ and FlgM have no obvious genetic or functional similarities, it is interesting how two systems involved in the biosynthesis of cell surface organelles have adopted a similar regulatory mechanism to coordinate gene expression.

As discussed above, the ability to inject effector proteins is functionally conserved in different species. Thus, not only the secretion system but also the ability to inject the effectors across the plasma membrane of the eukaryotic cell is mediated through a similar process in different species. Functional conservation of the injection process was first shown for the *Yersinia* YopE protein, which was secreted and translocated into the eukaryotic cytosol by the type III secretion system of *Salmonella*, even though *Salmonella* does not possess proteins that bear any obvious similarities to the *Yersinia* YopB and YopD proteins. The SipB protein of *Salmonella* does have a limited degree of similarity to YopB, which may indicate a similar function. SipB is encoded by the *sipBCDA* operon; all proteins encoded by this operon except SipA are involved in translocation of effector proteins. Interestingly, like YopD, the Sips are injected themselves, which could indicate that they are active within eukaryotic cells.

The injection apparatuses of *Pseudomonas* and *Yersinia* are probably similar since *Pseudomonas* encodes proteins that are very similar to YopB and YopD. It follows that *yopB* and *yopD* mutants of *Yersinia* can be functionally complemented by the corresponding *popB* and *popD* genes of *Pseudomonas*. It is generally accepted that the overall mechanism used by various gram-negative bacteria to deliver effector proteins into eukaryotic cells via the type III secretion system will turn out to be highly conserved. Nevertheless, it should not come as a surprise that variations would exist between the various species since the needs of each species, in terms of a protein delivery system, would be influenced by such factors as the preferred site of infection and the local host defense system.

Effectors

As discussed above, gram-negative pathogens possessing type III secretion systems secrete proteins that directly affect eukaryotic cells. These bacterial proteins probably exert their activities by physically interacting with key eukaryotic proteins involved in host defense functions. Our knowledge of what these proteins are doing varies greatly depending on the protein. For some of the injected proteins, nothing is really known about their effects on eukaryotic cells, although in many cases it is known that the protein is required to cause disease in the animal model system. At the other extreme, for a few proteins it is known what cellular process is affected as well as the eukaryotic proteins that physically interact with the bacterial effector protein. Some of the better-characterized effector proteins and what is known about their function are discussed in this section.

Yersinia

The three species of *Yersinia* that are recognized as human pathogens (*Y. pestis*, *Y. pseudotuberculosis*, and *Y. enterocolitica*) have a common tropism for lymphatic tissue and are able to resist the naive immune system by blocking both phagocytosis and the induction of proinflammatory cytokines. Both these processes are dependent on the type III secretion system and the directional transfer of the Yop effector proteins into the cytosol of the infected cell (Figure 13.6). Since both these activities could be assayed by infecting cultured eukaryotic cells with mutant strains of *Yersinia*, it was relatively simple to determine which of the Yops were required to block phagocytosis and inductive cytokine expression.

The full antiphagocytic activity of *Yersinia* requires the additive effects of the YopH and YopE proteins. In order to engulf a bacterial cell, eukaryotic cells must reorganize their cytoskeletal system (the actin-based component that confers the structural integrity of a cell). This reorganization encloses the bacterium in a membrane-bound vacuole. It appears that YopH and YopE are able to interfere with the structural rearrangements required for the normal phagocytic process to occur. YopH is a protein tyrosine phosphatase (PTPase) that is similar to eukaryotic PTPases (Figure 13.4 and Box 13.1). *Yersinia* strains mutated in YopH are avirulent in the mouse model and are rapidly phagocytosed by cultured eukaryotic cells. Following injection into the cytosol, YopH rapidly dephosphorylates specific target cell proteins. In cultured HeLa cells, these dephosphorylated proteins include p130^{cas} and focal adhesion kinase (FAK). Both of these proteins are involved in the regulation and assembly of new focal adhesion complexes, a cellular process that accompanies phagocytosis. Focal adhesion complexes, located on the inner surface of the plasma membrane, are thought to link the cellular cytoskeletal system to extracellular structures such as matrix proteins. Interestingly, YopH was found to localize to focal adhesion structures after

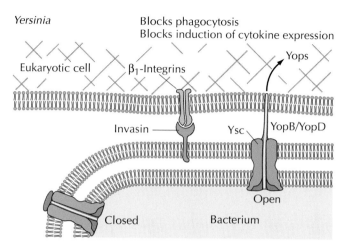

Figure 13.6 Proposed model for *Yersinia* interaction with target cells. Upon intimate association between the pathogen and its target cell, the type III secretion system is opened and the Yops are secreted and subsequently injected into the target cell. This results in inhibition of both phagocytosis and induction of cytokine expression.

infection of HeLa cells, and YopH activity was associated with the disassembly of these structures. Additionally, YopH was found to physically interact with FAK and p130^{cas} in vivo, further suggesting that YopH targets these proteins during the infection cycle. Thus, these observations have led to the suggestion that YopH specifically targets and destroys focal complexes as a means of blocking phagocytosis. It is also reasonable to expect that p130^{cas} and FAK, whose transient tyrosine phosphorylation is required for normal phagocytosis, are the actual in vivo substrates for the YopH PTPase activity.

YopE works together with YopH to block phagocytosis, but unlike YopH, no eukaryotic protein has so far been identified that is specifically targeted by YopE. Both the phagocytic process and cell motility require a substantial amount of cytoskeletal reorganization, which involves a balance of actin polymerization and depolymerization. After being injected into the cytosol of the infected cell, YopE disrupts the cellular cytoskeletal system by inducing a massive depolymerization of actin stress fibers. Although very little is known about the actual mechanism of how YopE exerts its effect, YopE does have similarity to the *Pseudomonas* protein exoenzyme S (ExoS). Like YopE, ExoS is both injected into eukaryotic cells via a type III secretion system and blocks the phagocytic process by a mechanism that includes actin depolymerization. However, unlike YopE, which has not been observed to have any enzymatic activity, ExoS possesses an ADP-ribosyltransferase activity. In vitro studies have shown that ExoS can inactivate eukaryotic proteins by covalently linking ADP-ribose to the target proteins. The preferred targets appear to be members of the H-Ras and K-Ras families, small GTP-binding proteins involved in intracellular signaling. Although YopE is similar only to the N-terminal part of ExoS, it has been suggested that YopE also targets small GTP-binding signaling proteins.

In addition to extracellular growth in eukaryotic tissues (due to the antiphagocytic activity of YopH and YopE), another characteristic of *Yersinia* infections is that the pathogen is able to proliferate in tissues without provoking an inflammatory-like response. For all three human pathogenic *Yersinia* species, blocking of the host inflammatory response is dependent on the 70.5-kb virulence plasmid encoding the type III-associated genes (Figure 13.2). For example, in mice infected with *Y. pestis*, visceral organs such as the liver become heavily colonized with the pathogen. In contrast, in tissues of mice infected with plasmid-cured strains of *Y. pestis*, inflammatory cells are recruited to the site of infection, leading to the formation of protective granulomas. The lack of an inflammatory response at the tissue level is thought to be due to the ability of *Yersinia* to repress, at the cellular level, the expression of proinflammatory cytokines. Inducible cytokine secretion is a result of a multistep process: a danger signal is perceived at the cell surface; this signal has to be transmitted to the nucleus, where the appropriate gene must be transcribed; the transcript must be exported to the cytoplasm and translated; and, finally, the resulting protein must be secreted from the cell. Recent work has indicated that *Yersinia* most probably blocks inducible cytokine secretion by somehow blocking the activation of signaling molecules involved in some of the early events that occur when eukaryotic cells encounter bacteria and other forms of stress such as UV radiation.

The *yopJ* locus of *Y. pseudotuberculosis* (and the homologous locus in *Y. enterocolitica, yopP*) is required to block inducible expression of the cytokines tumor necrosis factor alpha (TNF-α) and interleukin-8 (IL-8). The TNF-α and IL-8 gene promoter regions contain a number of binding sites for various families of transcription factors that control their basal and/or inductive expression levels. Binding sites for one such family, NF-κB, are important for the inducible expression of these genes. The NF-κB transcription factors are retained in the cytoplasm of uninduced cells by anchor proteins. Upon the appropriate stimulus, such as lipopolysaccharide, the anchor proteins release NF-κB which subsequently translocate to the nucleus, where they bind to the promoter regions of several genes such as the TNF-α and IL-8 genes and increase their expression. Not surprisingly, it was found that *Yersinia* blocked the activation of NF-κB and that this activity is dependent on YopJ/YopP. It was also found that YopJ/YopP was required for the *Yersinia*-mediated blockage of the mitogen-activated protein kinases (MAPKs), another family of signaling proteins involved in eukaryotic stress responses. Currently, it is not known whether YopJ/YopP blocks the activation of the MAPKs and NF-κB by a common mechanism. Whatever the mode of action of YopJ/YopP, it is impressive that it is capable of shutting down very sensitive stress response signaling pathways in cells undergoing what must be an extreme amount of distress.

In addition to blocking phagocytosis and inductive cytokine expression, *Yersinia* induces or triggers apoptosis (programmed cell death) in macrophage cell lines. Whether *Yersinia*-mediated apoptosis occurs during an actual infection remains to be determined. Apoptosis of cultured macrophages, like that of repressing cytokine expression, is dependent on the *yopJ* locus. Triggering apoptosis is also dependent on the *Yersinia* type III secretion system, suggesting that, like apoptosis-inducing proteins of plant-interacting bacteria, YopJ is recognized, or its activity occurs, inside the eukaryotic cell and that this recognition or activity triggers apoptosis. What is the significance of apoptosis in animal bacterial pathogenesis? Currently, there is disagreement about whether apoptosis is in the best interest of the bacterium or, alternatively, occurs as a host defense strategy. For plants, bacterium-induced apoptosis by proteins delivered by type III secretion machines is clearly a host defense response that serves to cordon off the bacterial pathogen to the initial site of infection (Box 13.2). Alternatively, inducing apoptosis in macrophage cells (and perhaps other cell types) may benefit *Yersinia* by preventing those cells from performing important immune functions.

Salmonella

Salmonella spp. cause a broad spectrum of different diseases, usually originating from oral infections. These infections can be life-threatening, such as typhoid fever, or relatively mild, such as gastroenteritis. As mentioned above, *Salmonella* possesses two different gene clusters, designated SPI-1 and SPI-2, each encoding apparently independent type III secretion systems. *Salmonella* organisms grown in culture secrete more than 20 proteins via a type III-dependent secretion mechanism. SPI-1-encoded secretion complexes secrete most of these proteins. *Salmonella* proteins secreted by SPI-2-encoding complexes have not yet been identified. Remarkably, neither mutants with mutations in the SPI-1 region nor mutants with mutations in the associated SPI-1-secreted proteins have any gross effect on virulence in the mouse model. In contrast, mutants with mutations in the SPI-2 region are strongly attenuated in virulence. Although the virulence effects are moderate (as measured in a mouse model), the *inv/spa* secretion system (as the SPI-1 is also known) of *Salmonella* is essential for the ability of the pathogen to invade epithelial cells and to induce cytotoxicity as well as apoptosis. This clearly indicates the importance of the SPI-1-associated secretion system in virulence. After infection of cultured epithelial cells, *Salmonella* induces membrane ruffling that is associated with rearrangements of the actin cytoskeleton of the eukaryotic cell. *Salmonella* can induce similar ruffles of the M cells of the Peyer's patches through which the bacterium induces its uptake into the M cells. The actual molecular mechanisms by which *Salmonella* induces these events are not understood.

When wild-type strains of *Salmonella* are grown in culture, the SipA, SipB, SipC, and SipD proteins, encoded by the *sipBCDA* operon, are secreted to high levels in the culture supernatant. Mutants with mutations in either *sipB*, *sipC*, or *sipD* are unable to translocate other effector proteins, showing that the Sips are involved in the actual translocation process. This suggests that they perform a similar function to YopB and YopD of *Yersinia*. Additionally, extracellular and intracellular *Salmonella* organisms translocate SipB and SipC into the eukaryotic cell cytosol (and probably SipA and SipD, although this has not yet been demonstrated). These findings may indicate that Sip proteins per se have effector functions. The Sips have been associated with all *Salmonella*-induced effects on cultured eukaryotic cells. Whether the Sips directly mediate these effects or whether other effector proteins translocated via the Sips are essential remains to be determined.

In addition to the Sips, *Salmonella* secretes at least five other proteins denoted SopA, SopB, SopC, SopD, and SopE. The Sops are injected into the eukaryotic cell by a process requiring at least the SipB protein (and perhaps other Sips). The SopE protein, when introduced into eukaryotic cells by the *Salmonella* type III secretion system (as occurs during an infection) or experimentally microinjected, induces cytoskeletal rearrangements that accompany membrane ruffling and eventual internalization of the bacterium. It is now known that SopE exerts its effects by physically interacting with and activating small GTP-binding signaling proteins of the Rho family.

Rho-type signaling proteins are normally activated when eukaryotic cells are exposed to hormones or growth factors. This activates cellular responses that typically include membrane ruffling. Apparently, *Salmonella* has evolved a way to manipulate a cellular signaling pathway to serve its own means.

Another interesting type III secreted and translocated protein of *Salmonella* is SptP (alternatively referred to as SopD). The N-terminal half of this protein is very similar to the N-terminal half of YopE and ExoS, while the C-terminal half is very similar to YopH, including the tyrosine phosphatase catalytic domain (Figure 13.4). As expected from the sequence, SptP possesses a protein tyrosine phosphatase activity and induces a "YopE-like" cytotoxic effect on cultured eukaryotic cells (disruption of the cytoskeletal system). Thus, SptP exhibits two apparently independent activities. It is therefore surprising that *sptP* mutants, in contrast to *yopE* and *yopH* mutants of *Yersinia*, are only moderately attenuated in virulence in the mouse model. It is important to stress that the mouse model does not reflect all aspects of virulence, and it seems very likely that SptA will be found to play a significant role in the overall pathogenicity of *Salmonella* disease.

Shigella

Shigella is the causative agent of bacillary dysentery and has a host range limited to humans and primates. Following oral infection, the pathogen colonizes the epithelial cell layer of the colonic mucosa, giving rise to the clinical symptoms. The infectious process can be divided into a number of discrete steps including transcytosis through the M-cells of the Peyer's patches, induction of a strong inflammatory response, invasion of epithelial cells, and multiplication and spread into adjacent cells (Figure 13.7). All these steps are dependent on the *Shigella* type III secretion system commonly referred to as *mxi/spa*. The proteins encoded by the *ipaBCDA* operon are secreted by a type III-dependent secretion complex and are very similar to the corresponding SipA, SipB, SipC, and SipD proteins of *Salmonella*.

The Ipa proteins are essential components for triggering bacterial entry into epithelial cells. The entry processes of *Salmonella* and *Shigella* appear to be very similar, both involving cytoskeletal rearrangements and the formation of plasma membrane projections which lead to the uptake of the pathogens into membrane-bound vacuoles. However, *Salmonella* enters eukaryotic cells at their apical surface, while *Shigella* enters at the basolateral side of the infected cell. An initial step in the uptake seems to be the binding between a complex consisting of the IpaB, IpaC, and IpaD proteins with the $\alpha_5\beta_1$-integrins of the target cell. Clustering of integrins is known to induce tyrosine phosphorylation of host proteins, and in *Yersinia*, ligand-induced clustering of integrins induces uptake of the pathogen. Since tyrosine phosphorylation is a prerequisite for the induced uptake of *Shigella*, it is possible that the Ipa-integrin interaction initiates the uptake process.

The entry process of *Shigella* leads to the accumulation of a number of host proteins to the zone of invasion. These proteins include cytoskeletal proteins, such as myosin and cortactin, as well as proteins associated with focal adhesions, such as paxcillin, talin, and vinculin. The recruitment of vinculin is particularly interesting since IpaA interacts with this protein. Although this interaction is not an absolute requirement for bacterial invasion, it is evident that the interaction facilitates the entry process. Thus, it seems likely that secretion of the Ipa proteins per se via the type III secretion system is sufficient to trigger uptake of *Shigella* into epithelial cells. Unlike *Yersinia* and *Salmonella* effector proteins, however, no *Shigella* pro-

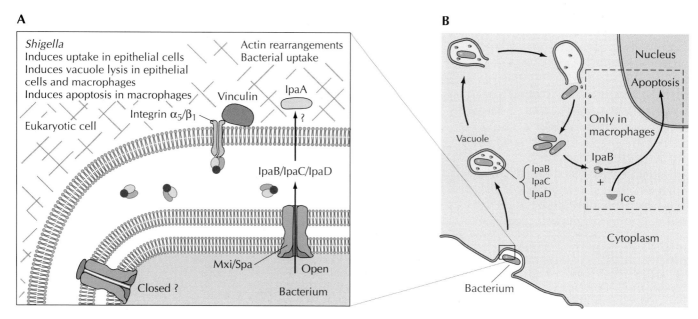

Figure 13.7 Proposed model for *Shigella* invasion of epithelial cells. **(A)** Upon contact with the eukaryotic cell, *Shigella* secretes several Ipa proteins. A protein complex of IpaB, IpaC, and IpaD interacts with $\alpha_5\beta_1$-integrins, generating a signal that induces membrane ruffling and subsequent uptake of the pathogen into a membrane-bound vacuole. IpaA is presumed to be translocated since it has been shown to interact with the eukaryotic protein vinculin. **(B)** Thereafter, the activity of the Ipas leads to lysis of the vacuole membrane, allowing release and proliferation of *Shigella* in the cytosol. Secretion of IpaB into the cytosolic compartment induces apoptosis of macrophages via the interaction of IpaB with ICE.

teins are known to be translocated into the cytosol of the target cells. Nevertheless, the vinculin-IpaA interaction suggests that at least IpaA is translocated. It is surprising that translocation of *Shigella* effector proteins has not been demonstrated, especially considering the similarity between the *ipa* and *sip* operons of *Shigella* and *Salmonella,* respectively. It is likely, however, that some translocated effector proteins remain to be discovered.

IpaB is involved in two important intracellular functions for *Shigella*-vacuole escape and induction of apoptosis. Immediately following the invasion step, *Shigella* appears in a membrane-bound vacuole. This vacuole is lysed, allowing the bacterium access to the cytosol. This lysis step requires the type III secretion machine as well as the Ipa proteins. IpaB appears to be especially important in this process, although the actual molecular events leading to vacuolar lysis remain elusive. Additionally, *Shigella* induces apoptosis after being released from the vacuole and gaining access to the cytoplasm. The induction of apoptosis is also dependent on IpaB, which binds to and activates the IL-Iβ-converting enzyme ICE. Activation of ICE leads to apoptotic death of the macrophage and the initiation of an acute inflammatory response.

EPEC

EPEC is one of several different *E. coli* strains that cause diarrhea after oral infection. These pathogens use a type III secretion system through which at least three proteins, called EspA, EspB, and EspC, are secreted. The secretion machine and the secreted proteins are important for EPEC to cause disease. At the tissue level, an EPEC infection results in localized destruction of intestinal brush border microvilli. Interestingly, bacterial binding to intestinal epithelial cells results in the formation of a pedestal-like structure, known as the attaching and effacing (A/E) lesion, underneath the bacterial cell (Figure 13.8).

The intimate interaction between the EPEC and the target cell as well as the subsequent formation of pedestals can experimentally be divided into three discrete steps: (i) initial adherence of the bacterium to the eukaryotic cell, (ii) signal transduction occurring within the eukaryotic cell,

Figure 13.8 Proposed stages in the generation of tight adherence between EPEC and its target cell. After initial interaction between EPEC and the eukaryotic cell has been established (independent of the type III secretion system), the EspA organelle, which is secreted by the type III system, bridges the small gap between the pathogen and the eukaryotic cell surface. This results in translocation of EspB and the insertion of Tir into the cell membrane. Following this, pedestal-like structures are formed as a consequence of dramatic actin rearrangements within the host cell. Concomitantly, EPEC binds very tightly at the tip of these pedestals via the interaction between the bacterial outer membrane protein intimin and Tir, which in this case works as a bacterially induced "eukaryotic cell receptor." During this process the ability to visualize EspA is lost.

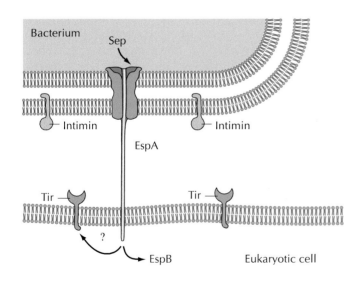

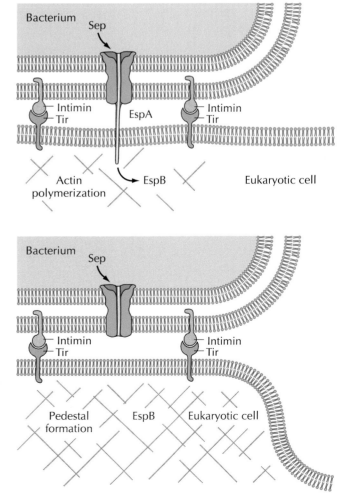

and (iii) intimate adherence of the bacterium to the surface of the eukaryotic cell. The first step, which is not dependent on the type III secretion system, involves bundle-forming pili which mediate the initial adherence of the bacterium to the eukaryotic cell. After this initial attachment, EPEC induces the activation of signaling pathways within the eukaryotic cell. This activity is induced by two proteins, EspA and EspB, secreted by the EPEC type III secretion system and results in increased inositol triphosphate levels and Ca^{2+} fluxes, two events involved in intracellular signaling within eukaryotic cells. Interestingly, EspA was recently shown to form filamentous pilus-like structures observed at the surface of the bacterium. These structures apparently serve as a link between the bacterium and the eukaryotic cell surface and are required for translocation of EspB into the target cell. Both *espA* and *espB* mutants fail to induce A/E lesion formation. A/E lesion formation is followed by a step in which EPEC inserts its own receptor, a 78-kDa protein called Tir (Translocation Intimin Receptor), into the eukaryotic cell membrane. Interestingly, Tir becomes phosphorylated by host kinases following its insertion into the infected cell membrane and functions as a receptor for the bacterial surface-located adhesion molecule intimin. In contrast to Tir, intimin is not secreted by the type III secretion machine. Hence, EPEC encodes both an adhesin and its cognate receptor, which together promote intimate attachment.

Conclusion

Only recently have we become aware of the subtle ways in which bacteria interact with eukaryotic cells. As discussed in this chapter, one very sophisticated approach that gram-negative bacteria have evolved to manipulate eukaryotic cells involves direct injection of proteins into the cytosol of the infected cell. These bacterial proteins injected by the type III secretion system either impede or activate particular eukaryotic cellular responses depending on what is beneficial for the bacterium. The variety of species that possess type III secretion systems, ranging from animal pathogens to plant symbionts, suggest that animal- and plant-interacting bacteria have common strategies to overcome, or at least survive, the eukaryotic defense system.

Knowledge of the type III secretion system and the proteins it injects will undoubtedly allow the design of novel therapeutic strategies for use against microbial pathogens. For example, the *Salmonella* type III system has been used to inject viral proteins into eukaryotic cells that are eventually displayed on the eukaryotic cell surface in association with MHC proteins. The resulting surface presentation of MHC-viral peptides promotes the development of the adaptive immune response that serves to protect the animal from subsequent viral challenge. Similarly, understanding how the injected effector proteins function within eukaryotic cells will certainly give us a better grasp of the animal and plant defense systems.

Selected Readings

Anderson, D. M., and O. Schneewind. 1997. A mRNA signal for the type III secretion of Yop proteins by *Yersinia enterocolitica*. *Science* **278:**1140–1143.

Hardt, W.-D., L.-M. Chen, K. E. Schuebel, X. R. Bustelo, and J. E. Galán. 1998. *S. typhimurium* encodes an activator of Rho GTPases that induces membrane ruffling and nuclear responses in host cells. *Cell* **93:**815–826.

Hueck, C. J. 1998. Type III protein secretion systems in bacterial pathogens of animals and plants. *Microbiol. Mol. Biol. Rev.* **62:**379–433.

Kenny, B., R. DeVinney, M. Stein, D. J. Reinsheid, E. A. Frey, and B. B. Finlay. 1997. Enteropathogenic *E. coli* (EPEC) transfers its receptor for intimate adherence into mammalian cells. *Cell* **91:**511–520.

Knutton, S., L. Rosenshine, M. J. Pallen, L. Nisan, B. C. Neves, C. Bain, C. Wolff, G. Dougan, and G. Frankel. 1998. A novel EspA-associated surface organelle of enteropathogenic *Escherichia coli* involved in protein translocation into epithelial cells. *EMBO J.* **17:**2166–2176.

Kubori, T., Y. Matsushima, D. Nakamura, J. Uralil, J. M. Lara-Tejero, A. Sukhan, J. E. Galán, and S.-I. Aizawa. 1998. Supramolecular structure of the *Salmonella typhimurium* type III protein secretion system. *Science* **280:**602–605.

Rosqvist, R., S. Håkansson, Å. Forsberg, and H. Wolf-Watz. 1995. Functional conservation of the secretion and translocation machinery for virulence proteins of yersiniae, salmonellae and shigellae. *EMBO J.* **14:**4187–4195.

Rosqvist, R., K.-E. Magnusson, and H. Wolf-Watz. 1994. Target cell contact triggers expression and polarized transfer of *Yersinia* YopE cytotoxin into mammalian cells. *EMBO J.* **13:**964–972.

Van den Ackerveken, G., E. Marois, and U. Bonas. 1996. Recognition of the bacterial avirulence protein AvrBs3 occurs inside the host plant cell. *Cell* **87:**1307–1316.

Woestyn, S., M.-P. Sory, A. Boland, O. Lequenne, and G. R. Cornelis. 1996. The cytosolic SycE and SycH chaperones of *Yersinia* protect the region of YopE and YopH involved in translocation across eukaryotic cell membranes. *Mol. Microbiol.* **20:**1261–1271.

14

Bacterial Type IV Secretion Systems: DNA Conjugation Machines Adapted for Export of Virulence Factors

PETER J. CHRISTIE AND ANTONELLO COVACCI

Bacteria have evolved transport systems that until recently were thought to function specifically for the conjugative transfer of plasmid DNA between cells. Our early understanding of conjugation was developed with studies of the F plasmid of *Escherichia coli*. Extensive genetic experiments showed that transfer initiates upon pilus-mediated contact between donor and recipient cells. The pilus then retracts, bringing donor and recipient cells into direct physical contact. Stabilization of the mating pair triggers the processing of F plasmid into a translocation-competent complex of single-stranded DNA and protein. This nucleoprotein particle is delivered across the donor cell envelope via a conjugal or "mating" pore to the recipient cell, and this step is followed by second-strand synthesis and dissociation of the mating pair. By definition, conjugation is a contact-dependent process. Thus, conjugation bears a strong mechanistic resemblance to the type III secretion systems (see chapter 13). Pathogenic bacteria of humans and plants have coopted conjugation systems to export virulence factors to eukaryotic host cells. Although this is a functionally diverse family, there are some unifying themes: (i) exporters are assembled at least in part from subunits of DNA conjugation systems, and (ii) the known substrates recognized by these transporters are large macromolecules such as nucleoprotein particles and multisubunit toxins.

Members of the Type IV Secretion Family

The best-characterized type IV system is that used by *Agrobacterium tumefaciens* for exporting oncogenic DNA (T-DNA) to plant cells. The result of DNA transfer is the proliferation of plant cells and ultimately the formation of tumorous tissues termed crown galls (Figure 14.1). T-DNA transfer requires approximately 20 virulence (Vir) proteins, which are encoded by six operons. Some of the Vir proteins, notably the VirD2 endonuclease and the VirE2 single-stranded-DNA-binding protein (SSB), interact with single-stranded T-DNA (the T strand) to form the transfer intermediate (the T complex). Other Vir proteins, including the 11 *virB* gene products and the *virD4* gene product, assemble into a gated channel-pilus complex. The channel serves as a transenvelope conduit through which the T complex passes, while the pilus is proposed to mediate the physical contact between *A. tumefaciens* and recipient plant cells. Figure 14.2 shows that the VirB components of the T-DNA transport system are highly related to Tra protein components of transfer systems encoded by several broad-host-range plasmids. The most extensive similarities are to Tra proteins encoded by the IncN plasmid, pKM101, and by the IncW plasmid, R388. Both of these Tra systems code for homologs of all 11 VirB proteins. In addition, the genes coding for related proteins are colinear in the respective *tra* and *virB* operons, providing further evidence of a common ancestry. A subset of the Tra proteins comprising the transfer systems of other broad-host-range plasmids, including RP4 (IncP), and the narrow-host-range plasmid F (IncF) also are related to VirB proteins.

The type IV secretion family also is composed of toxin exporters used by several bacterial pathogens of humans (Table 14.1). *Bordetella pertussis*, the causative agent of whooping cough, uses the Ptl transporter to export the six-subunit pertussis toxin across the bacterial envelope. All nine Ptl proteins are related to VirB proteins, and, as with several of the plasmid Tra systems, *ptl* genes are colinear with the corresponding *virB* genes in the respective operons (Figure 14.2). Type 1 strains of *Helicobacter pylori*, the causative agent of peptic ulcer disease and a risk factor for the development of gastric adenocarcinoma, contain the 40-kb *cag* pathogenicity island that encodes for several virulence factors. Among these are five Cag proteins related to VirB proteins and one Cag protein related to VirD4, another pro-

Figure 14.1 Model of the *Agrobacterium* infection process. Specific classes of phenolic and sugar molecules released from wounded plant cells induce the Vir regulon, resulting in the interkingdom transfer of the oncogenic T complex to plant nuclei.

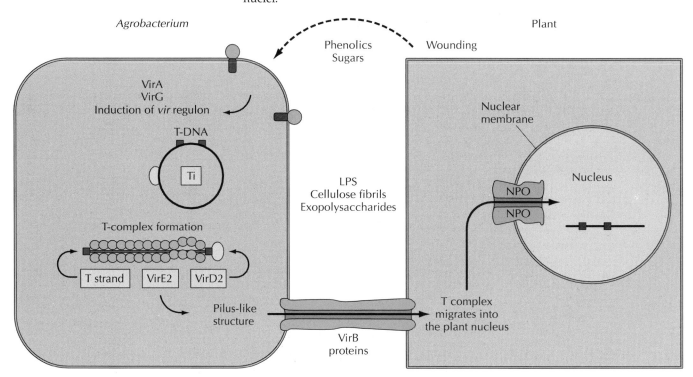

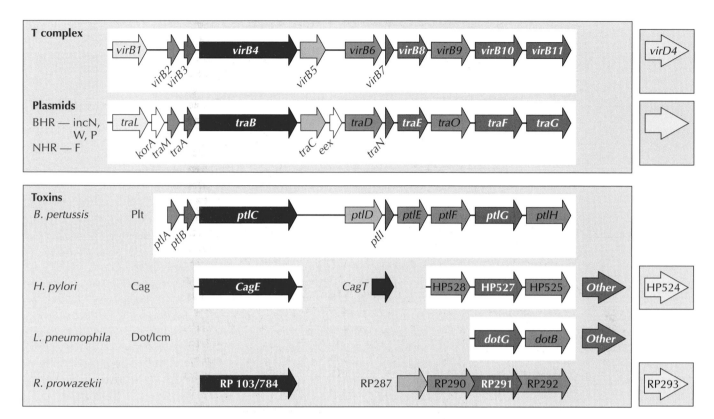

Figure 14.2 Alignment of genes encoding related components of the type IV transport systems. Of the 11 VirB proteins, those encoded by *virB2* through *virB11* are essential for T-complex transport to plant cells. The broad-host-range (BHR) plasmid pKM101 encodes a conjugation apparatus composed of the products of the *tra* genes shown. Other broad-host-range plasmids of the IncW and IncP incompatibility groups and the narrow-host-range (NHR) F plasmid code for Tra proteins related to most or all of the VirB genes. A second subfamily of type IV transporters found in bacterial pathogens of humans export toxins or other protein effectors to human cells.

posed component of the T-DNA transfer system. Deletion of the *cag* PAI and several insertion mutations results in a block in induction of synthesis of the proinflammatory cytokine interleukin-8 (IL-8) in gastric epithelial cells. The Cag homologs of the VirB, VirD4, and Tra proteins are thought to inject into the host cell an as yet unidentified factor involved in promoting IL-8 synthesis. Very recently, another member of the type IV family was shown to be important for the virulence of *Legionella pneumophila*, the causative agent of Legionnaires' disease and Pontiac fever. Mutational studies aimed at identifying genes involved in intracellular growth and in macrophage killing resulted in the identification of the *icm*/*dot* genes. Products of two of these genes, *dotG* and *dotB*, are related to VirB proteins, and products of several other *icm* or *dot* genes have homologs in other bacterial conjugation systems. As with the related Cag proteins of *H. pylori*, the Icm/Dot proteins act as an exporter for virulence factor(s) targeted to the intracellular environment to enhance survival and to kill host macrophages. Two homologues of VirB4 and one homologue each of VirB8, VirB9, VirB10, VirB11, and VirD4 were detected in the complete nucleotide sequence of *Rickettsia prowazekii*, suggesting that this microorganism encodes a type IV transport system similar to *H. pylori* and *Legionella*. In this chapter, we assume that some of the conclusions about *Helicobacter*, an extracellular pathogen, and *Legionella*, an intracellular pathogen, are also valid for the intracellular pathogen *Rickettsia*.

The *A. tumefaciens* T-Complex Transfer System as a Paradigm for Assembly of Type IV Secretion Systems

Extensive studies of the *A. tumefaciens* T-complex transporter components have generated the following information. VirB1, a probable *trans*-glycosylase, is a nonessential virulence factor, but a truncated, exported form of the protein (VirB1*) has been postulated to play a role in mediating specific contacts with recipient cells. There is compelling evidence that VirB2 is the major pilin subunit for the pilus (T pilus). VirB2 localizes at the inner membrane, and, upon an unknown signal, a processed form is recruited to the outer membrane for pilus polymerization. The VirB proteins are required for pilus assembly, but it is not known whether these proteins participate directly in pilus morphogenesis or simply provide an anchor at the membrane for pilus attachment. VirB4 and VirB11 contain consensus Walker "A" nucleotide-binding motifs, and each hydrolyzes ATP in vitro. These proteins utilize the energy of ATP hydrolysis to drive transporter assembly or substrate translocation. Each of these ATPases forms a homodimer early during assembly with other transporter components, although the final oligomeric structure of these proteins in the fully assembled transporter is unknown. Both ATPases are tightly associated with the inner membrane. VirB4 is an integral membrane protein with two domains either embedded into or through the membrane into the periplasm. VirB11 is tightly bound

Table 14.1 Biological effects mediated by type IV secretion systems in various bacterial pathogens

Bacterium	Effects
Helicobacter pylori	IL-8 induction on epithelial cells
	Tyrosine phosphorylation of a 145-kDa molecule present in gastric epithelial cells
	Pedestal formation on gastric epithelial cells
	NF-κB activation after exposure of epithelial cells to type I strains
	Colonization and bacterial density in animal models and human biopsy specimens
	DNA transfer(?)
Agrobacterium tumefaciens	Export of T-DNA–protein complex to plant cells
Escherichia coli	Intra- and interspecies conjugative DNA transfer to other bacteria and yeasts
Bordetella pertussis	Pertussis toxin export in the extracellular compartment
Legionella pneumophila	Phagolysosome fusion in macrophages
	Host cytotoxicity by pore formation
	DNA transfer(?)
Rickettsia prowazekii	Phagolysosome fusion in eukaryotic cells(?)
	DNA transfer(?)

to the membrane and most probably is localized exclusively on the cytoplasmic face of the membrane. The functions of the remaining transporter components have not been defined. Several integral inner membrane protein components such as VirB6 and VirB10 are likely candidates as structural components of an inner membrane pore. Other outer membrane components such as VirB7 and VirB9 are likely candidates as components of an outer membrane pore. The remaining VirB proteins may be structural subunits of the putative transenvelope channel, or they may transiently assist in the assembly of this structure. Figure 14.3 depicts a model of the T-complex transporter-pilus with the likely positions of the VirB and VirD4 proteins.

The possible roles of individual VirB proteins in transporter assembly have been analyzed in part by construction of nonpolar virB mutants. Studies of these mutants revealed that some of the VirB proteins, most notably VirB6, VirB7, and VirB9, provide important stabilizing functions for other VirB proteins. VirB7, an outer membrane lipoprotein, was shown to stabilize itself as well as VirB9 through the assembly of a disulfide cross-linked VirB7-VirB9 heterodimer. Several lines of evidence suggest that the correct positioning of the VirB7-VirB9 heterodimer at the outer membrane is required for assembly of a functional transporter. Thus, it has been proposed that this heterodimer functions as a nucleation center for transporter assembly. Of further interest, VirB6, a polytopic inner membrane protein, recently has been shown to facilitate VirB7 dimer formation or stabilization, as well as the stabilization of other VirB proteins including the VirB4 and VirB11 ATPases. Taken together, these biochemical results have led to the hypothetical model shown in Figure 14.3, depicting several early stabilizing interactions that must occur before the biogenesis of a functional T-complex transporter can take place. Interestingly, the PtlI and PtlF homologs of VirB7 and VirB9 also assemble as disulfide cross-linked dimers, and dimer formation also is important for Ptl protein stabilization. Furthermore, the PtlD homolog of VirB6 has a stabilizing effect on other Ptl transporter components. Finally, the PtlH homolog of VirB11 requires an intact ATP-binding domain for function, supporting the sequence-based prediction that PtlH provides energy from ATP hydrolysis for transporter assembly or function. Thus, even at this early stage of analysis of these systems, it is clear that several common mechanistic principles are applied to build functional T-complex and pertussis toxin exporters.

The existence of a subset of VirB homologues in the *H. pylori cag* and *L. pneumophila icm/dot* systems underscores the functional importance of these types of proteins in macromolecular export. Of particular note, the *cag* PAI codes for homologs of the two VirB ATPases, VirB4 and VirB11, and proposed structural components of the transfer channel, VirB7, VirB9, VirB10, and VirD4. Together, these findings raise the intriguing possibility that this subset of proteins corresponds to a minimal ancestral protein subassembly that bacteria have built upon to construct transporters designed for novel purposes ranging from intercellular DNA transfer, toxin export, and, possibly, direct injection of virulence factors into eukaryotic cells.

Substrate Recognition and the Export Route

During conjugation, the donor cell generates a transfer intermediate consisting of a single strand of DNA covalently bound at its 5' end with the processing endonuclease and coated along its length with a single-

I
Lipoprotein export
and processing

II
Formation of
disulfide-linked VirB7
dimers

III
Dimer sorting

IV
B7/B9 nucleation
center for complex
assembly

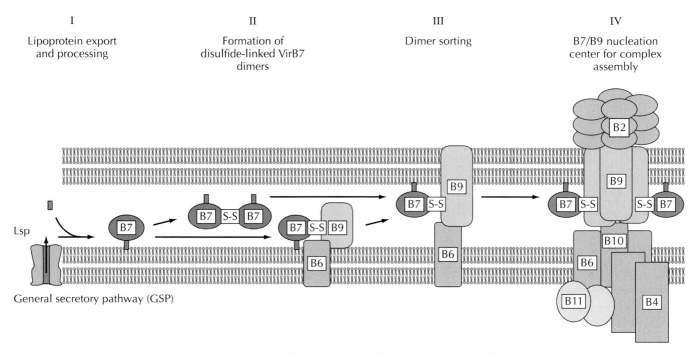

Figure 14.3 Early steps in a proposed assembly pathway for the T-complex transporter. VirB proteins, including VirB7 and VirB9, are exported across the cytoplasmic membrane via the general secretory pathway. Lsp, signal peptidase II, which processes the prelipoproteins. The VirB7 lipoprotein (fatty acid modification denoted by lollipop stick) assembles as homodimers and heterodimers with VirB9, possibly facilitated by VirB6. The VirB6-VirB7-VirB9 protein subcomplex is a proposed nucleation center for recruitment and stabilization of other VirB proteins, leading to assembly of the proposed gated channel-T pilus shown on the right.

stranded-DNA-binding protein (SSB). How this nucleoprotein particle is recognized by the conjugation machine is a subject of great interest. There are several lines of evidence suggesting that substrate recognition is conferred not by the DNA but, rather, by proteins associated with the DNA. Although the nature of these signals is unknown, it is likely the exported protein components contain a conserved peptide sequence or a recognition motif generated upon protein folding. It is therefore tempting to speculate that conjugation machines are in fact protein exporters that evolved to accommodate "hitchhiker" DNA during transport.

The *A. tumefaciens* T-DNA-processing reaction resembles the conjugative DNA processing reaction, resulting in formation of the T-strand/VirD2/VirE2 nucleoprotein particle or T complex. Perhaps the most compelling evidence that conjugation machines recognize proteins as translocation-competent substrates is that VirE2 SSB can be exported to plant cells independently of the T-strand/VirD2 complex. Conjugal DNA transfer intermediates are thought to be delivered across the donor cell envelope in a single step through a proteinaceous channel (mating pore). In striking contast, the highly related Ptl exporter is proposed to function quite differently. Pertussis toxin is a multisubunit toxin composed of five different subunits, each possessing characteristic signal sequences. These subunits are proposed to be delivered across the inner membrane via the general secretory pathway. Pertussis toxin assembles in the periplasm, where it is exported by the Ptl system across the outer membrane.

In view of the proposed one-step translocation model for conjugal DNA transfer and the two-step model for export of pertussis toxin, it is of considerable interest to recall that the two transporters are assembled from almost identical sets of protein homologs (Figure 14.1). Among the essential VirB proteins, only VirB5 has no counterpart in the Ptl system. This observation raises the interesting possibility that VirB5 and related Tra proteins supply a function(s) specifically for assembly of DNA transporters. In addition, in contrast to the contact-dependence of conjugal DNA transfer, *B. pertussis* excretes functional pertussis toxin into the extracellular milieu, which has led to the hypothesis that the Ptl proteins do not elaborate a pilus. However, it is notable that PtlA is related to the VirB2 pilin subunit. This could mean that in fact the Ptl system does assemble at least a vestige of a pilus or that VirB2-type proteins supply a function which is critical to transporter assembly but unrelated to pilus assembly.

The proposed toxin substrates exported by the *H. pylori cag* and *L. pneumophila icm/dot* systems have not yet been identified. However, one known substrate for the *icm/dot* system is the IncQ plasmid RSF1010. This is a non-self-transmissible plasmid that possesses the origin of transfer (*oriT*) and cognate mobilization (*mob*) proteins for processing the transfer intermediate (here referred to as the R complex). The proposed R complex resembles the T complex as a linearized single strand of RSF1010 DNA covalently associated at its 5′ end with MobA endonuclease and coated along its length with SSB. Many conjugative plasmid transfer systems, as well as the T-complex transporter, recognize and export the R complex to recipient cells. Indeed, wild-type *A. tumefaciens* cells carrying RSF1010 show a reduced ability to export T-DNA, possibly as a result of substrate competition for a limited number of transporters.

Similarly, the *L. pneumophila icm/dot* system directs the movement of plasmid RSF1010 between bacteria. Cells carrying RSF1010 display a reduction in intracellular multiplication and human macrophage killing. Both of these features of the infection process are thought to result from export of a toxin effector via the *icm/dot* system. These observations suggest that the *icm/dot* system has retained a functional vestige of the ancestral DNA conjugation system from which it evolved. Whether other type IV toxin export systems, including the *B. pertussis* Ptl system and the *H. pylori* Cag system, also are capable of directing conjugative DNA transfer to recipient bacteria or even to eukaryotic cells remains to be tested.

Molecular Cell Biology of Type IV Secretion Systems

Genomes increase in complexity through processes like gene duplication. The flagellar apparatus, or a simplified version of it, was proposed as a common ancestor of the type III secretion machines. These systems eventually specialized as extracellular tubular protrusions or as an intracellular gated complex involved in substrate transfer between different subcellular compartments. This analogy can be extended to the type IV family, which was proposed to originate from a conjugative apparatus that later associated with other classes of genes, providing, over the evolutionary timescale, an entirely new range of functions. We speculate that both the types III and IV systems have a progenitor in the filamentous-phage assembly and secretion processes.

Type IV systems are involved in a variety of physiopathological effects (Figure 14.4). These systems deliver toxins and nucleoprotein particles that interact with discrete receptors and, for all of the known systems, are in-

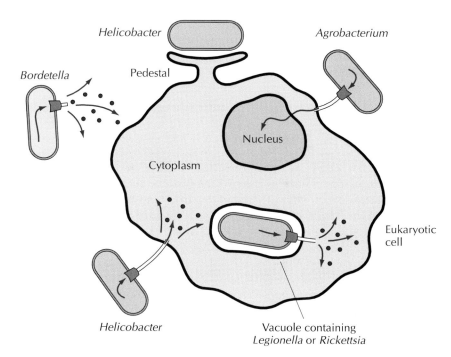

Figure 14.4 As with the type III systems, type IV systems are present in bacteria with different pathogenic behaviors. *Helicobacter* attachment to the eukaryotic-cell surface is mediated by pedestal induction. *Agrobacterium* and *Helicobacter* likely establish contact with eukaryotic host cells via a pilus structure prior to intracellular delivery of effector molecules. *Bordetella* secretes toxin via the Ptl system. Intracellular microorganisms, like *Legionella* and *Rickettsia*, confined in vacuoles, cross talk with the cytoplasmic compartment with a Ptl-like system, presumably deprived of piliated protrusions.

ternalized into cells, where they mediate a variety of intracellular responses. Within the eukaryotic cell, the effector molecules may perturb cellular functions through direct structural interactions. The liberation of pertussis toxin is a well-known example: the toxin has an ADP-ribosylating activity specific for GTP-binding proteins, causing cell intoxication. Alternatively, the transferred molecule may be DNA, in which case the expression of genes alters the cellular physiology. The example highlighted in this article is *Agrobacterium*-mediated T-DNA transfer, which ultimately disturbs plant hormone balances, leading to loss of cell growth control and to tumor formation. Similarly, the conjugal transfer of plasmids between bacterial cells often induces changes in cell physiology that aid in cell survival in harsh environments or increased virulence.

In the other type IV secretion systems, the variety of effects is extremely rich even though the exported substrate and its precise mode of action are not known. In *Helicobacter*, the *cag* system induces the epithelial secretion of IL-8, a mediator of chronic inflammation, and this process depends on NF-κB activation. In addition, a cellular protein of 145 kDa is tyrosine phosphorylated, and pedestals are formed after actin rearrangements and vasodilator-stimulated phosphoprotein (VASP) accumulation. Interestingly, similar events are induced by type III secretion-mediated effector export by enteropathogenic strains of *Escherichia coli*. The *Legionella icm/dot* system transfers DNA to other bacterial cells, although DNA is not thought to be the substrate involved in *Legionella* virulence. Rather, in mammalian hosts, the *dot/icm* products export a toxin that promotes intracellular multiplication and killing of human macrophages and prevention of phagosome-lysosome fusion. This latter response suggests the *dot/icm* system plays a direct role in altering communication processes between distinct subcellular compartments.

Conclusion

The evolution of a family of secretion systems from ancestral DNA conjugation machines raises many interesting questions and exciting new research directions. For example, it is of considerable biomedical importance to identify the toxin effectors exported by the *H. pylori* Cag and *L. pneumophila* Icm/Dot transporters. Equally important are detailed mechanistic and comparative studies of the type IV transporters with respect to the definition of (i) the basis for substrate recognition, (ii) the architectural arrangement and assembly pathway(s), (iii) the energy requirements for transporter assembly and function, and (iv) molecular interactions between these systems and eukaryotic host cells. Such information ultimately will be useful for applied studies aimed, for example, at designing drugs for selective inactivation of type IV secretion systems, or, even more enticing, at testing the potential of these systems for delivery of therapeutic protein or DNA macromolecules directly to eukaryotic cells.

Selected Readings

Akopyants, N. S., S. W. Clifton, D. Kersulyte, J. E. Crabtree, B. E. Youree, C. A. Reece, N. O. Bukanov, E. S. Drazek, B. A. Roe, and D. E. Berg. 1998. Analyses of the *cag* pathogenicity island of *Helicobacter pylori. Mol. Microbiol.* **28:**37–53.

Censini, S., C. Lange, Z. Xiang, J. E. Crabtree, P. Ghiara, M. Borodovsky, R. Rappuoli, and A. Covacci. 1996. *cag,* a pathogenicity island of *Helicobacter pylori,* encodes type I-specific and disease-associated virulence factors. *Proc. Natl. Acad. Sci. USA* **93:**14648–14653.

Christie, P. J. 1997. The *Agrobacterium tumefaciens* T-complex transport apparatus: a paradigm for a new family of multifunctional transporters in eubacteria. *J. Bacteriol.* **179:**3085–3094.

Covacci, A., S. Falkow, D. E. Berg, and R. Rappuoli. 1997. Did the inheritance of a pathogenicity island modify the virulence of *Helicobacter pylori? Trends Microbiol.* **5:**205–208.

Covacci, A., and R. Rappuoli. 1998. *Helicobacter pylori:* molecular evolution of a bacterial quasi-species. *Curr. Opin. Microbiol.* **1:**96–102.

Segal, E. D., C. Lange, A. Covacci, L. S. Tompkins, and S. Falkow. 1997. Induction of host signal transduction pathways by *Helicobacter pylori. Proc. Natl. Acad. Sci. USA* **94:**7595–7599.

Segal, G., and H. A. Shuman. 1998. How is the intracellular fate of the *Legionella pneumophila* phagosome determined? *Trends Microbiol.* **26:**253–255.

Vogel, J. P., H. L. Andrews, S. K. Wong, and R. R. Isberg. 1998. Conjugative transfer by the virulence system of *Legionella pneumophila. Science* **279:**873–876.

Winans, S. C., D. L. Burns, and P. J. Christie. 1996. Adaptation of a conjugal transfer system for the export of pathogenic macromolecules. *Trends Microbiol.* **4:**64–68.

15

Induction of Apoptosis by Microbial Pathogens

JEREMY E. MOSS, ILONA IDANPAAN-HEIKKILA, AND ARTURO ZYCHLINSKY

There are two distinct mechanisms of cell death: apoptosis and necrosis. The most significant difference between them is that in apoptosis the cell's own molecules are ultimately responsible for its death while in necrosis the cell is a victim of molecules synthesized by other cells. For example, apoptosis can be initiated by a cytokine binding to its receptor, which initiates a signaling cascade that activates a cell death program. Necrosis results from the effects of toxic molecules on a cell. Activation of complement results in the formation of a pore on the membrane of the target cell called the membrane attack complex. This pore disrupts the integrity of the cell membrane, causing a rapid exchange of ions and, eventually, osmotic lysis.

In multicellular organisms both development and homeostasis are achieved through a balance between cell death and cell growth. Programmed cell death, or apoptosis, is activated in multicellular organisms when death is desirable for the well-being of the whole organism. A classic and graphic example of the importance of apoptosis in development can be seen during the morphogenesis of the limb. During embryogenesis, the digits are interconnected by a web of cells. At a precise time point, the cells that constitute the interdigital web undergo programmed cell death, allowing the formation of independent fingers.

There are two criteria to distinguish apoptosis from necrosis: morphology and DNA fragmentation. Typical morphological changes (Figure 15.1) that occur during apoptosis include cell shrinkage and loss of normal cell-to-cell contacts, blebbing at the cell surface, and intense cytoplasmic vacuolization. Conservation of organelle structure, condensation of the chromatin (often at the perinuclear region), and loss of normal nuclear architecture also occur. In contrast, during necrotic cell death, the organelles are critically damaged, the plasma membrane is ruptured, the cytoplasmic elements are dispersed into the extracellular space, and the shape of the nucleus is normally conserved, although flocculation of chromatin may be detected.

DNA fragmentation is a biochemical characteristic that was detected early in the study of cell death. Apoptotic cells break up their DNA into

275

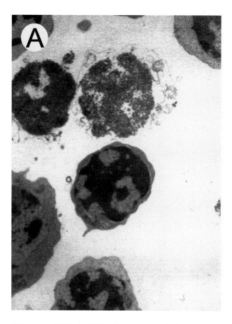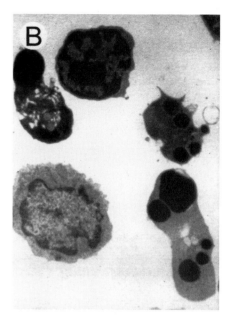

Figure 15.1   Morphology of thymocytes undergoing necrosis and apoptosis. **(A)** Thymocytes undergoing necrosis. Organelles are critically damaged, the plasma membrane is ruptured, the cytoplasmic elements are dispersed, the shape of the nucleus is normally conserved, but flocculation of chromatin can be detected. **(B)** Thymocytes undergoing apoptosis. Cell shrinkage, blebbing at the cell surface, intense cytoplasmic vacuolization, conservation of organelle structure, condensation of the chromatin at the perinuclear region, and loss of nuclear morphology, which appears like segmentation of the nucleus, are all apparent.

multimers of approximately 200 bp, correlating roughly with the size of a nucleosome. This fragmentation of DNA can be quantified and used as a marker for programmed cell death. In contrast, during necrosis, the DNA of the dying cell remains intact initially although it is degraded eventually.

Another characteristic of apoptotic cells is that they express specific markers on their surface, such as vitronectin and phosphatidylserine, which can be recognized by professional phagocytes. These cells can rapidly engulf the apoptotic bodies before their intracellular contents are spilled into the tissue, and therefore they prevent inflammation.

Interestingly, the morphological changes, the fragmentation of DNA, and the expression of markers for recognition by phagocytes are very similar across different cell types and species. This reflects remarkable conservation of the mechanism of apoptosis, which is further underscored by the homology across species of the genes involved in this process.

Apoptosis Is Genetically Programmed

Initially, the demonstration that there was genetic information for cell death, i.e., that cell death was "programmed," came from now classical studies of the nematode *Caenorhabditis elegans*. Two features of *C. elegans* make it an ideal organism for the study of development and consequently of cell death. First, these worms are transparent, and so the fate of each cell can be microscopically observed in living animals. Second, of 1,090 somatic cells generated during hermaphrodite development, 131 naturally undergo programmed cell death. Furthermore, each cell in the worm has been mapped in both genealogy and position.

In mutagenesis studies, two cell death (*ced*) genes, *ced-3* and *ced-4*, were shown to be required for cell death. If either of these genes was inactivated, death failed to occur and mutant adult worms ended up with "extra" cells (Table 15.1). A third gene involved in the regulation of cell death is *ced-9*. The product of this gene inhibits cell death and functions upstream of *ced-3* and *ced-4*. The discovery of these genes demonstrated unequivocally that cell death was genetically programmed. Amazingly, the apoptotic program described in *C. elegans* is conserved in mammalian cells since not only are there mammalian homologues of *ced-3*, *ced-4*, and *ced-9*, but also, in some cases, mammalian genes can complement mutations in worms and worm genes are functional in mammalian cells.

In mammals, the program for cell death is more complex than that in worms and the induction of apoptosis is highly regulated. Furthermore, there are many homologues of certain cell death genes, suggesting both redundancy and multiple pathways of apoptosis. There has been a recent explosion in the identification of genes that are involved in apoptosis; however, a clear picture of how all of the pieces to the puzzle fit together has not been established.

Triggering of Apoptosis

Signal Transduction of a Cell Death Signal

Many apoptotic signals are received by the cell through surface receptors. Several mechanisms exist to transduce either the pro- or the antiapoptotic signal from the cell surface to the cell death machinery. The initial signal frequently comes from the binding of a ligand to a specific receptor at the cell membrane. This signal is transduced within the cell and can initiate several pathways, including generation of second messengers like an increase in intracellular cyclic AMP (cAMP) or calcium concentrations, activation of specific kinases that start the apoptotic program, and interaction with molecular adapters that directly connect with cell death effectors.

An illustrative and relatively simple example of this process is the case of Fas ligand-induced apoptosis. Fas ligand is a molecule related to the cytokine tumor necrosis factor alpha (TNF-α) and is important in T-lymphocyte cytotoxicity. Fas ligand binds to Fas, its receptor, on the target cell. This binding leads to an active receptor complex which transduces a death signal via the cytosolic adapter molecule Fas-associating protein with death domain (FADD). FADD connects the receptor and an effector protease of the caspase family. The function of caspases is discussed below.

Table 15.1   *C. elegans* genes involved in apoptosis

Type of gene	Gene in:	
	C. elegans	Mammals
Proapoptotic	*ced-3*	Caspases
	ced-4	*apaf-1* family
Antiapoptotic	*ced-9*	*bcl-2* family

Apoptosis and the Cell Cycle

Apoptosis can also be triggered from within, when cells receive conflicting signals for cell cycle progression and arrest or after irreparable damage to DNA. p53 is proposed to be instrumental in blocking cell cycle progression and activating apoptosis. p53 is a DNA-binding protein that can transactivate genes. Inactivation of p53 prevents cells from undergoing apoptosis after specific stimuli, supposedly because it controls the expression of cell death effector genes at a crucial cell cycle checkpoint. This tumor suppressor gene might prevent malignant transformation by activating programmed cell death. In fact, it is the most commonly mutated gene in human cancer.

Effector Molecules of Apoptosis

Caspases

Caspases are a family of cysteine proteases that play a central role in the apoptotic pathway. Among cysteine proteases, caspases are unique in requiring an aspartate at the cleavage site. The first caspase to be isolated, interleukin-1β (IL-1β)-converting enzyme (ICE; caspase 1), was identified by classical biochemical studies, using the limited proteolysis of IL-1β as an assay. IL-1β is a cytokine, i.e., a protein that signals to other cells, with significant proinflammatory activity. IL-1β is synthesized as a biologically inactive 30-kDa protein which is cleaved to a mature form of 17 kDa. The initial link between ICE and apoptosis was made by sequence comparison; the *ICE* gene is homologous to the *C. elegans* cell death gene *ced-3*. Furthermore, both *ICE* and *ced-3* induce apoptosis when overexpressed in mammalian cells in tissue culture and *ICE* complements the *ced-3* mutation in *C. elegans*.

More than 10 caspases have been identified in humans. Most of these induce apoptosis when overexpressed in tissue culture cells, but not all are required for different apoptosis pathways, as shown in gene disruption experiments with mice. For example, ICE knockout mice develop normally and have no overt phenotype in apoptosis. This suggests either that ICE plays no role in developmental or homeostatic cell death or that there is redundancy in caspase function. Caspase 3 is one of the most commonly activated caspases in apoptosis. In contrast to ICE knockout mice, caspase 3 knockout mice have profound developmental alterations, although some of the mice are viable. These mice usually die young due to severe brain malformations.

The regulation of caspase activation is one of the key steps in the control of the apoptotic process (Figure 15.2). The caspase precursors are abundant in the cell cytosol. In response to diverse apoptotic stimuli, they can be proteolytically cleaved into active enzymes. Caspase precursors contain an amino-terminal end that varies in size depending on the caspase and is thought to regulate the cleavage of the precursor. The activation of some caspase precursors is autocatalytic and can lead to cleavage and activation of other members of the family. It has not been clearly established whether caspases are organized in a proteolytic cascade or whether there are caspases that can cleave their relevant substrates and induce apoptosis independently of all other caspases.

The key caspase substrates are beginning to be discovered. Caspase targets appear to be crucial in maintaining the cell architecture, RNA splic-

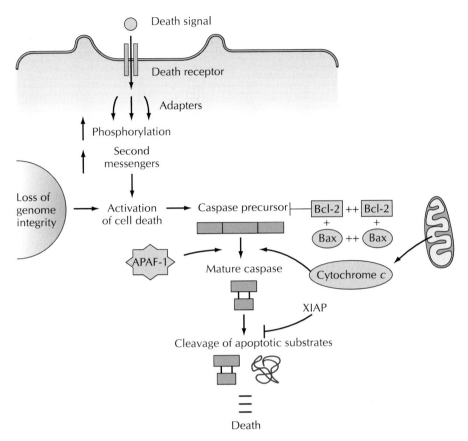

Figure 15.2   Models for induction of apoptosis. Receptor binding or DNA damage activates a signal cascade that culminates in caspase activation and/or disruption of binding of members of the Bcl-2 family and release of mitochondrial proteins required for cell death.

ing, and DNA repair. These substrates include nuclear lamins, gelsolin, poly(ADP-ribose) polymerase, and the retinoblastoma protein. The hallmark of apoptosis, degradation of DNA into nucleosomal fragments, results from caspase cleavage of a substrate called DNA fragmentation factor. It is still unclear whether all the aforementioned substrates are required for the induction of apoptosis or whether a coordinate cleavage of substrates is needed. Therefore, although caspase activation results in apoptosis, the cascade of programmed cell death that leads from these proteases to apoptosis is not yet understood. It is interesting that although all caspases require an aspartate at the cleavage site, each caspase recognizes a different amino acid sequence, suggesting that they cleave different substrates.

Caspase activity can be blocked by specific inhibitors, like the cellular inhibitors of apoptosis (IAPs). Viruses, like baculovirus and poxvirus, which are strict intracellular pathogens and can benefit from preventing host death encode other caspase inhibitors.

Bcl-2 Family

bcl-2, the first gene identified in the Bcl-2 family, maps to a conserved chromosomal translocation site in B-cell lymphomas. Several Bcl-2 family members have been identified in mammals and viruses. Members of the *bcl-2* family encode proteins that either inhibit (like Bcl-2 and Bcl-xL) or activate (like Bax, Bad, Bik, and Bak) apoptosis. Bcl-2 is homologous to the C. *elegans* apoptosis inhibitor gene *ced-9*. *bcl-2* complements the *ced-9* mutation in worms and inhibits apoptosis in many different instances when over-expressed in mammalian cells. The highest homology between members of this large family is found in two specific regions called Bcl-2 homology domains 1 and 2 (BH-1 and BH-2). Both of these domains are crucial for binding to other Bcl-2 homologues.

The Bcl-2 family of proteins form homo- and heterodimers which antagonize or enhance the functions of the two partners (Figure 15.2). For example, overexpression of Bad, which results in apoptosis, leads to both formation of Bad-Bad homodimers and disruption of Bcl-2–Bcl-2 homo-dimers to form Bad–Bcl-2 heterodimers. In this case, it is an open question whether the proapoptotic activity of Bad is the result of the activity of Bad-Bad homodimers or of the disruption of the protective Bcl-2–Bcl-2 homo-dimer or both. It appears that the fate of the cell depends on the relative amounts of Bcl-2 inhibitors and activators present.

Mutating both the BH-1 and BH-2 domains of Bcl-xL does not abrogate the antiapoptotic activity of this molecule, indicating that it has important functions other than binding Bcl-2 family members. Bcl-xL also binds to Ced-4, the product of a gene important in the induction of apoptosis in C. *elegans*. Ced-4 can bind to and activate caspases. Recently, Apaf-1, the first human protein with sequence similarity to Ced-4, was isolated and shown to activate caspase 3 in a cytochrome *c*-dependent manner. Thus, it appears that members of the Bcl-2 family might modulate caspase activity through Ced-4 homologues.

Bcl-2 also prevents cytochrome *c* release from mitochondria. Cyto-chrome *c* is a mitochondrial protein localized in the intermembrane space and is involved in cellular respiration. Intact cells undergo apoptosis after release of cytochrome *c* in the cytosol, indicating that this protein, in ad-dition to its function in respiration, has a function in apoptosis. Thus, it has been postulated that Bcl-2 could act in situ on mitochondria by inhibiting cytochrome *c* release.

The mechanism of action of this family of proteins remains elusive. Bcl-2 localizes to the outer mitochondrial membrane and the endoplasmic reticulum and might regulate the redox potential of the cell. Interestingly, the three-dimensional structure of Bcl-xL has revealed striking similarity to the pore-forming subunits of bacterial proteins, such as diphtheria toxin and colicins. The structural similarity led to experiments that indicate that Bcl-xL forms channels in lipid membranes. This activity allows for altera-tions in mitochondrial permeability by the Bcl-2 family of proteins.

Apoptosis in Diseases

The inappropriate induction of cell death is involved in the pathogenesis of a number of diseases. The untimely activation or inhibition of apoptosis plays a role in cancer, viral latency, autoimmune diseases, and microbial infections.

Infectious agents have evolved ways to affect the host cell suicide pro-gram. Some viruses and bacteria are able to activate the apoptotic program of the infected cell as part of their means of causing disease. Several mi-crobes modulate apoptotic pathways to survive host defense mechanisms or otherwise guarantee optimal living conditions. Another goal might be to facilitate the initiation of infection and inflammation, thus securing efficient microbial spread. Finally, other microbes might inhibit apoptosis to guarantee the life of a host cell that is useful for their survival or persistence.

Induction of Apoptosis by Microorganisms

Many bacterial pathogens induce apoptosis in host cells (Table 15.2). This review groups microbial pathogens by the mechanisms they use to induce apoptosis. The proapoptotic strategies used include activation of cell surface receptors, mimicry of second messengers, regulation of caspase function, inhibition of protein synthesis, disruption of the host cell membrane, and, finally, unknown mechanisms.

Activation of Host Cell Receptors That Signal for Apoptosis

Staphylococcus aureus is a gram-positive coccus that can be part of the normal flora of the skin and mucosa. This microorganism is also the etiological agent of a variety of diseases including skin lesions, food poisoning, toxic shock syndrome, endocarditis, and osteomyelitis. Many virulence factors implicated in staphylococcal virulence, including superantigens, can cause apoptosis (discussed below). *Streptococcus pyogenes* is another gram-positive coccus; it is an important cause of pharyngitis and is sometimes associated with serious sequelae. It has also been associated with toxic shock-like syndrome through exotoxins that act as superantigens. Superantigens are proteins that activate T cells by directly binding both to major histocompatibility complex class II molecules on antigen-presenting cells and to the T-cell receptor (TCR) on T cells. Normally, T cells are activated only when the TCR recognizes a peptide in the context of the major histocompatibility complex. The superantigens of streptococci and staphylococci recognize specific TCRs of the Vß family.

Engaging the TCR on T cells can activate programmed cell death. Activation of the TCR through superantigens like staphylococcus exotoxin B

Table 15.2 Bacteria that cause apoptosis

Actinobacillus actinomycetemcomitans
Bordetella pertussis
Clostridium difficile
Corynebacterium diphtheriae
Escherichia coli
Helicobacter pylori
Legionella pneumophila
Leptospira interrogans
Listeria monocytogenes
Mycobacterium spp.
Pasteurella haemolytica
Pseudomonas aeruginosa
Salmonella spp.
Shigella spp.
Staphylococcus aureus
Streptococcus pyogenes
Yersinia spp.

(SEB), SEA, SED, and SEE induce apoptosis in T cells and thymocytes both in vitro and in vivo. Although it is likely that superantigens activate apoptosis by engaging the TCR, the precise pathway by which superantigen activation results in programmed cell death is not yet known. It is clear, however, that T cells have to be active and proliferating in order to die. SEB-induced cell death is mediated by both protein kinase C and an ATP-gated ion channel.

Vß cells are depleted in patients infected with toxigenic *S. pyogenes*. Furthermore, peripheral blood mononuclear cells isolated from these toxic shock patients undergo apoptosis in vitro. This suggests that induction of apoptosis plays an active role in superantigen-induced toxic shock syndrome.

It has yet to be established whether superantigen induction of T-cell apoptosis plays a role favorable or harmful to the host in the pathogenesis of microbial infections. On the one hand, it can be argued that the reduction in the number of T cells after superantigen exposure and the consequent decrease of immune system function may be important for pathogen survival. On the other hand, the deletion of T cells by superantigens might control the unregulated proliferation and activation of T cells during an infection and thus prevent an exaggerated immune response.

Induction of Second Messengers

Bordetella pertussis, a gram-negative rod, is the etiological agent of whooping cough, an upper respiratory tract infection characterized by a cough with an inspiratory "whoop." *Bordetella* is highly contagious and is transmitted through air droplets. In the course of the infection, the bacteria remain localized to the respiratory tract, where they evoke an acute inflammation.

B. pertussis kills macrophages by apoptosis in vitro by secreting adenylate cyclase-hemolysin (AcHly) toxin. This toxin has two domains: (i) a potent adenylate cyclase activity, which is activated by calmodulin, and (ii) a hemolysin activity, which is a pore-forming protein that is thought to allow the translocation of the cyclase into the host cell cytoplasm. AcHly kills macrophages only when both parts of the toxin are functional. This indicates that the disruption of the host cell cytoplasmic membrane through the pore-forming domain of this toxin is not sufficient to kill the cell.

An increase in the intracellular concentration of cAMP triggers pathways that lead to apoptosis, and curtailing the production of this second messenger can prevent programmed cell death. Hence, it is interesting to speculate that *Bordetella* initiates apoptosis by sharply increasing the intracellular concentration of cAMP to activate a program for cell death.

Bordetella encodes another toxin, pertussis toxin (PT), which also increases the intracellular cAMP concentration. PT is a member of the A/B family of toxins, in which the B subunits allow the translocation of the enzymatically active A subunit into the cytoplasm of the target cell. PT ADP-ribosylates a G protein, inhibiting the inhibitory subunit which acts on the host adenylate cyclase. Thus, indirectly, by inhibiting the downregulator of the host cell endogenous cyclase, PT activity leads to an increase in the intracellular cAMP concentration. Interestingly, PT is not necessary for *Bordetella*-mediated induction of macrophage cytotoxicity. Thus, it appears that the kinetics and/or magnitude of the increase in the cAMP concentration may be the key determinant of apoptosis activation.

The importance of *Bordetella*-induced apoptosis is still not understood. AcHly is produced early in the infection and is then down-regulated during the later, chronic phase of the disease. Therefore, it is possible that *Bordetella* kills alveolar macrophages to eliminate the first line of defense that it encounters. Alternatively, early macrophage apoptosis might be important in triggering an inflammatory response (see the next section).

Regulation of Caspase Activity

Activation of Caspases

Shigella is a gram-negative rod that causes dysentery, a severe form of diarrhea that often contains blood and mucus. *Shigella* is transmitted through the fecal-oral route and is an extremely infectious agent.

Shigella is an invasive organism that induces macrophage apoptosis. This pathogen is phagocytosed by macrophages but escapes from the phagocytic vacuole. Inside the cytoplasm of the macrophage, *Shigella* secretes, among other proteins, invasion plasmid antigen B (IpaB). IpaB disseminates throughout the cytoplasm, binding to and activating ICE (caspase 1 [see above]). The activity of mature ICE is responsible for both macrophage apoptosis and maturation of IL-1β. *Shigella* induces classical apoptosis, as determined by morphological changes and the fragmentation of the host cell DNA. Since IpaB has to be delivered by the bacterium into the compartment where ICE resides, *Shigella* is able to induce apoptosis only from within the cytoplasm.

All clinical isolates of *Shigella* tested thus far induce macrophage cell death. Apoptosis is up-regulated in tissues of animals experimentally infected with *Shigella* and in patients suffering from shigellosis. Taken together, these data indicate that apoptosis is activated in *Shigella* infections and is not an in vitro artifact.

Recent findings strongly suggest that macrophage apoptosis is an important step in *Shigella* pathogenesis. This bacterium has a very specific tropism for the colon, where it penetrates through M cells (Figure 15.3) into lymphoid follicles. After M-cell translocation, *Shigella* encounters resident macrophages. Colonic macrophages are usually activated, probably because of the constant sampling of bacterial products from the lumen of the colon. The infected macrophages undergo apoptosis, and because *Shigella*-induced apoptosis is dependent on ICE activation, the proinflammatory cytokine IL-1β is processed to its mature form. IL-1 is likely to be the first signal to initiate the inflammatory response to this invasive organism. An acute inflammation, rich in polymorphonuclear lymphocytes (PMN), ensues. These inflammatory cells compromise the integrity of the intestinal barrier as they migrate toward the lumen of the intestine and allow for the entry of more bacteria. The disruption of the epithelial cell barrier is a necessary step in the pathogenesis of *Shigella*.

Salmonella, like *Shigella*, is a gram-negative enteric pathogen that is transmitted orally. Depending on the bacterial serovar and host specificity, *Salmonella* either remains localized in the gut and produces gastroenteritis or is dispersed hematogenously to other organs such as the spleen and the liver, as in typhoid fever.

Salmonella can also kill macrophages by apoptosis, although the mechanism is unclear. It has been demonstrated that during *Salmonella*-induced apoptosis there is no autocrine induction of a death signal by macrophages. It appears, however, that *Salmonella*, like *Shigella*, induces apoptosis only when it is inside macrophages. Only bacteria competent for invasion and capable of generating characteristic ruffles in the host cell membrane kill

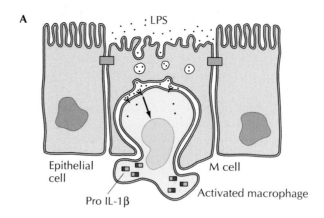

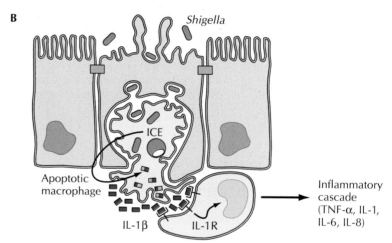

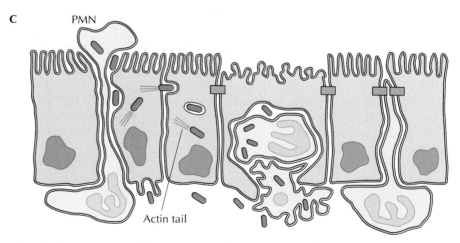

Figure 15.3 Model of *Shigella* pathogenesis. **(A)** Colonic macrophages are activated to synthesize IL-1β by exposure to bacterial products of the normal microflora, like lipopolysaccharide (LPS). **(B)** *Shigella* translocates through M cells, infects resident macrophages, and escapes the phagocytic vacuoles. It secretes IpaB, which binds to ICE, which is activated. ICE activation leads to macrophage apoptosis and IL-1β maturation. IL-1β initiates an acute inflammatory response. **(C)** PMN break the epithelial barrier, allowing further bacterial invasion. Bacteria invade epithelial cells through the basolateral side and then spread from cell to cell. Redrawn from A. Zychlinsky and P. J. Sansonetti, *Trends Microbiol.* 5:201–204, 1997.

macrophages; noninvasive mutants do not activate programmed cell death. *Salmonella*, unlike *Shigella*, induces a significant increase in the concentration of intracellular calcium when it invades host cells. As mentioned above, increases in the intracellular calcium concentrations can act as a second messenger to activate apoptosis. Whether the changes in intracellular calcium concentrations are important in the activation of *Salmonella*-induced apoptosis remains to be determined.

Salmonella-induced apoptosis requires an intact type III secretion apparatus. Interestingly, one of the proteins that is secreted by the *Salmonella* type III system is *Salmonella* invasion protein B (SipB), which has sequence homology to IpaB of *Shigella*. It is possible that SipB also activates the cell death machinery of macrophages by binding to a caspase.

Salmonella typhimurium, which causes a syndrome similar to typhoid fever in mice, requires a functional type III secretion apparatus to cause disease when it is administered orally. If the intestinal infection is bypassed and *Salmonella* is inoculated directly into the peritoneum, type III secretion is no longer required for virulence. These data suggest that if apoptosis played a role in these diseases, it would be important for the intestinal phase of the infection. Based on the parallels between *Salmonella* and *Shigella* infections, it is possible that apoptosis is proinflammatory in salmonellosis. Apoptosis has also been observed in the liver, a target organ of *S. typhimurium* in its systemic phase. Future research will have to address this issue.

Inhibition of Caspases

Several viral proteins are known to inhibit caspase activity, preventing the host cell from undergoing apoptosis. An interesting example is that of the poxviruses, a family of DNA viruses that replicate in the eukaryotic cytoplasm. Cowpox virus, a member of this family, causes skin lesions that are the result of both viral replication and the host inflammatory response. This virus contains the *crmA* gene, encoding cytokine response modifier A (CrmA). CrmA is a caspase competitive inhibitor of the serpin family. This inhibitor is quite selective in its ability to block caspases, showing the highest affinity for ICE. In vitro, CrmA inhibits the enzymatic activity of ICE, and when it is ectopically expressed in mammalian cells, it prevents programmed cell death initiated by a variety of stimuli.

Infections with poxvirus *crmA* mutants result in larger skin lesions than those from wild-type viruses. This suggests that in infections with wild-type virus, inhibition of ICE results in a lower production of IL-β and that this cytokine is a key mediator of the inflammatory response to cowpox virus. The viral inhibition of ICE prevents the intense inflammation, possibly allowing further viral replication. As with activation of caspase in *Shigella* infections, it remains to be determined whether the poxvirus modulation of inflammation through ICE is linked to the proapoptotic potential of this enzyme.

Inhibition of Apoptosis via Bcl-2

Until recently, there has been no description of a bacterium that inhibits apoptosis of its host cell (Box 15.1). This is surprising since some intracellular microbial pathogens could benefit from prolonging the life of their host. Many viruses, however, use homologues of Bcl-2 or other members

BOX 15.1

Bacteria also inhibit apoptosis

Many bacterial species can induce apoptosis of host cells. Nevertheless, for obligate intracellular pathogens, inducing apoptosis would translate into losing their host. Recent work on the obligate intracellular bacterium

Rickettsia rickettsii, the etiological agent of Rocky Mountain spotted fever, shows that this bacterium inhibits cell death. *R. rickettsii* infects endothelial cells and activates the transcription factor NF-κB, which turns on genes that block apoptosis. Inactivation of NF-κB during *R. rickettsii* infections results in apoptosis. These studies

suggest that *R. rickettsii* can modulate the host cell apoptotic response to determine its fate in the host.

Clifton, R. D., R. A. Goss, S. K. Sahni, D. van Antwerp, R. B. Baggs, V. J. Marderm, D. J. Silverman, and L. A. Sporn. NF-κB-dependent inhibition of apoptosis is essential for host cell survival during *R. rickettsii* infection. Submitted for publication.

---

of that gene family to prolong the life of their host cell. For example, adenovirus encodes the protein E1B19-kD, which is a Bcl-2 homologue. Infections with adenovirus carrying mutations in E1B-19kD induce apoptosis in infected cells. Bcl-2 can complement this mutation, suggesting that Bcl-2 and E1B-19kD have an analogous function. Although the sequence similarity between E1B-19kD and Bcl-2 is not significantly high over the entire protein, the domains involved in the inhibition of apoptosis are conserved. Furthermore, it has been shown that E1B-19kD and Bcl-2 interact with a similar set of cell death proteins, explaining the analogy in function between these two proteins.

A number of herpesviruses including Epstein-Barr virus, herpesvirus saimiri, and Kaposi's sarcoma-associated virus contain homologues of *bcl-2.* Most of these viruses can cause persistent infections, and the products of the *bcl-2* homologues might allow the extended survival of the host cell. Epstein-Barr virus encodes another gene, latent membrane protein 1 (LMP1), which regulates the expression of genes that control apoptosis, including *bcl-2.*

Inhibition of Protein Synthesis

Corynebacterium diphtheriae, a non-spore-forming gram-positive rod, and the gram-negative bacteria *Pseudomonas aeruginosa, Shigella dysenteriae,* and enterohemorrhagic *Escherichia coli* (EHEC) secrete toxins that inhibit eukaryotic translation and activate apoptosis in host cells. All these toxins are of the A/B type, like PT. They inhibit translation through different mechanisms. Inhibition of eukaryotic translation or transcription by other substances, like cycloheximide or actinomycin, also eventually activates apoptosis in many cell types.

The symptoms of diphtheria are sore throat, fever, and a characteristic gray, adherent pseudomembrane at the back of the pharynx. This fibrillar pseudomembrane contains bacteria and infiltrating inflammatory cells. A dangerous complication of diphtheria is obstruction of the airways by an enlarged pseudomembrane.

C. diphtheriae lives extracellularly and secretes diphtheria toxin (DT). The B subunit of DT binds to an extracellular glycoprotein receptor, allowing the toxin to be endocytosed by several different cell types including epithelial and myeloid lines. The acidic environment of the phagolysosome exposes a specific domain and creates an aqueous pore that allows the translocation of the A subunit into the host cell cytoplasm. The A subunit is an enzyme that ADP-ribosylates elongation factor 2 (EF-2), an essential component of the translation machinery. The ribosylation inhibits EF-2 activity and blocks protein synthesis, eventually leading to apoptosis. Three lines of evidence strongly support the hypothesis that DT induces apoptosis by inhibiting translation: (i) the levels of inhibition of protein synthesis and cytotoxicity correlate very tightly in DTX-treated cells; (ii) host cells carrying mutations in EF-2 which prevent ADP ribosylation are insensitive to DTX-induced apoptosis; and (iii) inhibition of ADP-ribosylation by DT blocks both cytotoxicity and the inhibition of protein synthesis.

Although inhibition of protein synthesis is clearly necessary in DT-mediated apoptosis, it may not be sufficient. It has been proposed that DT might have another proapoptotic activity. Apparently, this toxin has a calcium- and magnesium-sensitive nuclease activity in the A subunit. It is interesting to speculate that DT might initially damage DNA and that the cell, in the absence of protein synthesis, might be incapable of making the proteins that would repair the damaged DNA. The notion that DT-mediated inhibition of translation is not sufficient to kill the host cell is supported by the cloning of the cellular apoptosis susceptibility (cas) gene. CAS is homologous to a yeast protein involved in cell division. Overexpression of antisense cas makes susceptible cells resistant to DT cytotoxicity, although protein synthesis is still blocked in these cells. Whether apoptosis plays a role in diphtheria pathogenesis is still an open question. Cell death might help the bacteria evade the host response by killing macrophages. Alternatively, the killing of cells may facilitate bacterial spread and colonization.

P. aeruginosa is found in soil and water and sometimes in the flora of the gut. It can cause sepsis, urinary tract infection, and pneumonia, particularly in immunocompromised persons and cystic fibrosis patients. One of the key virulence factors of P. aeruginosa is exotoxin A (ExoA), which, like DT, ADP-ribosylates EF-2 and inhibits protein synthesis. ExoA and DT do not have significant sequence homology. However, ExoA also induces apoptosis in a human monoblastoid cell line. Mutations in EF-2 confer resistance to ExoA, suggesting that, as with DT, inhibition of translation is necessary for ExoA-mediated apoptosis. Interestingly, it has been reported that ExoA, like DT, has a nuclease activity. The role of ExoA in Pseudomonas infections has not been thoroughly analyzed.

S. dysenteriae and EHEC cause a dysenteric syndrome and can be associated with sequelae involving the kidneys and the central nervous system (CNS). S. dysenteriae and EHEC both produce toxins, Shiga toxin (ST) or Shiga-like toxins (SLT), respectively, which are almost identical. The B subunits of these A/B toxins mediate binding to globotriaosylceramide (Gb$_3$), the receptor on the host cell membrane. The A subunit cleaves eukaryotic rRNA and disrupts ribosomal function and hence inhibits protein synthesis. Induction of apoptosis by these toxins has been documented both in vitro and in vivo. The purified B subunit of the toxin is sufficient to kill epithelial cells, suggesting that Gb$_3$ binding might activate a signal transduction cascade that culminates in apoptosis. The inhibition of protein synthesis might synergize with the Gb$_3$-dependent pathway, preventing the synthesis of inhibitors of apoptosis.

Although these toxins are cytotoxic in vitro, the significance of the induction of apoptosis in vivo is still unclear. Cells in both the kidneys and the CNS are enriched in Gb$_3$ receptors. This distribution might explain the specific localization of the sequelae of infections with bacteria that secrete either ST or SLTs. Significantly, macrophages do not express the Gb$_3$ receptor and are not susceptible to ST or SLT cytotoxicity.

Disruption of the Cytoplasmic Membrane

Several pore-forming proteins (PFP) are made by microbes. These toxins include *Staphylococcus aureus* alpha-toxin, *Actinobacillus actinomycetemcomitans* leukotoxin, *Listeria monocytogenes* listeriolysin, and *E. coli* hemolysin. It appears that large doses of PFP massively disrupt the integrity of the cell membrane and cause necrosis. At lower doses, a more delicate modification of the host cell membrane results in apoptosis. It is still unclear how disruption of the cell membrane can lead to apoptosis. Nevertheless, mutations in PFP that lower their hemolytic activity also decrease their apoptotic potential, demonstrating that the pore-forming ability is directly linked to apoptosis.

It is still not known whether bacterial pore-forming proteins bind to a specific host cell receptor. It has been postulated that low doses of *S. aureus* alpha-toxin bind to a host cell receptor and allow the influx of Na^+. This Na^+ influx might be cytotoxic because of the ionic imbalance it generates. Hyperpolarization of the cytoplasmic membrane has been previously shown to lead to apoptosis. Alternatively, the Na^+ influx might indirectly lead to an increase in the intracellular concentration of Ca^{2+}, which serves as a second messenger to activate apoptosis. High doses of alpha-toxin generate larger holes in the host cell membrane. These larger holes allow the free flow of Ca^{2+} and are rapidly cytotoxic to cells. Alpha-toxin appears to be an important staphylococcal virulence factor; however, the relevance of alpha-toxin-induced apoptosis in any of the diseases that *S. aureus* can cause (see "Activation of host cell receptors that signal for apoptosis," above) is not clear.

A. actinomycetemcomitans is a gram-negative coccobacillus that is associated with periodontitis as well as meningitis and endocarditis. It produces leukotoxin which causes cytotoxicity in lymphoid cells but spares fibroblasts, platelets, and endothelial and epithelial cells. Destruction of immune system cells may be important in the development of disease. In addition, *Actinobacillus* induces apoptosis independently of leukotoxin. There are *Actinobacillus* strains that do not make leukotoxin but still kill macrophages. This second pathway requires intracellular bacteria and involves a protein kinase C pathway but not a cAMP-dependent protein kinase pathway.

L. monocytogenes is a gram-positive rod that can cause meningitis and sepsis in infants and immunosuppressed patients. It is transmitted orally, and after invading the gastrointestinal tract, it can spread systemically. *Listeria* invades cells and escapes from the phagolysosome into the cytoplasm of the host cell. It can kill cells, however, without invading them. *Listeria* induces apoptosis in dendritic cells and hepatocytes, but it is not cytotoxic to macrophages. Dendritic cells are antigen-presenting cells located in lymphoid aggregates throughout the body including the gut. Killing of dendritic cells might be an important way of preventing the immune system from mounting a timely immune response against this pathogen.

The liver receives much of the blood drainage from the lower gastrointestinal tract and is one of the first organs to be infected with *Listeria* in a systemic infection. When *Listeria* infects the liver, it induces hepatocyte apoptosis. Hepatocyte apoptosis causes the release of a yet to be defined PMN chemoattractant. Consequently, these important immune system cells are brought into an infected liver, probably to promote bacterial clearance. Thus, it is likely that, like *Shigella* infections of macrophages (see above), *Listeria* infections of hepatocytes are proinflammatory.

E. coli hemolysin can kill stimulated but not unstimulated T-lymphocytes, suggesting that a factor that allows apoptosis is up-regulated (or an inhibitor is down-regulated) upon stimulation. Hemolysin is important in the pathogenesis of *E. coli* infections of the urinary tract.

Unknown Mechanisms

Yersinia

Yersinia, like *Salmonella* and *Shigella*, to which it is closely related, has recently been shown to cause apoptosis of murine and human macrophages. *Y. enterocolitica* and *Y. psuedotuberculosis* are acquired by ingestion of contaminated food. *Yersinia* gains access to the intestinal epithelium and replicates in Peyer's patches. Unlike *Salmonella* and *Shigella*, however, *Yersinia* does not persist in macrophages; instead, it exerts cytotoxic effects from outside the cell. Several bacterial genes are associated with the ability of *Yersinia* to induce macrophage apoptosis. Mutants which cannot attach to host cells are not cytotoxic. Also, *yopJ* and *yopP* mutants are not cytotoxic, suggesting that the products of these genes are important in induction of apoptosis. Interestingly, these *Yersinia* genes are homologous to genes in cytotoxic plant pathogens.

Mycobacterium

Mycobacterium tuberculosis is an acid-fast bacterium that causes the respiratory tract disease tuberculosis. *M. tuberculosis* is transmitted by respiratory droplets and invades the lungs, where it infects alveolar macrophages, which serve as its residence for the duration of the infection. The macrophages elicit an immune response against the bacteria, leading to the characteristic histological caseating granuloma of tuberculosis. *M. avium-M. intracellulare* causes a similar disease but only in immunocompromised individuals. *M. avium-M. intracellulare*, like *M. tuberculosis*, also resides in alveolar macrophages. Mycobacteria induce apoptosis in macrophages in vitro by an as yet undefined mechanism that possibly involves TNF-α. In tissue sections of mycobacterially infected lungs, apoptotic cells have been detected by electron microscopy and by in situ demonstration of DNA fragmentation. Since mycobacteria require the host cell to replicate and survive, induction of macrophage apoptosis would probably be beneficial to the host, since it would deprive the pathogen of a place to live. Indeed, it has

---

BOX 15.2

Can bacteria survive host cell apoptosis?

Mycobacteria live in a special compartment in the macrophages, where they are protected from the bactericidal mechanisms of the host. Mycobacteria are killed to a greater extent when the host macrophage is stimulated to undergo apoptosis than when the host cell dies by necro-

sis. In a series of experiments, macrophages infected with bacillus Calmette-Guérin (BCG) were treated either with ATP, a known inducer of apoptosis in macrophages, or H_2O_2, which causes necrosis, and the surviving bacteria were counted. There were significantly fewer colony-forming units in macrophages treated with ATP than in those incubated with H_2O_2. The way in which apoptotic

macrophages kill BCG is still not understood. Whether induction of host cell apoptosis plays a role in mycobacterial infections remains to be determined.

Molloy, A., P. Laochumroonvorapong, and G. Kaplan. 1994. Apoptosis, but not necrosis, of infected monocytes is coupled with killing of intracellular bacillus Calmette-Guérin. *J. Exp. Med.* **180:**1499–1509.

been shown that more virulent *Mycobacterium* strains induce less apoptosis than avirulent strains do (Box 15.2).

Helicobacter pylori

The gram-negative rod *Helicobacter pylori* causes gastritis, peptic ulcers, gastric atrophy, and carcinoma. This bacterium contains several virulence factors, but none of them has been shown to induce apoptosis. There is a significant increase in the number of apoptotic cells in patients infected with *H. pylori* compared to controls, and eradication of the infection after treatment reduces the number of apoptotic cells to control levels. The apoptosis-inducing ability of *H. pylori* may be the key to understanding the gastric atrophy in infected persons.

Selected Readings

Chen, Y., M. R. Smith, K. Thirumalai, and A. Zychlinsky. 1996. A bacterial invasin induces macrophage apoptosis by binding directly to ICE. *EMBO J.* **15:**3853–3860.

Ellis, H. M., and H. R. Horvitz. 1986. Genetic control of programmed cell death in the nematode *C. elegans. Cell* **44:**817–829.

Hengartner, M. O., R. E. Ellis, and H. R. Horvitz. 1992 *Caenorhabditis elegans* gene ced-9 protects cells from programmed cell death. *Nature* **356:**494–499.

Lazebnik, Y. A., S. Cole, C. A. Cooke, W. G. Nelson, and W. C. Earnshaw. 1993. Nuclear events of apoptosis *in vitro* in cell-free mitotic extracts: a model system for analysis of the active phase of apoptosis. *J. Cell Biol.* **123:**7–22.

Vaux, D. L., S. Cory, and J. M. Adams. 1988. Bcl-2 gene promotes haemopoietic cell survival and cooperates with c-myc to immortalize pre-B cells. *Nature* **335:**440–442.

Vaux, D. L., I. L. Weissman, and S. K. Kim. 1992. Prevention of programmed cell death in *Caenorhabditis elegans* by human bcl-2. *Science* **258:**1955–1957.

Wyllie, A. H. 1980. Glucocorticoid-induced thymocyte apoptosis is associated with endogenous endonuclease activation. *Nature* **284:**555–556.

Yuan, J., S. Shaham, S. Ledoux, H. M. Ellis, and H. R. Horvitz. 1993. The *C. elegans* cell death gene ced-3 encodes a protein similar to mammalian interleukin-1β-converting enzyme. *Cell* **75:**641–652.

Yonish-Rouach, E., D. Resnitzky, J. Lotem, L. Sachs, A. Kimchi, and M. Oren. 1991. Wild-type p53 induces apoptosis of myeloid leukaemic cells that is inhibited by interleukin-6. *Nature* **352:**345–347.

Zychlinsky, A., M. C. Prevost, and P. J. Sansonetti. 1992. *Shigella flexneri* induces apoptosis in infected macrophages. *Nature* **358:**167–169.

16

Interaction of Pathogens with the Innate and Adaptive Immune System

Emil R. Unanue

The host response to pathogenic microorganisms is extraordinarily diverse. The extent and degree of the host response depend upon the nature of the pathogen itself and the route and extent of the infection. Some general features of host-pathogen interactions are discussed in this chapter.

Innate and Adaptive Immunity

Microbial infections can be controlled by one or more effector systems that are brought into play during the infection. Two broad kinds of responses or interactions, distinguished by the extent of involvement of the lymphocyte, take place between the host and the pathogen. The responses that are independent of lymphocytes have been termed "natural immunity," "innate immunity," or "T-independent" responses. The responses that involve lymphocytes are the adaptive responses. The innate response involves various leukocytes, in particular the cells of the mononuclear phagocyte system (macrophages), the cells of the dendritic cell lineage, the granulocytes, and the natural killer (NK) cells, which can be mobilized and activated without the direct participation of the T cell. Their response to pathogens is fast and immediate. These findings have led to the concept that the innate cellular response is the initial step in the host response. The cells of the innate response also cooperate with the lymphocytes, in part by presenting antigens in the form of peptide fragments and in part by releasing a number of modulatory molecules. It is safe to say that the cells of the innate system serve to regulate and support the adaptive response, when it comes into operation.

Much of the adaptive response centrally involves the T cell. T cells are the cells derived from the thymus, which form part of the recirculating pool of lymphocytes (those that migrate from the blood to the lymphoid tissues, to the lymph, and back to the blood). Upon activation, T cells rapidly produce and release mediators, particularly cytokines, that regulate and activate the cellular response. T cells also respond by direct cellular interac-

291

tions. Thus, T cells are involved not only in activation of macrophages, for example, but also in killing of infected cells and in regulating B cells for antibody formation. All these different cellular expressions are centrally focused on an activated T lymphocyte, i.e., T lymphocytes of either the CD4 and/or CD8 subset that have responded to antigen. Finally, the B cell plays a major role in the antimicrobial response by producing antibodies. The antibody response plays a major role by neutralizing many microbes and/or their products. The growth of many microbes is controlled by antibody molecules, while the growth of others is resistant (pus-forming gram-positive bacteria are rapidly eliminated if bound to antibodies and engulfed by neutrophils, but this does not happen with intracellular facultative bacteria). Antibody molecules are usually produced by B cells interacting with T cells, both recognizing microbial antigens (Box 16.1).

Innate Response

Probably the best experimental approach to the analysis of the innate response is to examine mutant mice that can use only the innate system, by virtue of the absence of lymphocytes. The first strains of mice used for this purpose were the SCID mice, discovered by the Bosma's when examining the antibody response of the CB.17 strain of inbred mice. SCID mice have a defect in the formation of the T-cell and B-cell antigen receptor, resulting from a defect in the enzyme DNA-PK involved in the recombination of the V gene segments required to form the antigen receptor of T and B cells. Mice which lack other enzymes in this recombination process, like Rag and

---

BOX 16.1

Functions of B cells and T cells

B Cells

1. Recognition of antigen molecules occurs by way of surface Ig (IgM and IgD). The Ig assemble by recombination of the different gene segments (V, D, and J). Each B cell has a unique receptor (clonal selection).
2. Recognition of multimeric repeating epitopes can partially trigger B-cell activation. (This is the response to bacterial polysaccharides, the "T-independent response.")
3. Recognition of proteins requires the helper function of T cells.
4. B-cell–T-cell interaction involves binding of protein by B cells, internalization, and processing with generation of peptides bound to class II MHC molecules. CD4 T cells recognize the peptide-MHC complex and activate B cells.

5. Activation involves switching of Ig constant heavy-chain genes, point mutations of Ig genes resulting in selection of high-affinity Ig molecules, and generation of B-cell memory.

T Cells

1. There are two major subsets of T cells, the CD4 and CD8 T cells, each with receptors for peptide-MHC complexes displayed by APC; i.e., their receptors have dual specificity (self-MHC and peptides).
2. The thymus is the main source of T cells. These undergo a selection process in the thymus in which the thymocytes first express both CD4 and CD8 molecules together with the TCR, maturing to express either one or the other. For positive selection, if the TCR has affinity for self-MHC but weak reactivity to the self-peptides, the T cells mature to CD4 or CD8 and exit the thymus.

Each T cell will react with a unique peptide-MHC combination usually produced by a "foreign" peptide. For negative selection, if the TCR also recognizes self-peptides at high affinity, the T cell dies. This results in the death of many self-reactive T cells. For death by negligence, if the TCR does not show specificity for self-MHC, the T cells die by apoptosis.
3. CD4 T cells that recognize peptide-MHC complexes are activated, secrete cytokines, proliferate, and differentiate into distinct sets of cytokine-secreting cells. CD4 T cells are involved in macrophage activation and in B-cell–T-cell interaction. Each CD4 T cell recognizes a unique complex of peptide with class II MHC molecules.
4. CD8 T cells are activated after recognition of peptides presented by class I MHC molecules. CD8 T cells can kill cells presenting the peptide-MHC complex.

Ku proteins, and which, like the spontaneous SCID strain, exhibit a selective absence of lymphocytes have now been produced by gene ablation techniques.

The examination of SCID mice infected with a variety of viruses, bacteria, and parasites has allowed the response to be examined, unencumbered by lymphocytes. By reconstituting these mice with lymphocytes, the influence of the adaptive response on the infection could be compared with that in SCID mice. SCID mice exhibited a surprising first line of defense toward many pathogens. Leukocytes were mobilized, some sets of cytokines were produced, and the infection was restricted or delayed. The three cell types that permitted the innate response in the SCID mice to operate were the macrophages, neutrophils, and NK cells. Two results need to be emphasized regarding the response of SCID mice. First, SCID mice showed no sterilizing immunity; thus, the growth of the pathogens could be partially controlled but they were not eliminated. Second, as predicted, SCID mice did not develop an anamnestic or secondary response following a primary infection. The state of cellular activation was finite and could not be perpetuated unless lymphocytes were present.

Macrophage Response—the Activated Macrophage

The mononuclear phagocyte system comprises the circulating monocytes and their products of differentiation, the various tissue macrophages, and the dendritic cell lineage. Macrophages are found widely among tissues, usually near epithelial surfaces and blood vessels. Such macrophages constitute a first barrier to the dissemination of exogenous material. Mononuclear phagocytes are also involved in the reorganization of tissue during inflammation (i.e., in wound healing) and in the removal of tissue debris and apoptotic cells. Macrophages function in part by releasing mediators that affect the surrounding cells. The early release of cytokines by macrophages is an important response to microbes. Macrophages also express histocompatibility molecules and participate in interactions with CD4 and CD8 T cells (Table 16.1).

The production and differentiation of macrophages are controlled by colony-stimulating factors (CSF), of which CSF-1 is the major regulatory factor. Indeed, the number of circulating monocytes is much influenced by CSF-1, a protein elaborated by many cells including mesenchymal cells. The absence of CSF-1 translates into a deficit of mature macrophages in some organs.

The interaction between pathogens and macrophages that results in their uptake and internalization is mediated through a variety of cell surface receptors, which recognize a range of proteins and polysaccharides of bacteria, viruses, and parasites. Among the cell surface receptors identified are the family of scavenger receptors, several of which have been cloned (types A, I, II, and B, CD36, and others). These receptors are of broad specificity and interact with gram-positive and gram-negative bacteria, participating in their clearance from the circulation. Scavenger receptors bind to negatively charged molecules, low-density lipoproteins, polynucleotides, and anionic polysaccharides and phospholipids. Scavenger receptors have homologies to structures found in hemocytes of insects, presumably the early evolutionary counterparts of phagocytes. Macrophages also have a variety of other receptors for polysaccharides, including macrosialin (CD68) and the mannose receptor, for microbial lipoproteins (CD14), and

Table 16.1 Properties of macrophages

Membrane receptors for diverse chemical structures
    Scavenger receptor
    Complement receptors
    Fcγ receptors
    Sialoadhesin
    Mannose receptors
    Macrosialin
    Cytokine receptors (IFN-γ and TNF)
    CD14-LPS receptor

Production of cytokines
    IL-1α and β
    TNF-α
    IL-12
    IL-10
    IL-6
    Fibroblast growth factor

Antigen presentation

Production of enzymes involved in antimicrobial responses and/or acting on
    connective tissue proteins or cells
    Collagenase
    Elastase
    Lysozyme
    Lysosomal enzymes

Production of bioactive lipid and small radicals
    Prostaglandins
    Platelet-activating factor
    Reactive oxygen intermediates
    Reactive nitrogen intermediates

for immunoglobulin G (IgG) and complement (C) proteins. The Fc and C receptors are of major importance in that they promote, by several thousandfold, the uptake of microbes containing bound antibody and/or C proteins. Various Fc receptors for IgG (Fc-RI, Fc-RII, and Fc-RIII) have been identified and cloned. The same holds true for CR1, CR2, and CR3, the receptors for degradation intermediates of the C3 protein, the major C opsonin in blood. These receptors are also expressed in neutrophils, B cells, and the follicular dendritic cells of germinal centers. Since the original description of its phenomenon, opsonization is recognized as a major step in the clearance and elimination from the blood and extracellular fluids of many microbes and proteins. Some further comments on general features of this phenomenon of opsonization should be made.

    1. C proteins can directly interact with many microbial surfaces in the absence of antibodies (and without the binding of C1, C4, and C2, the first three components of the C activation cascade). This alternative pathway of complement activation is thought to be vital in

the clearance of many encapsulated organisms. Indeed, genetic deficiency of C3, the key opsonic protein deposited on microbe surfaces, results in pronounced infections with diverse microorganisms but particularly with extracellular bacteria (Box 16.2).

2. Blood contains a pool of antibodies that arise spontaneously, in the absence of overt antigenic stimulation. Many of these natural antibodies arise from a subset of B cells called B-1 and have low affinity and broad binding specificities, some of which are directed to polysaccharides. It is still unknown whether these natural antibodies act in an early recognition step in infections.

3. Other proteins have been identified that interact with polysaccharides and help in the clearance of abnormal glycoproteins. Collectins are proteins that also constitute a first defense element by interacting with polysaccharides and glycoproteins. These proteins have general structural features: they are polymers made of subunits, each with a carbohydrate recognition unit and a collagen-like stalk. Among the collectins are the mannose-binding protein found in blood, the serum bovine protein conglutinin, and the lung surfactant proteins A and D. The surfactant proteins are presumed to be involved in the rapid clearance of inhaled bacteria at the level of lung alveoli.

Macrophages can respond to external stimuli and become activated. An activated macrophage exhibits distinct features. Morphologically it is a larger cell than a normal macrophage and has more vacuoles and more pinocytic activity. The activated macrophage expresses a number of cyto-

---

BOX 16.2

Complement cascade

Classical Pathway

Antibody binds to antigen, to which C1, the first component of C, binds to initiate the cascade of interactions. C1 is composed of three subunits: C1q, C1r, and C1s. C1q binds to the Fc portion of Ig molecules, C1r is a serine proteinase that cleaves C1s, and C1s is another serine protease that cleaves C4 and C2. C4 binds to the activating surfaces after its partial cleavage by C1 (C4b is the cleavage product). C2, upon cleavage, assembles with C4 to form the "C3 convertase," a complex that cleaves C3. C3, the major blood C protein, is the major serum opsonin. It binds to activating surfaces upon cleavage by the C3 convertase. The complex of C4bC2b cleaves C5.

Alternative Pathway
C3 binding to activating molecules, including those from microbes, results in cleavage. Unique proteins of the alternative pathway are factor D, which cleaves factor B to generate an active fragment (Db); factor B, which is a serine protease that upon its cleavage by factor D assembles with C3 to form the "C3 alternative pathway convertase," which cleaves C3; and properdin, a protein that assembles with the C3 convertase to produce a more stable enzyme.

Terminal Components
C5, C6, C7, C8, and C9 form the terminal components. After the cleavage of C5, the C5b fragment induces the assembly of the membrane attack complex. C6 to C9 assemble on membranes and form a transmembrane pore.

This box outlines the major components of the classical and alternative C activation pathways. (This summary is based on information in A. K. Abbas, A. M. Lichtman, and J. S. Pober, *Cellular and Molecular Immunology*, 1994, The W. B. Saunders Co.) Aside from the proteins involved in the cascade, other soluble and membrane proteins participate in regulating the activity of the C proteins. Among the soluble regulatory proteins are the C1 inhibitor (which controls the activity of C1), factor I (an enzyme that cleaves the C4b and C3b components of the C3 convertase), and factor H (a cofactor for factor I). Among the membrane proteins are CR1, a receptor-cleaved C3b, which results in the opsonization of cells and can also act to dissociate the C3 convertase. CD46 and decay accelerating factor are two proteins that control the activity of the C3 convertase.

cidal molecules, including reactive oxygen and nitrogen metabolites, that enable them to control the growth of intracellular pathogens. Highly reactive oxygen derivatives are formed during the consumption of oxygen by the macrophages. The reactions involve the assembly of a phagocyte oxidase that utilizes NADPH. The oxidase contains three cytosolic proteins and a unique membrane cytochrome b, which utilizes NADPH to transfer oxygen. Oxidants that are formed include the superoxide anion (O^{2-}), the perhydroxyl radical (HO_2), and the hydroxyl radical ($OH^{.}$). The production of $NO^{.}$ by the activated macrophage results from the expression of an inducible enzyme, the nitric oxide synthetase. (There are two constitutive isoforms of the enzyme, the endothelial and neuronal forms.) $NO^{.}$ is produced from the metabolism of arginine to citrulline. Nitration of a number of enzymes, including RNase reductase, results in the impaired growth of cells. Activated macrophages that are infected with pathogens have restricted growth, and part or all of this restriction can be attributed to $NO^{.}$ production. For example, inhibition of $NO^{.}$ production in mice infected with a number of intracellular pathogenic bacteria results in uncontrolled infection. The same effects were found in mice lacking the gene for the inducible nitric oxide synthetase.

Activated macrophages are the hallmark of the cellular response to intracellular pathogens. They were first detected in the initial clinical and experimental studies of the tuberculous granuloma by Koch. The tuberculous granuloma consists of a mass of activated macrophages, some of which contain the bacilli. Some of the macrophages fuse to form multinucleated giant cells. (This capacity of macrophages to organize in infective foci and restrict the spread of microbes may be a primitive evolutionary behavior. A similar behavior is noted in invertebrates as a response to exogenous stimuli, for example, an accumulation of hemocytes that restrict the dissemination of the inflammation-inciting material.)

There is a relationship between activated macrophages and control of infection with intracellular pathogenic bacteria. Lurie and collaborators first showed that activated macrophages curbed the growth of the tubercule bacilli whereas serum antibodies did not. This relationship was subsequently established by Mackaness and coworkers in studies of murine infections with *Listeria monocytogenes*. The results of their observations established that antibody molecules were not the major effector molecules that restricted the growth of intracellular pathogens but, rather, that a particular cellular change in the phagocytes, characteristic of the activated phagocyte, restricted the growth. In marked contrast, the antibody molecules controlled the changes induced by extracellular toxins, the infections with many extracellular pyogenic bacteria, and the blood and extracellular stages of some viral infections.

The activation of macrophages results, to a major extent, from the interaction of macrophages with the cytokine gamma interferon (IFN-γ). IFN-γ binds to specific receptors found in many different cells, including those of the mononuclear phagocyte lineage. Neutralizing IFN-γ with monoclonal antibodies or infecting mice that do not produce IFN-γ or do not respond to it because of lack of the IFN-γ receptor (i.e., due to gene ablation techniques) results in overwhelming infection, with the absence of activated macrophages. IFN-γ regulates a number of cellular effector pathways by activating a number of important genes (including the LMP2 and LMP7 genes, involved in proteasome activation, and the NO synthase gene, involved in NO production). The effect of IFN-γ on macrophages is markedly potentiated by second stimuli that include interactions with bacteria or their products (lipopolysaccharides are the most notable) or with cytokines, particularly tumor necrosis factor (TNF). Macrophage activation, on the other hand, can be inhibited, in particular by cytokines like interleukin-4 (IL-4), IL-10, and transforming growth factor β (TGF-β), which are anti-inflammatory. The way in which the balance of IFN-γ versus IL-4/TGF/IL-10 takes place during an infectious process can be critical and needs to be evaluated, particularly for chronic infections.

The activated macrophage is found as a result of activation of either the innate immune system or the T-cell system (Table 16.2). The innate system is activated when macrophages interact with microbes and release early cytokines that induce NK cells to produce IFN-γ. The adaptive T-cell system is activated in a clonal manner when the antigen-reactive cloned T cells recognize the specific peptides from pathogens presented by histocompatibility molecules.

Interactions Involving Neutrophils

Neutrophils are the other sets of essential effector cells that control the growth of a number of infectious organisms. Their importance is clearly indicated by the susceptibility to infections in the neutropenic individual. Neutrophils are essential in infections with extracellular bacteria, which are rapidly eliminated by the oxidative and nonoxidative neutrophil microbicidal mechanisms. The oxidative burst, as described for monocytes, involves activation of the NADPH oxidase with the generation of superoxide anion. In the neutrophil a great part of O_2^- is converted to hydrogen peroxide, which in turn reacts with Cl^- ions to generate hypochlorous acid (HOCl) in a reaction catalyzed by the enzyme myeloperoxidase. HOCl is a short-lived compound but can also react with amines to form the more stable microbicidal N-chloramines. The importance of the NADPH oxidase became evident in studies of the clinical disease chronic granulomatous disease, where defects in the elimination of gram-positive organisms are

Table 16.2 Generation of activated macrophages

Pathway I (T-cell independent)
    Microbe activates macrophages
    Macrophages release early cytokines: IL-12, TNF-α, IL-1β, IL-6
    IL-12 + TNF activates NK cells
    NK cells produce IFN-γ

Pathway II (T-cell dependent)
    Microbe activates macrophages
    Macrophages release cytokines, as above
    Macrophages present microbial peptides to T cells
    Macrophages up-regulate B7-1/B7-2
    T-cell activation takes place
        TCR is engaged by the peptide-MHC complex of the macrophage
        T-cell–APC interaction is fostered by adhesion/costimulatory molecules/
            CD40-CD40L molecules
    T cells differentiate to Th1 pattern of differentiation through IL-12 release

evident. The nonoxidative mechanisms involve cationic peptides of the neutrophil granule. The defensins comprise a family of antimicrobial peptides, some of which are located in the primary granules of neutrophils. (However, defensins and related molecules are also produced in epithelial cells and could serve as major microbicidal molecules in infections of respiratory and gastrointestinal epithelia. Similar molecules have been identified in invertebrates and could represent the most primitive line of defense.)

The migration of neutrophils to sites of inflammation is critical. Directed migration or chemotaxis takes place when the neutrophil recognizes a gradient of the chemoattractant. Among the chemoattractants are products derived from C activation (C5a), small lipids (LTB$_4$, from the leukotriene cascade of mediators), microbial products, and a large family of small polypeptides, the chemokine family. Chemokines comprise about a dozen or more small proteins, now classified into three groups, CXC, CC, and C chemokines, depending on the presence of terminal cysteine residues. The CXC chemokines (i.e., two terminal cysteines with a residue in between) are powerful neutrophil chemoattractants, while the CC and C chemokines attract monocytes and lymphocytes. Chemokines are induced strongly by interaction with a range of bacteria. Particularly prominent is the release of CXC family molecules by streptococci and staphylococci.

Neutrophils are activated by interactions with microbes, particularly if these are opsonized. Cytokines also contribute to part of the activation. Neutrophils are also found in tuberculous granulomas, but their role is not clear. For some intracellular pathogens, neutrophils reduce the microbial load.

In summary, antibody molecules opsonize extracellular pyogenic bacteria, together with complement protein C3; the bacteria opsonized by neutrophils in particular are killed in a process involving oxygen intermediates. In contrast, the many facultative and obligate intracellular pathogens are not entirely or partially eliminated by neutrophils, even if opsonized. They are eliminated by macrophages activated by IFN-γ released by NK cells during innate stages of infections and/or CD4 and CD8 T cells during the adaptive phase.

Interaction Involving the NK-Cell Response

NK cells represent about 10% of circulating leukocytes, are derived from stem cells of the bone marrow, and mature in the absence of lymphocytes. NK cells participate in the early response to viruses and some bacteria (e.g., *Listeria*) and in the control of some parasitic infections (i.e., *Toxoplasma gondii*). Tumor growth and tumor metastasis can be influenced by NK cells.

For many of these responses, NK cells function through their cytolytic properties, involving the pore-forming protein perforin. In this process, NK cells establish contact with the target cell, and this contact results in their degranulation and the release of perforin. Other granule-associated proteins, like the granzymes, are involved. The granzymes enter the target cell through the pore made by perforin and cause cell death via apoptosis. However, NK cells are also involved in the response to intracellular pathogens by virtue of their production of large amounts of cytokines, particularly IFN-γ, as described above. The production of IFN-γ can be extensive, placing the NK cell as a central cell in the regulation of the early stages of an infectious process, particularly with intracellular pathogens.

Although NK cells can produce cytokines after their interaction with target cells, the main stimuli for their cytokine production derive from macrophages. A pathway of stimulation from macrophages to NK cells has been identified whereby interaction of macrophages with pathogenic organisms results in release of the cytokines IL-12 and TNF, which in turn result in activation of NK cells for IFN-γ release (Table 16.2). Other cytokines (IL-1α, IL-1β, and IL-6) are also released from the macrophages, and each has a predominant target of action. This cytokine cascade pathway (i.e., pathogen $\rightarrow$ macrophages $\rightarrow$ IL-12 + TNF $\rightarrow$ NK cells $\rightarrow$ IFN-γ) operates in a variety of infections, particularly with intracellular pathogenic bacteria, parasites, and some viruses. Upon production of IFN-γ, the macrophage system is activated and primed to become a highly cytocidal cell. NK cells are also regulated by a number of cytokines, including IL-2 and IL-15 molecules, as well as IFN-α/β, that promote their growth and activation.

NK cells recognize their target cells by using two sets of surface receptors: the activation receptors, which permit recognition and killing of the target cells, and, importantly, the inhibitory receptors, which inhibit the activation of the NK cell. The nature of all the activating receptors is not entirely known. One activating receptor is FcγRIII (CD16), which allows the interaction of NK cells with target cells bound to antibody molecules. There are examples of receptors (see below) that can be inhibitory or activating depending on the presence or absence, respectively, of inhibitory signals in their cytosolic domains. The inhibitory receptors interact with various forms of class I major histocompatibility complex (MHC) molecules. An NK cell that interacts with a target cell bearing class I MHC molecules may be inhibited from killing the target cell. This inhibition can be released by blocking and/or removing the class I MHC molecules (this is the "missing-self hypothesis" proposed by Karre). Thus, the NK cells may favor the recognition of target cells bearing low levels or abnormal forms of class I MHC molecules, as can occur with some virally infected or tumor cells. The inhibitory receptors are diverse and are represented by two sets of molecules, C-type lectin receptors and Ig-like receptors. The former are disulfide-linked type 2 integral membrane dimers (e.g., human CD94 and mouse Ly49A). The Ig-like receptors, which are termed KIR, are diverse membrane proteins that vary in their expression of Ig domains. NK cells appear to vary in the extent and diversity of NK-cell-inhibiting receptors. Different allelic forms of receptors exist. All these express cytoplasmic tyrosine inhibitory products (ITIM), which upon phosphorylation associate with tyrosine phosphatases, antagonizing the activation of tyrosine protein kinases. It is noteworthy that some viruses express a class I-like MHC molecule which can have biological consequences by engaging and neutralizing the NK-cell receptors. This has been documented for murine cytomegalovirus.

Macrophage–NK-Cell Interaction: Lessons from Listeriosis

The most extensively studied infection that has led to insights into the activation of the innate immune system is *Listeria monocytogenes* infection in the SCID mouse (Figure 16.1; Table 16.3). Results similar to those initially found in *Listeria* were also found in infections involving other intracellular pathogens. In *Listeria* infection in SCID mice, exponential growth of *Listeria* takes place after systemic infection. After a few days, growth of *Listeria* stops but the numbers of *Listeria* organisms are maintained at a steady state

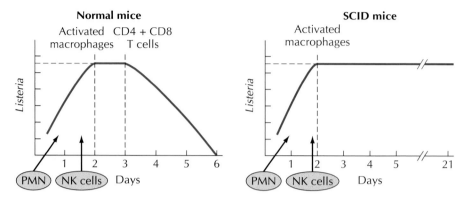

Figure 16.1   Dynamics of *Listeria* infection in normal and SCID mice. In normal mice there is exponential growth and the infection is controlled. The cells involved at different stages are indicated. In SCID mice, the absence of lymphocytes results in persistence of the infection. The early components of the infection involve neutrophils and NK cells, while lymphocytes are essential for clearance of the infection.

for prolonged periods. Examination of the macrophage discloses a heightened expression of class II MHC molecules and of their antigen-presenting capabilities. The macrophages show enhanced microbicidal activity and all the properties of activated macrophages. The activated macrophages restrict the growth of *Listeria* in the granulomas, particularly from the release of NO.

In *Listeria* infection of normal mice, the macrophages are activated by both CD4 and CD8 T cells, which recognize antigens from the microorganism. However, interestingly, both normal and SCID mice exhibit the same number and distribution of activated macrophages after infection. The difference in the infection between the two strains is marked: the normal strain develops sterilizing immunity, whereas the SCID strain becomes a chronic carrier of the infection. Thus, activated macrophages by themselves are not capable of entirely eliminating an infection with intracellular pathogens, and so lymphocytes must play a role in producing sterilizing immunity

Table 16.3   Properties of SCID mice

Selective absence of B and T lymphocytes

Acellular thymus with only stromal cells; small lymph nodes and spleen

Normal number of granulocytes, macrophages, and NK cells

Leukocytes can be mobilized upon sterile or infectious inflammation.

Normal antigen-presenting function: in culture, their antigen-presenting cells stimulate allogeneic or syngeneic T-cell responses.

Spleen cells do not release cytokines upon addition of mitogens (like concanavalin A); cytokines are released by addition of some microbes; macrophage-NK interaction can induce IFN-γ release from NK cells.

Partial resistance to microbial infections; reduced microbial growth, development of a carrier state of resistance, no sterilizing immunity

Different cytokines produced from macrophages and NK cells; these early cytokines include IL-1α, 1L-1β, TNF-α, TNF-β, IL-6, IL-10, IL-12 (all from macrophages), and IFN-γ (from NK cells); no production of IL-2, IL-4, or IL-5.

Addition of T cells (or bone marrow stem cells) reconstitutes normal immune system function.

beyond that of activating macrophages by producing IFN-γ. Indeed, sterilizing immunity in *Listeria* infection results from CD4 and/or CD8 T cells that are generated during presentation of *Listeria* antigens by the macrophages or dendritic cells. *Listeria* organisms can reach the cytosol as a consequence of their production of a pore-forming enzyme, listeriolysin O. This cytosolic stage allows *Listeria* to disseminate within cells in tissues with an extracellular phase. The vacuolar and cytosolic stages of *Listeria* generate peptides for either class I or II MHC molecules, which trigger T-cell activation (see the next section).

The crux in *Listeria* infection is that this organism is a powerful agonist for the release of cytokines by macrophages. The elements in *Listeria* that induce this response probably involve more than one chemical entity. Not all microbes induce this powerful macrophage response, and thus the extent to which SCID mice mobilize the innate system will vary. The SCID mouse responds to *Listeria* uptake via the release of IL-12 and TNF by macrophages that have taken up the microbe. Both of these cytokines play seminal roles in the regulation of the innate system, as well as the lymphocytes. IL-12 and TNF both bind to specific receptors and drive NK cells to express IFN-γ. As with IFN-γ, neutralization of these cytokines by specific antibodies or in mice with ablation of the IL-12 gene or TNF p55 receptors results in uncontrolled infection. The role of TNF is, however, much broader than that of IL-12. TNF acts not only on various leukocytes but also in the vascular endothelium. TNF-activated endothelia express adhesion molecules (E-selectin, intercellular cell adhesion molecule [ICAM]) that promote the adhesion of leukocytes and their migration to the extracellular milieu.

Another issue that has become very prominent in *Listeria* infection is the influence of different effector mechanisms during the various stages of the infective process. The early-exponential growth of *Listeria* requires control by an early infiltration by neutrophils within a few hours of the infection. The activated macrophages do not become apparent until 2 or 3 days later. The neutrophil role is not only to curb the dissemination of the microbe but also to control the liver infection. *Listeria* infects hepatocytes, where it grows exuberantly unless neutrophils restrict it. The last stage of the infection is that ultimately mediated by CD4 and CD8 T cells, producing sterilizing immunity. The CD8 T-cell response can be very prominent and persists for long periods, ensuring a state of immunological memory.

Symbiosis between the Innate Cellular System and the T-Cell Response—the MHC System

T-cell responses depend on the recognition by T cells of peptides derived from the intracellular processing of protein antigens. These peptides are bound to the MHC molecules. Initially discovered as transplantation antigens, the MHC molecules were later shown to regulate all the cellular interactions involving T cells. MHC molecules are peptide-binding molecules that rescue peptides from intracellular digestion and transport them to the cell surface. The peptide-MHC molecular complex represents the molecular substrate that engages the T-cell receptor (TCR) for antigen. The composition of the peptide-MHC complex reflects the intracellular milieu of the antigen-presenting cells (APC). Thus, it is via the MHC molecules that the T-cell system is informed whether an abnormal or previously unrecognized molecule has entered the APC. The MHC molecules thus connect the innate cellular response to the T-cell response.

The presentation of a peptide-MHC complex by the APC initiates activation of the T cells; with T-cell activation, a series of cellular events profoundly change the cellular environment. These events result from the different expressions of activated T cells: first, the release of cytokines like IL-2, IL-4, IL-6, lymphotoxin, and IFNs that alter the environment, and second, direct cellular interactions that can activate or kill cells. Prominent among these interactions are those of CD4 T cells with B cells, resulting in antibody formation; those of CD4 or CD8 T cells with APC, releasing IFN-γ and activating the macrophage system; and those of CD8 T cells with APC or other MHC-bearing cells, causing the death of these cells.

The large MHC gene segments encode a series of proteins involved in host defense, particularly two major sets of proteins, the class I and II MHC molecules. Several loci encode distinct class I and II MHC molecules with common structural and biological features. The class I MHC molecules are made of a heavy chain (of ~44 kDa) and a small (12-kDa) polypeptide, β_2-microglobulin. (The gene encoding β_2-microglobulin is not included in the MHC gene segment.) The class II MHC molecules are made up of two chains (α and β), which are noncovalently linked. Both class I and II MHC molecules are transmembrane proteins with a small segment in the cytosol (Figure 16.2).

The MHC genes are highly polymorphic. The sites of allelic differences are the amino acid residues located in the peptide-binding site. Thus, the evolutionary selective pressure for allelic diversity resides in the peptide-binding property of these molecules. The greater the allelic diversity, the more capable is the species of binding peptides from different pathogens.

Both class I and II MHC molecules bind peptides: their binding site lies at the most distal end of the molecule. The overall structure of the peptide-binding site is similar in both sets of molecules. In the class I MHC molecules, both the α_1 and α_2 domains, the two most distal domains, at the amino end of the protein, contribute to the binding site (the third domain, the α_3 domain, is an Ig-like domain). In the class II MHC molecules, each of the two distal domains of each chain contributes to peptide binding (α_1 and β_1 domains) (Figures 16.2 and 16.3; Table 16.4).

The combining site is made of a platform of pleated chains bounded by two helices which leave an open cleft or groove where the peptide binds (Figure 16.3). Although both class I and II binding sites are similar, there are some important differences. At a given time, the combining site holds a single peptide, usually of 8 to 10 residues for class I MHC molecules but 10 or more for the class II molecules. The binding specificity for peptides by MHC molecules tends to be broad. A given MHC allele can interact with a wide range of peptides. This broad, "promiscuous," binding ensures that many different peptides can be recognized. Peptides bind to MHC molecules with different affinities. For a given protein, some segments are preferentially displayed on the plasma membrane; these give rise to the dominant peptides that preferentially stimulate T-cell activation.

Several features responsible for peptide binding are well recognized since the initial analysis of the X-ray crystal structure. Peptides interact via amino acid side chains that establish hydrophobic and polar interactions with sites or pockets in the combining site. These sites are usually those having allelic specificity. Contributing to the binding affinity is an extensive network of hydrogen bonding between the peptide main carbon chain and many of the conserved residues of the helices and platform of the combining site. Some of the amino acid side chains are solvent exposed and serve

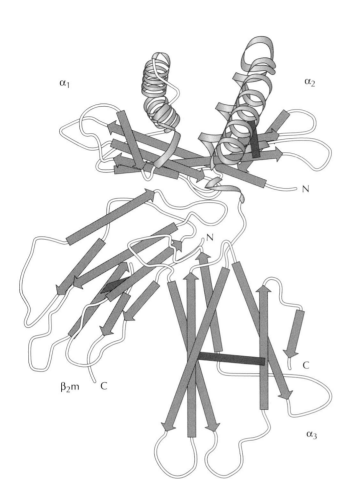

Figure 16.2 General features of a class I MHC molecule. The α_1 and α_2 domains create the peptide-binding site, which is on top of the molecule. The α_3 domain is an Ig-like domain. A small polypeptide, β_2-microglobulin, forms part of the complex. (In the class II MHC molecule, the combining site is formed by the two most external domains of the α and β chains, which make up a structure similar to the combining site of class I molecules.)

to contact the TCR. The receptor establishes contact both with peptide residues and with the residues in the α-helices. This dual specificity is the basis for the phenomenon of MHC restriction; i.e., the recognition of a foreign peptide is always linked to recognition of self MHC molecules.

From the very early studies, it became clearly apparent that class I MHC molecules were sampling peptides from proteins that localized to the cytosol. In contrast, the peptides that bound to class II MHC molecules were derived mainly from proteins taken by the vesicular system of the APC. Regardless of the source, it is important to note that the MHC system does not discriminate between autologous and foreign proteins. Peptides derived from foreign proteins or from self-proteins bind similarly; i.e., some bind well, others bind weakly, and some do not bind at all, depending on their sequence of amino acids. This issue is vital both for our understanding of how T-cell differentiation takes place and for placing the recognition of foreign and autoimmunity in the correct perspective.

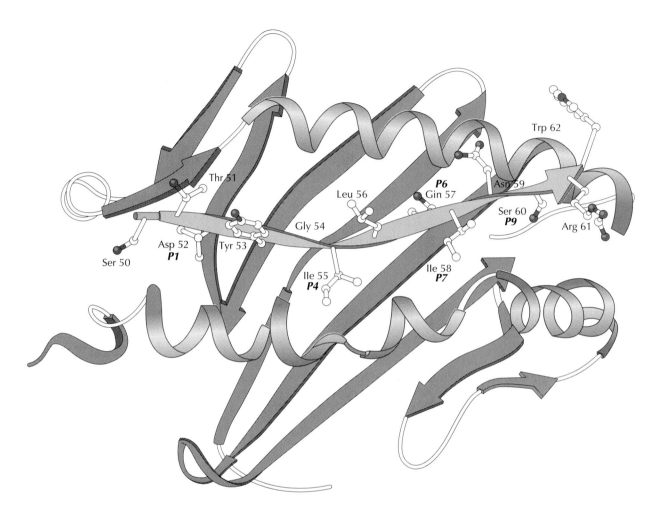

Figure 16.3 Structure of a peptide bound to the combining site of a class II MHC molecule. This specific example shows a lysozyme peptide bound to the murine *I-A^k* molecule. The peptide is stretched out and slightly twisted. The TCR contacts some of the solvent-exposed residues as well as the helices of the α1 (top) or β1 (below) domains. Adapted from D. H. Fremont, D. Monnaie, C. A. Nelson, W. A. Hendrickson, and E. R. Unanue, *Immunity* **8**:305, 1998.

We now understand that because the MHC binds to self-peptides, this allows, in the thymus, the elimination of T cells that spontaneously express an autoreactive receptor. The T cells go through a process of selection during their maturation in the thymus gland, as they interact with epithelial cells and thymic APC, both of which express MHC molecules. Only developing T cells that express a receptor that interacts with self-MHC molecules but poorly with the self-peptides contained in them are selected to mature, and they eventually spread to seed the lymphoid tissues. In contrast, the T cells that have a TCR directed to self-MHC containing an autologous peptide are eliminated and will die. As a result, part of the self-reactive repertoire of receptors is purged in the thymus.

The result of thymic selection is that the TCR from T cells that peripheralize to secondary lymphoid organs will recognize many nonself (for-

Table 16.4   Properties of class I and II MHC molecules

Property	Class I	Class II
Chemistry	Two chains, a heavy chain that forms the binding site and α_2-microglobulin	α and β chains that pair to form the binding site
Peptide binding	Peptides of 8–10 residues	Peptides usually of >10 residues
Site of peptide binding	Primarily in ER	Vesicular system
Derivation of peptides	From cytosolic proteins, but not exclusively	From proteins in the vesicular system
Assembly pathway	Proteins in cytosol $\rightarrow$ proteasome catabolism $\rightarrow$ ER transport via TAP $\rightarrow$ assembly involving auxiliary molecules (tapasin, calnexin)	α and β chains transported from ER with invariant chain $\rightarrow$ release of invariant chain $\rightarrow$ peptide binding
Expression	All cells; high in hematopoietic cells	Mostly in hematopoietic cells

eign) peptides if they are presented by their own syngeneic APC. This "MHC restriction" was discovered, to a great extent, in the context of microbial peptides. Thus, in the APC–T-cell interaction, it was first noted by Rosenthal and Shevach in inbred guinea pigs immunized with dead tubercule bacilli. Importantly, the antiviral response to lymphocytic choriomeningitis virus was found by Zinkernagel and Doherty to be restricted by the class I MHC alleles. Thus, in the latter experiments, CD8 T cells from mice from the $H\text{-}2^k$ strain would lyse target cells infected with this virus but only if they were of the $H\text{-}2^k$ strain. Either the virus was modifying the structure of the H-2 molecules (altered self) or the virus and the H-2 protein in some way contributed to an interaction between these molecules and the T cells.

The class I MHC molecules are central for presentation of peptides derived from viruses or bacteria that reach the cytosol. The CD8 T-cell response is the major cellular system responsible for the antiviral response. CD8 T cells have cytolytic properties and will lyse infected cells. Thus, by lysing infected cells and releasing cytokines with antiviral effects, like IFN-α/β and IFN-γ, the CD8 T cell restricts the growth and spread of the virus. An important point is that class I MHC molecules are expressed in most cells. Thus, most infected cells can signal their infection through class I MHC molecules.

The pathway of presentation of peptides from cytosolic proteins is complex, involving a number of critical steps. The catabolism of the protein is carried out by the multicatalytic proteasome, a structure of about 1,500 kDa, made up of several protein subunits. The proteasome is a barrel-shaped structure with a central channel where the unfolded peptide interacts. Two of these proteins, LMP2 and LMP7, are encoded within the MHC gene complex. The expression of both proteins is also enhanced by IFN-γ. The presence of the different subunits influences the subsets of peptides that are generated. Indeed, mice engineered to have mutations in either LMP2 or LMP7 have susceptibilities to different viruses. The importance of proteasomal catabolism in the generation of peptides has been shown by the use of aldehyde inhibitors like lactacystine.

The peptides generated from proteasomal catabolism are then transported into the endoplasmic reticulum (ER) by two peptide transporters, which form a bimolecular molecular complex, the TAP-1 and TAP-2 molecules (for "transports of antigen processing"). There is some degree of specificity in the peptides transported by way of TAP-1 and TAP-2. Once in the ER, the peptide itself forms part of the assembly of the complex. The heavy chain associates with several accessory molecules like tapasin, calnexin, and calreticulin, which maintain it in an unfolded state. The exact sequence of assembly is still under evaluation. Once assembled, the peptide-MHC complex moves out of the ER into the Golgi and the plasma membrane.

The class II MHC system samples proteins found in the vesicular system, i.e., proteins that are internalized from the extracellular milieu by either receptor-mediated or fluid-phase endocytosis. Thus, the class II molecules are essential for presentation of many of the microbes that are taken into the APC through phagocytosis. CD4 T cells recognize the class II MHC-peptide complex and are the central cells in the response to intracellular pathogens.

The assembly of the class II MHC-peptide complex is also intricate. The class II MHC molecules are synthesized in the ER as separate chains which pair and associate with a third chain, the invariant chain (called Ii because it does not show allelic polymorphism like the MHC molecules). Ii, also a transmembrane protein, plays two distinct roles: it has cytosolic residues that govern its transport out of the ER and Golgi, and it assembles with the class II MHC molecules, covering the combining site and thus blocking it from binding peptides. Once out of the ER-Golgi the complex reaches a proteolytic environment, where the Ii is degraded by cathepsins. A small peptide from the Ii chain is left in the combining site. At this stage, another auxiliary molecule, HLA-DM/H2M, comes into play; it favors the release of the Ii-derived peptide and the association of peptides from surrounding proteins. The new peptide-MHC complex is now free to be transported to the plasma membrane. The foreign proteins in the vesicular compartment are transported to the proteolytically rich vesicles bearing the MHC molecules, where they are denatured and partially proteolysed. How the opened polypeptide chain binds to the class II MHC molecule is not entirely settled. The point is, however, that once a segment of the polypeptide chain becomes bound, it becomes protected from catabolism by the cathepsins that surround it.

Recent studies have indicated an important facet of the interaction between microbes and the host, i.e., interference with antigen processing and presentation. This interference has been found for several DNA viruses. An important finding has involved the inhibition of presentation of one of the nuclear antigens of Epstein-Barr virus (EBV). For example, EBNA-1, despite displaying a class I MHC epitope, is not presented. This is as a result of having a protein domain, made up of a series of Gly-Ala repeats, that inhibits processing. Manipulation of the EBNA-1 gene by deleting this domain results in the expression of the class I MHC epitope. This cycle could be important in the maintenance of the viral latency in EBV-infected B cells, where viral gene expression is restricted to EBNA-1.

Cytomegaloviruses are important for their effect on immunocompromised individuals. They contain a gene, U6, that inhibits class I MHC assembly, also by targeting the TAP transporters. Another gene, US11, also inhibits class I MHC expression by channeling these molecules from ER to the cytosol, where they are rapidly degraded. Herpes simplex viruses also interfere with antigen presentation; they do this through the expression of a protein, ICP47, which inhibits the assembly of class I MHC molecules by binding to TAP molecules and inhibiting their function of peptide transport.

Another example of how microbes alter the interaction with MHC molecules is that of superantigens. Superantigens are proteins, derived from bacteria or viruses, that are capable of binding to class II MHC molecules outside the combining site and to TCR of particular sets of T cells. This can result in extensive activation of T cells. Much of the pathology of infection with toxic shock syndrome toxin type A of staphylococci is caused by the sudden release of cytokines like TNF.

APC–T-Cell Interaction

The interaction of T cells with APC bearing the peptide-MHC complex involves two components. The first component consists of the TCR binding to the MHC-peptide complex. The second component involves auxiliary molecules in or from the APC and from the T cell that foster and modulate the cellular interactions and the differentiation of each cell. There are four sets of interactions dominated by auxiliary or non-antigen-specific proteins and cytokines.

Interactions with Adhesion Molecules

As the name implies, pairs of adhesion molecules foster the cell-to-cell contact. A number of coreceptors involving integrins and molecules of the Ig superfamily have been defined. (The Ig superfamily contains proteins with an Ig fold or domain, usually of 90 to 110 residues, made up of two antiparallel β-strands connected by a critical disulfide bond. Many molecules involved in cellular interactions bear Ig domains. Proteins of the Ig superfamily are probably derived from an early gene that diversified as required for regulation of multiple cellular interactions in higher vertebrates.) The integrins belong to various families that foster not only cell-cell interactions but also interactions with connective tissue proteins and between lymphocytes and various cells including endothelial, epithelial, and connective tissue cells. Important to cite here are the interactions between ICAM molecules on APC or target cells and the family of integrins on T cells (i.e., LFA-1 or CD11/CD18).

Interaction of CD40 with CD40L

CD40L is a molecule found on T cells that pairs with CD40 on APC including B cells. This interacting pair is crucial for B-cell activation and differentiation and is also important for activation of APC. The importance of CD40-CD40L was first appreciated during analysis of infants with the hyper-IgM syndrome. These infants show marked susceptibility to infections caused by mutations of the gene encoding CD40L. In these infants, the B cells produce IgM but do not class switch to produce IgG antibodies. The IgG is required for interaction with the opsonic receptors. CD40-CD40L interaction also has a profound effect on the biology of APC. APC will produce cytokines like IL-1, TNF, and IL-12 and also produce a burst of

NO· production. Thus, APC stimulation can develop not only directly as a result of microbes, as described above, but also during the stage of interaction with T cells. It follows that interactions with microorganisms or proteins that do not directly stimulate the macrophages depend highly on the CD40-CD40L system to induce this state of activation. The CD40-CD40L interaction importantly enhances the expression of B7-1 and B7-2 molecules involved in the third set of interactions (see below)

Pairing of B7-1/B7-2 with CD28/CTLA-4

B7-1 and B7-2 are a pair of molecules of the Ig superfamily that are expressed under basal conditions at low levels on APC. B7-1 and B7-2 have complementary molecules on T cells, i.e., CD28 and CTLA-4. Both are disulfide-linked homodimers. B7-1 and B7-2 are up-regulated during interactions of APC with a variety of microorganisms. CD28 is expressed in a constitutive way, while CTLA-4 is up-regulated during T-cell activation. The engagement of B7-1/B7-2 with CD28 favors T-cell activation and the expression of a number of T-cell cytokines. Also important is the up-regulation of a number of antiapoptotic molecules of the Bcl-2 family. Lack of CD28 engagement results in poor or limited T-cell stimulation. The CTLA-4 molecule appears to inhibit T-cell activation, an issue made very apparent by the phenotype of mice having null mutations of it, which show massive lymphoproliferation and activation of T cells. Thus, there is a balance between the two molecules, one favoring activation (CD28) and the other dampening it.

Role of Cytokines

Several conditions during the APC–T-cell interaction have a profound influence on the differentiation of T cells. T cells polarize into two subsets depending on their activation of particular sets of cytokine genes. The Th1 subset produces the cytokine IFN-γ as well as IL-2. Thus, Th1 cells, via IFN-γ, markedly influence the cellular response that eliminates many intracellular pathogens. The Th2 subset produces little if any IFN-γ but makes many of the B-cell-activating molecules, such as IL-4, IL-5, IL-6, and IL-10. Some of these cytokines have a negative effect on macrophage activation. Thus, the Th2 response favors antibody responses and is much less favorable for the macrophage activation pathway.

The decision by the T cells to polarize in a Th1 or Th2 direction rests on the early stage of interaction with APC and is influenced by many factors, but a major influence is the cytokine environment. Production of IL-12 is a major and important signal for Th1 differentiation, while IL-4 (resulting for mast cells or lymphocytes) favors Th2 polarization (Figure 16.4).

Many of the cellular responses are of mixed type at the start of the response, but persistence of antigenic stimulation will skew the response toward one or the other, and these are of major importance in determining the outcome of an infection. The importance of Th1-Th2 differentiation has been strikingly shown in the model of *Leishmania major* infection of mice. *Leishmania* infection is controlled by activated macrophages produced primarily by IFN-γ secreted particularly by CD4 Th1 cells. The activated macrophages restrict *Leishmania* growth because of the release of NO·. Indeed, infections where IFN-γ is not produced or is neutralized by antibodies or where NO· is likewise not produced because of ablation of the inducible NO synthase gene or NO· is neutralized by drugs all result in dissemination of the infection (Table 16.5). In brief, because of as yet unidentified suscep-

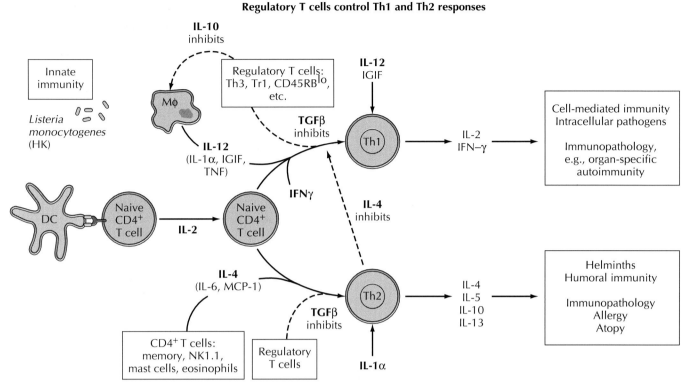

Figure 16.4 Regulation of CD4 T-cell differentiation. Reprinted from A. O'Garra, *Immunity* **8:**275–283, 1998.

tibility genes, the BALB/c strain responds early to the infection by showing a predominant IL-4 response over IL-12 response, resulting in a Th2 differentiation pattern. The IL-4 and IL-10 cytokines do not favor macrophage activation, resulting in dissemination of the parasite. Manipulations that favor a Th1 pattern (i.e., neutralization of IL-4 or addition of IL-12) produce the contrasting response, with macrophage activation and elimination of the parasite. One of the manipulations summarized in Table 16.5 is the one involving the early response to a *Leishmania* peptide, which is very fast. In the BALB/c susceptible strain the early response has a high IL-4 compo-

Table 16.5 Experimental manipulations that influence resistance to *Leishmania major* in the mouse

Favor resistance	Favor susceptibility
Injecting IFN-γ early in the infection	Neutralizing IFN-γ, IL-12, or TNF with antibodies
Neutralizing IL-4 with antibodies or infecting IL-4 null mice	Using mice with ablation of IFN-γ, IL-12, or TNF genes
Injecting IL-12 early in the infection	Neutralizing NO· or infecting mice lacking inducible nitric oxide
Abolishing early immune response to dominant epitope of *L. major* for CD4 T cells	Neutralizing early production of IFN-αβ

nent, which immediately skews the response to a Th2 pattern. If such a response is neutralized, this lack of IL-4 response can allow the IL-2 stimulation to the protective Th1 phenotype. The differences in Th1 and Th2 responses may be important in chronic infections and may have been indicated in the two polar extremes of leprosy infection: tuberculoid leprosy, with activated granulomas controlling the bacterial load, and lepromatous leprosy, with widespread bacillary infiltration.

Apoptosis of Lymphocytes

Finally, during an ongoing response to infection, lymphocyte growth is controlled largely by mechanisms that involve apoptosis. Apoptosis of lymphocytes is a component of the normal response as lymphocytes become activated. Activation-induced cell death involves a number of cell surface ligand pairs, the most extensively studied of which involves Fas and Fas ligand, both of which are expressed in lymphocytes and up-regulated during cell activation. The levels of these molecules also increase during infection, particularly with intracellular pathogens. During diverse viral infection, virally infected cells die; this process is mediated by CD8 T cells (or NK cells) but also not infrequently by direct viral infestation. Interestingly, a number of viral products influence apoptosis. These have been identified, particularly in DNA viruses, and include inhibitors of caspases, the enzymes involved in the apoptotic response, which can be produced by poxviruses (the CrmA inhibitor); the viral FLICE proteins, which interrupt the signaling pathways involved with Fas proteins; and the AIA protein from adenoviruses, which affects the DNA cycle of the infected cell.

Selected Readings

Babbitt, B. P., P. M. Allen, G. Matsueda, E. Haber, and E. R. Unanue. 1985. Binding of immunogenic peptides to Ia histocompatibility molecules. *Nature* **317:**359–363.

Bjorkman, P. J., B. Saper, B. Samraoui, W. S. Bennett, J. L. Strominger, and D. C. Wiley. 1987. The foreign antigen binding site and T cell recognition regions of class I histocompatibility regions. *Nature* **329:**512–515.

Buchmeier, N. A., and R. D. Schreiber. 1985. Requirement of endogenous interferon-γ production for resolution of Listeria monocytogenes infection. *Proc. Natl. Acad. Sci. USA* **82:**7404–7409.

Epstein, J., Q. Eichbaum, S. Sheriff, and A. B. Ezekowitz. 1996. The collectins in innate immunity. *Curr. Opin. Immunol.* **8:**29–35.

Freeman, M., J. Ashkenas, D. J. Rees, D. M. Kingsley, N. G. Copeland, N. A. Jenkins, and M. Krieger. 1991. An ancient, highly conserved family of cysteine-rich protein domains revealed by cloning type I and II murine macrophage scavenger receptors. *Proc. Natl. Acad. Sci. USA* **88:**4931–4935.

Gordon, S., S. Clark, D. Greaves, and A. Doyle. 1995. Molecular immunobiology of macrophages: recent progress. *Curr. Opin. Immunol.* **7:**24–33.

Grewal, I. G., and R. A. Flavell. 1998. CD40 and CD154 in cell-mediated immunity. *Annu. Rev. Immunol.* **16:**111–135.

Hengel, H., and U. H. Koszinowski. 1997. Interference with antigen processing by viruses. *Curr. Opin. Immunol.* **9:**470–476.

Hsieh, C. S., S. E. Macatonia, C. S. Tripp, S. F. Wolf, A. O'Garra, and A. M. Murphy. 1993. Development of Th1 CD4 T cells through IL-12 produced by *Listeria*-induced macrophages. *Science* **260:**547–549.

Lanier, L. L. 1997. Natural killer cells—from no receptor to too many. *Immunity* **6:**371–378.

Lurie, M. B. 1964. *Resistance to Tuberculosis: Experimental Studies in Native and Acquired Defensive Mechanisms.* Harvard University Press, Cambridge, Mass.

Mackaness, G. B. 1964. The immunological basis of acquired cellular resistance. *J. Exp. Med.* **120:**105–113.

Moore, K. V., A. O'Garra, R. de Waal Malefyt, P. Vieira, and T. R. Mosman. 1993. Interleukin 10. *Annu. Rev. Immunol.* **11:**165–184.

Nelson, C. A., N. J. Viner, and E. R. Unanue. 1996. Appreciating the complexity of MHC class II peptide binding: lysozyme peptide and I-Ak. *Immunol. Rev.* **151:**81–105.

Ravetch, J. V., and R. A. Clynes. 1998. Divergent roles for Fc receptors and complement *in vivo. Annu. Rev. Immunol.* **16:**421–432.

Rollins, B. J. 1997. Chemokines. *Blood* **90:**909–928.

Unanue, E. R. 1997. Studies in Listeriosis show the strong symbiosis between the innate cellular system and the T cell response. *Immunol. Rev.* **158:**11–25.

17

Microscopy

CHANTAL DE CHASTELLIER

Pathogens have evolved a wide variety of strategies to circumvent the host microbicidal activities and to use the cellular machinery to their own advantage. When pathogens invade the host, they immediately encounter professional phagocytes. Depending upon their surface properties, they will either be or not be phagocytosed by macrophages. In the latter case, they will either remain as extracellularly growing bacteria or invade nonphagocytic cells.

Each intracellular pathogen has evolved distinct strategies to manipulate host cell organelles and/or constituents, thus enabling it to find favorable conditions for survival and multiplication. Under normal conditions, the newly formed phagosomes, in which bacteria are enclosed after phagocytosis, intermingle their contents and membrane with the successive compartments of the endocytic pathway through a complex series of fusion and fission events. As they are processed into phagolysosomes, they undergo gradual modifications by specific addition and removal of membrane constituents. In addition, they become acidified due to the vacuolar proton pump ATPase located in the membrane and acquire toxic constituents, including hydrolases, that will destroy bacteria. One of the most important strategies, but by no means the only one, used by pathogens is to modulate these interactions. A wide variety of situations have been described. Pathogens can (i) lyse the phagosome membrane, escape from the phagosome, and invade the cytoplasm, in which they multiply; (ii) use the acidic and hydrolase-rich phagolysosomal environment to survive and multiply; or (iii) affect phagosome processing into phagolysosomes, in particular by inhibiting phagosome maturation and/or acidification. This strategy is of the utmost importance since it will profoundly affect drug targeting to the intracellular site of replication of pathogens and also antigen presentation.

Although some of the survival strategies have been identified for several pathogens, the underlying molecular mechanisms remain largely unknown. For cells, several molecules have been reported to be involved in the successive events of phagosome formation and processing. Among these, one can cite cell surface receptors involved in recognition and adhesion of bacteria, membrane constituents necessary for recognition, docking, fusion, and fission of endocytic compartments among themselves and with the maturing phagosomes, and the cytoskeletal network, i.e., actin

313

filaments, microtubules, and associated proteins involved in phagocytosis, organellar movement, and fusion events. More molecules will probably be discovered in the near future. For pathogens, several constituents important for their survival have been identified. However, our knowledge of the interactions between pathogen and cell constituents remains very restricted for most pathogens.

Our understanding of the survival strategies of pathogens depends a great deal on the tools and techniques we use to obtain information about the pathogen and the cellular machinery. Morphological methods, at the light and, especially, electron microscopic levels, particularly when combined with molecular biology tools (mutants, knockouts) or with drugs that modify the cellular machinery, are the most valuable tools to unravel the molecular mechanisms that enable pathogens to survive and multiply within the host cells.

This chapter reviews some of the major morphological methods, recent and less recent, of special interest to host-pathogen interplay. The detailed protocols are not given here because they are documented and explained at length in excellent handbooks and recent review articles. The emphasis is on some of the important aspects of the methods and their advantages, drawbacks, and necessary compromises. It is important to realize that no standard recipes can be given and that trials and errors with different methods are often necessary. This is particularly true when one wants to localize molecules, as explained in the sections on immunolocalization of bacterial and/or cell constituents by fluorescence or electron microscopy. A large part of the chapter is devoted to electron microscopy, which is the technique of choice for optimal resolution and precision. Given the wide range of methods and applications, it is not possible to cite the many references related to the methods and/or applications described here. The most pertinent ones are cited (and the others can be found in these references), and a few articles have been selected as examples to illustrate the methods.

Light Microscopy

Phase-Contrast, Differential Interference Contrast (Nomarski Optics), and Video Microscopy

The observation method depends on the size (cell or tissue), thickness, and contrast of the sample. Densely contrasted or stained cells will be observed by bright-field or phase-contrast microscopy, reflective ones will be observed under dark-field or reflected light, and birefringent cells will be observed under polarized light. Transparent cells can be observed by phase-contrast microscopy but also by differential interference contrast microscopy (DIC) with Nomarski optics. The use of Nomarski optics permits resolution and contrast of structural details that surpass those of other light microscopic methods, including phase-contrast microscopy. The images have a three-dimensional appearance and provide a shallow depth of field, which means that out-of-focus information will not interfere. It can be useful, for example, to study colocalization of particles within phagosomes after double-infection experiments, provided that the particles are large enough. However, the resolution will remain low for smaller bacteria such as *Coxiella* or *Mycobacterium* that will appear as small bumps, as well as for small intracellular organelles.

The use of video microscopy imaging has greatly extended the use, resolution, and contrast provided by the light microscope in addition to offering the possibility of looking at living cells. There are several methods for imaging particles in the video microscope. The first application of bright-field microscopy to the imaging of gold particles, called "nanovid" microscopy, has the advantage of enhancing the contrast of the particles over other structures in the cells. By using this approach, it is possible to observe small gold particles (10 to 40 nm in diameter) moving both on the surface of and within cells. For example, it can be used to observe the internalization and subsequent fate of receptors or ligands through the endocytic pathway. A combination of DIC with video microscopy is often preferred because the narrow depth of focus (less than 0.5 μm) allows a precise determination of the vertical location of the image within a cell. Furthermore, the contrast of the gold particles is greater than with bright-field microscopy. Even small latex particles (80 nm) are readily imaged by video-enhanced DIC but not by bright-field microscopy.

A third alternative is to combine fluorescence and video microscopy to localize fluorescent particles. This will be useful to detect particles in a cell that has many small intracellular vesicles that could be mistaken for gold or latex particles under bright-field microscopy or DIC.

Fluorescence Microscopy

Fluorescence microscopy has become an important tool for cell biologists because the sensitivity of fluorescence permits one to obtain an adequate signal from cells at relatively low concentrations of probe. Also, because brightly fluorescent molecules contrast strongly with the dark background, the fluorescence is visible over cellular structures too small to be resolved by light microscopy.

Fluorescent Probes

One of the major aims of fluorescence microscopy has been to analyze the distribution of antigens at the cellular level. This has been done essentially by applying fluorescent-antibody techniques to fixed and permeabilized material. The problems of using such methods are common to fluorescence and electron microscopy immunolabeling methods and are discussed in the section on localization of intracellular antigens by immunoelectron microscopy.

Many other fluorescent probes have been developed. Intracellular proteins can be labeled indirectly by using fluorescent ligands which bind noncovalently to the protein of interest. One example is the fluorescent derivative of the fungal toxin phalloidin, which selectively stains actin filaments. This approach has been especially valuable for elucidating the spatial organization of actin filaments during phagocytosis and also the in vivo dynamics and interactions of actin with pathogens such as *Listeria* or *Shigella* that become surrounded by an impressive network of actin filaments as they multiply and move about in the cytoplasm after lysis of the phagosome membrane (Figure 17.1).

▶ *For Figure 17.1, see color insert.*

A direct method for visualizing the intracellular distribution of proteins is to covalently label the protein of interest and microinject it back into the cell. This approach has been used to study a variety of cytoskeletal proteins. It is conceptually simple but depends on being able to obtain pure samples of a protein and to covalently label it with a fluorescent probe without destroying its function.

Covalent binding of fluorescent probes to cell surface molecules of pathogens has been performed with the aim of tracing bacteria intracellularly and characterizing the compartment in which they reside. One important drawback, which is often neglected, is that as the phagosome matures, it acquires hydrolytic enzymes that will eventually cleave off the fluorescent probe from the bacterial surface. Due to dynamic intermingling of contents between phagosomes and endocytic compartments, the location of the pathogen will not necessarily coincide with the distribution of the fluorescent probe. For the same reasons, the interval when reliable pH measurements can be performed with pH-sensitive fluorochromes coupled to bacteria depends largely upon the acquisition and activity of hydrolases while phagosomes are being processed into phagolysosomes.

Green fluorescent proteins, which are added as fluorescent tags to proteins by placing their genes behind the promoter of interest or by fusing them directly to the corresponding gene, is becoming a probe of choice to localize pathogens and to decipher the role of cell and bacterial proteins involved in pathogen survival. One of the interesting applications is to study the expression and function of constituents which are synthesized by pathogens only when they reside within the host cell. This method has also been successfully used to analyze the expression and recruitment of cytoskeletal proteins during endocytosis and/or phagocytosis.

Another interesting approach has been to use fluorescent fluid-phase or protein markers to analyze trafficking kinetics through endosomal and biosynthetic pathways. This provides a means of studying vesicle movements, sorting and delivery of vesicle contents, and changes in pH, to cite a few examples. These methods are useful for studying interactions with phagosomes, although they are less precise than electron microscopy approaches.

Finally, a number of fluorescent dyes that are sensitive to the concentrations of specific ions have been developed. Some of these dyes are acidotropic. They freely permeate cell membranes and remain trapped in acidic organelles upon protonation. This property can be used to determine whether phagosomes become acidified, but it does not give precise indications about the intraphagosomal pH.

Conventional Fluorescence Microscopy

In fluorescence microscopy, the specimen is illuminated by light of a certain range of wavelengths. The fluorescence microscope is designed to maximize the transmission of emitted light uncontaminated by the much more intense excitation light. Conventional fluorescence microscopy was initially applied only to fixed cells. Photographic techniques for examining the distribution of fluorescence in living cells is severely limited by the sensitivity of the emulsion to low light levels. The resulting long exposures result in poor temporal resolution, probe bleaching, and possible photodynamic damage to the cell and cell movement artifacts.

Coupling of the fluorescence microscope to video image-processing systems has overcome these problems and greatly extended the use of fluorescence to probe living cells. Due to the speed and sensitivity of video, rapid changes in fluorescence intensity and distribution can be observed and recorded in real time.

One of the drawbacks of conventional fluorescence microscopes is that they create images with a depth of field at high power of 2 to 3 μm. Since the resolving power of optical microscopy is 0.2 µm, superimposition of detail within this thick plane of focus obscures structural detail that might otherwise be resolved. For specimens more than 10 µm thick, light from out-of-focus planes also degrades the image, creating a diffuse halo around structures under study. Out-of-focus fluorescence reduces the contrast and the resolution. Combined use of Nomarski optics and ultra-low-light-level-sensitive video detectors has made it possible to image fluorescence in living cells for relatively long periods with minimal photodynamic injury and dye bleaching, and it improves the resolution. However, the best way of eliminating out-of-focus fluorescence is to use confocal microscopy.

Confocal Microscopy

Confocal microscopy creates images whose depth of field is less than 1 µm thick. In addition, it differs from conventional fluorescence microscopy in the mode of specimen illumination and the detection of the emitted signal. The specimen is illuminated by a point source, consisting of a laser beam focused on a small aperture. The combined effects of point illumination and point detection (as opposed to field detection in conventional microscopy) result in a strong rejection of out-of-focus light. As a result of these two characteristics, improvement of contrast and resolution is achieved.

Sample preparation is about the same as for conventional fluorescence microscopy. However, acquisition of images by confocal microscopy is much more time-consuming and complicated than by conventional fluorescence. It is therefore wise to first examine samples by conventional fluorescence microscopy. If fluorescence cannot be observed by the latter procedure, one should not expect it to become visible by confocal microscopy.

Three steps can be distinguished for acquisition of images. The first is image analysis. To examine an entire sample, one must first scan it horizontally point by point and line by line, thus defining a field. Each field is then analyzed by displaying the samples vertically with respect to the objective; the second step is image storage. The images obtained are optical sections corresponding to the fields that have been successively analyzed. The images are digitized, and the information is stored in files in the computer hard disk. The third step is image processing. Software can be used to increase contrast, to introduce pseudo-colors to enhance subtle differences, and to subtract background. This processing step must be handled with extreme care because it is rather easy to lose important information. Finally, confocal microscopes can be equipped with special chambers and video devices to analyze living cells.

Confocal microscopy presents several advantages over conventional fluorescence microscopy. First, it allows unambiguous discrimination between intracellular and surface labeling. This cannot always be achieved with conventional fluorescence microscopy. Second, the information on the relative location of different antigens by double fluorescence and the extent of colocalization obtained by comparing or superimposing images is much more precise than with conventional microscopy. In such analyses, however, it is necessary to balance the signals emitted from the two fluorophores. The most important advantage is that after the analysis of a sample by well-defined horizontal scans, computer-generated three-dimensional reconstructions of the sample can be obtained. An example of the localization of major histocompatibility complex (MHC) class II molecules in macrophages infected with *Leishmania mexicana* is given in Figure 17.2.

▶ *For Figure 17.2, see color insert.*

There are a few drawbacks to confocal microscopy. First, the lasers that are usually used as the source of illumination to excite the samples with light of a defined wavelength are available for only a limited number of wavelengths. The fluorochromes must therefore be chosen to obtain the highest possible emission. Second, it is not possible to use phase-contrast microscopy to define the contours of the cell. One possibility is to use one of the fluorescent labels to visualize the entire sample; alternatively, the background fluorescence can be increased to define the contours. However, some of the newer models of confocal microscopes have been equipped with nonconfocal transmission devices of a Nomarski type that permit such viewing.

Drawbacks of Fluorescence Microscopy

One of the important drawbacks of fluorescence microscopy is the instability of fluorochromes. Upon illumination, the fluorochromes are degraded, especially when long exposure times or high intensities of illumination are needed. This is particularly important when colocalizing different antigens by double fluorescence. More rapid photobleaching of one of the fluorochromes, overintense fluorescence of one of the two markers, or emission interference from one marker to the other can give rise to serious misinterpretations of colocalization.

Also, incorporation and illumination of fluorescent probes in living cells may not be without adverse physiological effects. Excited fluorescent probes may have diverse metabolic or toxic effects, including generation of toxic oxygen free radical species, which could be detrimental to pathogens. Two general strategies are used to minimize these effects: (i) load cells with a minimal amount of dye, and (ii) attenuate the excitation light intensity.

Finally, the most important problem is that conventional fluorescence microscopy, even when coupled to Nomarski optics or video imaging systems, and confocal microscopy do not allow high enough resolution for precise visualization of small intracellular compartments such as the endocytic compartments, which are of special importance to pathogens.

The best way of obtaining precise information on the morphological state of a pathogen under given conditions, the interaction of a pathogen with the phagosome in which it resides (which is of the utmost importance to its fate in the host cell), and the exact location of molecules involved in its survival and/or multiplication is to turn to electron microscopy.

Electron Microscopy

Electron microscopy techniques are the most suitable methods of obtaining maximum resolution and precision in investigations of host-pathogen interactions. A wide variety of cytochemical and immunoelectron microscopy methods can be used to characterize pathogens, analyze the intracellular compartment in which they reside, and localize molecules involved in their survival. The most pertinent ones are discussed here.

Basic Conventional Electron Microscopy

Before going into complicated and sophisticated methods, very basic conventional electron microscopy methods should be used to characterize the survival strategies of pathogens. These techniques give very valuable information, provided that optimal fixation and embedding protocols are used. It is vital to choose fixation conditions that preserve ultrastructure and avoid shrinkage of pathogens or swelling of organelles. Maintenance of membrane integrity is of the utmost importance, especially if it is desired to determine whether a pathogen resides permanently in a phagosome or whether it escapes from the phagosome after membrane lysis. The quality of the products, the method of preparation and storage of fixatives, and the fixation procedure itself (temperature, incubation time, buffer, osmolarity, and added ions) are most important. A very satisfactory procedure consists of a three-step fixation procedure with glutaraldehyde followed by osmium tetroxide and, finally, en bloc staining with uranyl acetate.

The choice of the embedding resin is also important. Some bacteria, such as mycobacteria, are difficult to embed. Upon sectioning, tearing occurs at the interface between the bacterial wall and the resin. This can be avoided by choosing more fluid resins such as Spurr. However, depending on the method used for embedding with Spurr, membranes can become nearly impossible to visualize. This can be very misleading, and it is advisable to compare Spurr with other resins, such as Epon, before drawing any conclusions about bacterial escape from the phagosome. Other particles such as latex beads, for which there is no lysis of the phagosome membrane, can be used as a control.

These simple methods will provide definite and precise answers to the following questions: (i) the morphological state of bacteria, i.e., whether they are intact or damaged (Figure 17.3A and B) (intact bacteria are usually live, and damaged ones are dead); (ii) whether stationary CFU counts are detected because bacteria have become dormant or because they are being degraded at the same rate as they multiply; (iii) whether bacteria divide in a single, increasingly larger phagosome or whether bacterial division induces separation of the phagosome; (iv) whether there is any specific interaction between the pathogen and the phagosomal membrane (loose versus tight contact, for example); (v) whether bacteria lyse the phagosome membrane (Figure 3C); and (vi) whether bacteria induce a reorganization of the actin network (Figure 3C and D). To answer this specific question, one possibility is to use other fixation conditions before and especially after cell permeabilization and decoration with subfragment 1 (S1) of myosin.

Methods To Study Intersection with the Endocytic Pathway

Analysis of the intracellular compartment in which pathogens reside and how it intersects with the compartments of the endocytic pathway is a necessary component of any research on the survival strategies of intracellular pathogens. Two main methods can be used. The first consists of analyzing the kinetics of acquisition of contents or membrane markers added either before or after pathogen uptake. This can be achieved only by use of electron-dense markers, markers tagged with electron-dense particles, autoradiography, or cytochemical staining for specific enzymes. The other method consists of labeling cells for specific contents or membrane markers by immunoelectron microscopy.

Acquisition of Content or Membrane Markers

Acquisition of newly internalized content or membrane markers by phagosomes. Horseradish peroxidase (HRP) has been widely used by endocytologists as a content marker because it is easily stained by cytochemical methods, it is resistant to standard fixation conditions, and the reaction

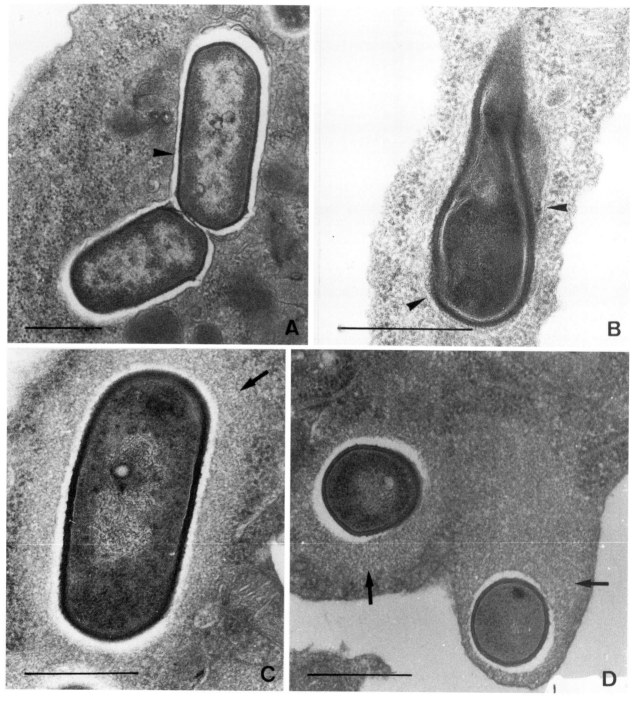

Figure 17.3 Morphological appearance of bacteria in conventional electron microscopy. Macrophages were infected with live *L. monocytogenes* LO-28 for 45 min. At 2 h after infection, the bacteria were found within phagosomes **(A and B)**, in which they were morphologically intact (A) or damaged (B), or had escaped from the phagosome **(C and D)** after lysis of the membrane (arrowheads). In the cytoplasm, the bacteria are surrounded by a thick network of actin filaments (arrows in panels C and D). Some bacteria are being extruded from the cell (D). Bar, 0.5 μm.

product forms a very dense insoluble deposit that is easy to visualize under the electron microscope. The great advantage of this marker is that early endosomes can be distinguished from prelysosomes and lysosomes in terms of two parameters. First, early endosomes differ from prelysosomes and lysosomes in their cytochemical staining pattern after HRP uptake. In early endosomes, the HRP reaction product only lines the inner face of the membrane (Figure 17.4A). In prelysosomes and lysosomes, the entire lumen is filled with the HRP reaction product (Figure 17.4B). Second, early endosomes acquire newly internalized endocytic marker (HRP) immediately whereas lysosomes display the marker only after a characteristic lag of 5 to 10 min after uptake in macrophages, and this time can reach 15 to 30 min for other cell types. The first parameter is used to observe whether phagosomes fuse with early endosomes or with lysosomes (Figure 17.5), and the second is used to classify phagosomes as resembling either early endosomes or lysosomes.

Many other endocytic content markers can be used, provided that they can be tagged with dense probes. One possibility is to use biotinylated ligands (dextran or albumin) as content markers and then exploit the biotin-streptavidin interaction, with streptavidin coupled to HRP or to gold particles, to localize the ligands intracellularly. A much simpler method consists of tagging ligands with gold particles (bovine serum albumin [BSA] tagged with gold is the most widely used). This method is quite useful to analyze interactions between compartments of the endocytic pathway by successive addition of ligands tagged with gold particles of different diameters (5, 10, and 15 nm). However, with all these ligands, it is not possible to morphologically distinguish early endosomes from late compartments,

Figure 17.4 Morphological appearance of early endosomes **(A)** and lysosomes (and prelysosomes) **(B).** Macrophages were given HRP (25 μg/ml) for 30 min, fixed, and stained for the endocytic content marker, HRP. In early endosomes, the reaction product only lines the inner face of the membrane, whereas in lysosomes, it entirely fills the lumen (B). Bar, 0.25 μm.

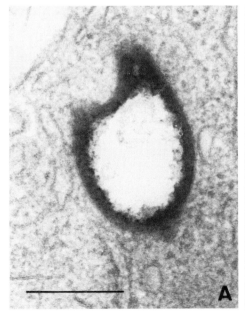
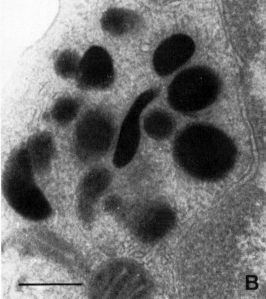

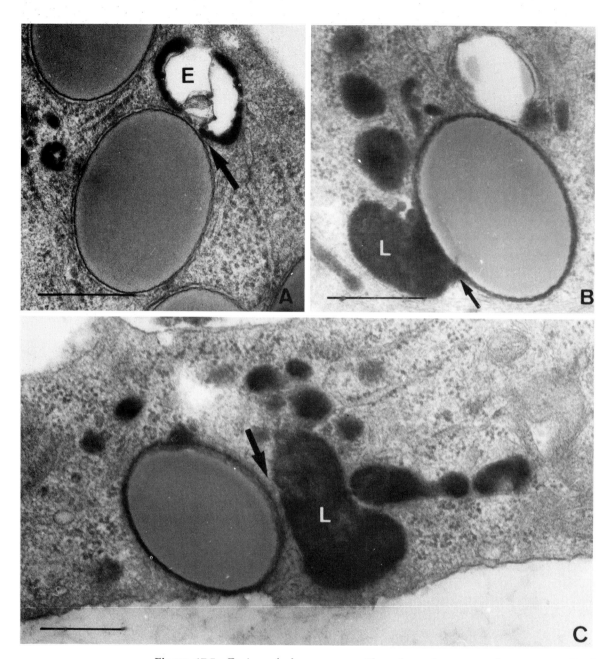

Figure 17.5 Fusion of phagosomes with early endosomes or lysosomes stained with HRP, as observed by electron microscopy. The cells underwent phagocytosis of different types of latex beads and were then given HRP as an endocytic content marker; they were then fixed and stained for HRP. The 1-μm-diameter hydrophobic bead-containing phagosomes fuse with early endosomes (E) (arrow in panel A) but not with lysosomes (L) (arrow in panel C). The 1-μm-diameter hydrophilic bead-containing phagosomes fuse with lysosomes (arrow in panel B). Bar, 0.5 μm.

as is the case with HRP. If one opts for BSA tagged with gold, it is advisable to prepare one's own, by the method of Slot and Geuze. Although it is tedious and sometimes tricky to prepare, the commercially available BSA tagged with gold does not have an adequate concentration of marker for studies of the acquisition of endocytosed material by phagosomes.

Another approach involves analyzing the acquisition of newly internalized membrane markers by phagosomes. After labeling of cell surface glycoconjugates with [^{3}H]galactose, this membrane marker redistributes to membranes of the different endocytic organelles with distinct kinetics. Early endosomes acquire membrane marker immediately, whereas lysosomes display the marker only after a characteristic lag of 5 to 10 min after uptake. When cell surface glycoconjugates are labeled with radioactive galactose, subsequent internalization and recycling lead to a redistribution of membrane marker between the plasma membrane and the membranes of the endocytic pathway. At steady state, early endosomes display a twofold-higher labeling intensity than lysosomes do. These parameters can be used to determine whether the phagosome membrane resembles the membrane of early endosomes or that of lysosomes. Such methods are difficult to apply because some expertise in autoradiography and morphometric methods is needed, but they give very precise kinetic information about the phagosomal membrane in relation to early endosomes and prelysosomes/lysosomes. Surface labeling with N-hydroxysuccinimide–sulfobiotin has also been used, but it is much less informative, because it cannot be used to estimate the labeling intensity of different endomembrane compartments.

Acquisition of lysosomal markers by phagosomes. Phagosomes do not fuse directly with lysosomes but, instead, interact with the successive compartments of the endocytic pathway in a very precise manner. To determine whether phagosomes have been processed into phagolysosomes, several laboratories have analyzed the acquisition of hydrolases at selected intervals after infection. However, hydrolases are concentrated in prelysosomes and lysosomes but are also present in small amounts in early compartments and can therefore be acquired by multiple fusion events with early endosomes.

Enzymes are not electron dense and so are not visible in electron microscopy unless enzyme cytochemistry or immunoelectron microscopy methods are applied. Cytochemistry does not visualize the enzyme itself but the product of the enzymatic reaction. The fact that it must be electron dense to be visualized has limited the number of enzymes that can be localized. Acid phosphatase can be studied since phosphate liberated during the enzymatic reaction will react with lead citrate to form insoluble and electron-dense lead phosphate precipitates. These precipitates remain in the organelles containing the enzyme, provided that the ultrastructure is preserved.

For this to be achieved, the cells must be fixed before the enzymatic reaction occurs and a compromise must be reached between preservation of ultrastructure and of enzymatic activity. Most enzymes are inactivated by chemical fixatives, and below a threshold concentration, they will no longer be detected. When studying the acquisition of hydrolases by phagosomes, it is advisable to try different fixation conditions and include controls such as phagocytosis of particles that do not inhibit phagosome-lysosome fusions and therefore acquisition of hydrolases by phagosomes. Staining for acid phosphatase is illustrated for *Mycobacterium avium*-containing phagosomes in Figure 17.6. The immunoelectron microscopy methods are discussed below and can be used to localize any enzyme, provided that antibodies are available.

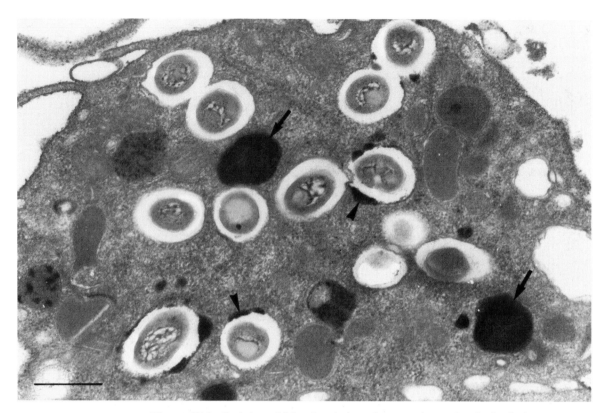

Figure 17.6 Staining of *M. avium*-infected macrophages for the hydrolytic enzyme acid phosphatase. Bone marrow-derived macrophages from BCG-susceptible mice (BALB/c) were infected with *M. avium* TMC 724. At 7 days later, the cells were fixed, stained for acid phosphatase, and processed for electron microscopy. Lysosomes were strongly labeled (arrow), but most phagosomes were not stained. Only a few of them displayed small deposits (arrowhead). Bar, 0.5 μm.

The most widely used method to study phagosome processing into phagolysosomes involves chasing an endocytic content marker to lysosomes prior to phagocytosis and then analyzing acquisition of the marker by phagosomes at selected intervals after phagocytosis. Several markers have been used, such as ferritin and thorotrast, to cite a few, but the most frequently used is BSA coupled to gold particles. This method is illustrated in Figure 17.7 to show fusion or no fusion of phagosomes with lysosomes. To obtain the most reliable data, the marker must be rapidly endocytosed by the cell and in sufficient amounts that it will label the entire lysosomal compartment when chased after uptake. However, the marker must not be too tightly packed within the lysosome, or it might form a sort of rigid gel or network. Under such conditions, the lysosomal contents would not be transferred to phagosomes upon fusion of lysosomes with phagosomes, and this would be misinterpreted as a "no-fusion" event. It is advisable to try different concentrations of marker and different incubation times in the presence of marker and to use control particles such as *Bacillus subtilis*, hydrophilic latex beads, or dead bacteria that do not inhibit phagosome-lysosome fusion events when defining the optimal conditions.

Finally, it is important to recall that lysosomal contents, especially nondegradable ones, are recycled out of the lysosomal compartment via small recycling vesicles. The latter can fuse with and deliver contents to early

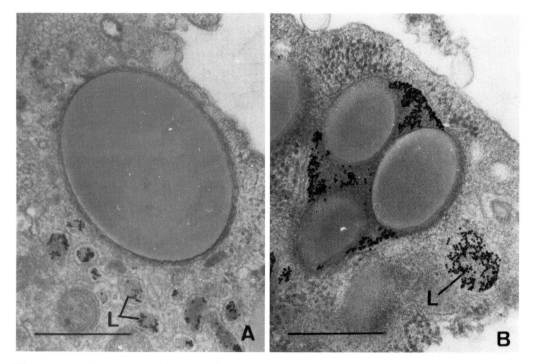

Figure 17.7   Acquisition of lysosomal marker by phagosomes. Macrophages were incubated for 30 min with BSA tagged with gold, washed, and incubated for 2 h in medium devoid of BSA tagged with gold to chase the marker to the lysosomes. The cells were then incubated in the presence of different types of latex beads. Phagosomes containing 1-μm-diameter hydrophobic beads do not fuse with lysosomes (L) **(A)**, but phagosomes containing smaller beads (0.1 to 0.5 μm in diameter) do **(B)**. Bar, 0.5 μm.

endosomes. This will eventually lead to excretion of the marker via the normal recycling pathway. This phenomenon is particularly important in the case of macrophages: BSA tagged with gold chased to lysosomes reappears as tiny patches of gold in early endosomes within a 2-h chase period and about 50% of the label is secreted within 20 h of the chase.

Phagosome Acidification

DAMP [3-(2,4-dinitroanilino)-3′-amino-*N*-methyldipropylamine] is a probe that is widely used to study phagosome acidification at the electron microscopy level. This weak base accumulates by diffusion within acidic compartments. Once protonated, it can no longer diffuse out. During chemical fixation with aldehydes, it becomes covalently linked to proteins, which allows it to be retained in acidic organelles during processing of samples for electron microscopy. DAMP is then localized with appropriate post-embedding immunolabeling methods (anti-DNP [dinitrophenol] antibodies followed by protein A coupled to gold) on Lowicryl or LR White thin sections.

An important advantage of this method is that the number of gold particles per phagosome can be converted to values for intraphagosomal acidification by the method described by Orci, provided that the phagosome contains sufficient protein for all the protonated DAMP molecules to be retained in the phagosome after fixation. An example of this labeling method is illustrated in Figure 17.8 for macrophages infected with *Listeria monocytogenes*.

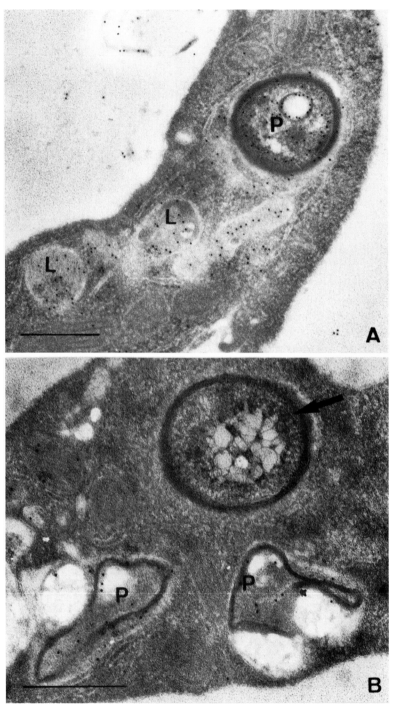

Figure 17.8 Morphological assessment of phagosome acidification. Macrophages were infected for 45 min with *L. monocytogenes*, washed, and reincubated in fresh medium. At 1 h later, DAMP (30 min, 60 μM) was added for passive accumulation in acidic organelles. The cells were fixed and embedded in Lowicryl. The probe was localized by a postembedding labeling method as follows. Lowicryl thin sections were sequentially incubated with rabbit anti-DNP antibodies and with protein A coupled to 10-nm-diameter gold particles. **(A)** A high abundance of gold particles was observed for lysosomes (L) and for phagosomes (P) due to their acidification. **(B)** Intracytoplasmic bacteria (arrow) were not labeled. Bar, 0.5 μm. From such pictures, one can estimate the intraphagosomal pH by the method described by Orci.

Localization of Membrane Markers

Immunolocalization of membrane proteins known to be located predominantly in the membrane of a specific compartment of the endocytic pathway is a widely used method to characterize the phagosome membrane as resembling the membrane of early endosomes, prelysosomes, or lysosomes. Some of the proteins most commonly used as markers are listed in Table 17.1. Proteins of the Rab family have become very useful, because these small GTPases, which are involved in the regulation of intracellular membrane traffic, are often associated with only one compartment.

However, pathways involved in targeting membrane proteins to lysosomes are very complex. Newly synthetized proteins in the endoplasmic reticulum are transported to the Golgi complex and upon arrival at the *trans*-Golgi network are often first targeted either to early endosomes or even to the cell surface, from where they can be rapidly internalized into early endosomes for final delivery to lysosomes. The presence of lysosomal markers therefore does not necessarily mean that a phagosome has been processed into a phagolysosome. While remaining immature, it may acquire membrane constituents that are considered to be lysosomal as a result of extensive fusion events with early endosomes. A typical example is the *Mycobacterium*-containing phagosome. It has been extensively shown that *Mycobacterium* inhibits phagosome maturation and therefore fusion with lysosomes, and yet the phagosome membrane has been shown to be lysosome-associated membrane protein (LAMP) positive. As a consequence, it is advisable to also label cells for specific early endosome membrane markers such as the transferrin receptor or Rab5, which are never found in the membrane of late compartments, before drawing any conclusions about the nature of the compartment in which a pathogen resides and how it modulates the interactions with the endocytic pathway. Different immunolabeling methods are described and discussed in the following section.

Immunoelectron Microscopy

Methods

Immunoelectron microscopy is the best choice to study the distribution of antigens at high resolution. Immunolabeling specialists face a perpetual dilemma, i.e., to find the best conditions for sample processing and labeling in order to achieve optimal preservation of ultrastructural details, maintenance of antigenic determinants, and immunoreactivity and precise im-

Table 17.1 Localization of endomembrane markers

Compartment	Markers
Early endosome	TfR,[a] Rab5, annexins I, II, and III[a]
	Immature cathepsin D
Late endosome	M6PR,[b] Rab7, LAMP1,[b] LAMP2,[b] CD63, CD68 (macrosialin)
	Hydrolases (acid phosphatase, cathepsin D, β-hexosaminidase)
Lysosome	LAMP1, LAMP2, CD63
	Hydrolases (acid phosphatase, cathepsin D, β-hexosaminidase)

[a]Tfr (transferrin receptor) and annexins are also present in large amounts on the plasma membrane.
[b]M6PR, mannose-6-phosphate receptor; LAMP, lysosome-associated membrane protein.

munolabeling with negligible background. It is equally important to ensure equal accessibility of all antigens, especially when they are located at different sites, and to find conditions where antigens and the markers used to localize them remain at their natural intracellular site.

A variety of different methods have been developed to detect antigens at the electron microscopy level. Most of them are depicted schematically in Figure 17.9. Because some antigens will be readily localized by one method and not by another, it is advisable to try several methods to optimize the labeling efficiency for a given antigen. The method of choice will also depend upon the material under investigation, the kind of information needed (qualitative versus quantitative), and the available equipment.

As shown in Figure 17.9, there are three main options: (i) preembedding ultrastructural immunocytochemistry, in which labeling is performed after fixation of the cell sample but before tissue dehydration, embedding, and sectioning; (ii) postembedding immunolabeling, in which labeling is performed on thin sections of fixed, dehydrated, and embedded (Epon, Lowicryl, LR Gold, or LR White) material; and (iii) immunolabeling of ultrathin cryosections, in which samples are fixed, cryoprotected, and frozen in liquid nitrogen. Samples are first cryosectioned, and labeling is performed on thawed cryosections which are then contrasted and stabilized before observation by electron microscopy. Postembedding immunolabeling and immunolabeling of ultrathin cryosections are illustrated in Figures 17.8 and 17.10, respectively. The detailed procedures, including recent improvements in ultracryomicrotomy, are not given here because they can be found in the handbook by Griffiths (1993), reviews, and articles as recommended at the end of this chapter. However, some of the crucial points are discussed here.

Figure 17.9 Schematic depiction of methods used for immunoelectron microscopy. See the text for details.

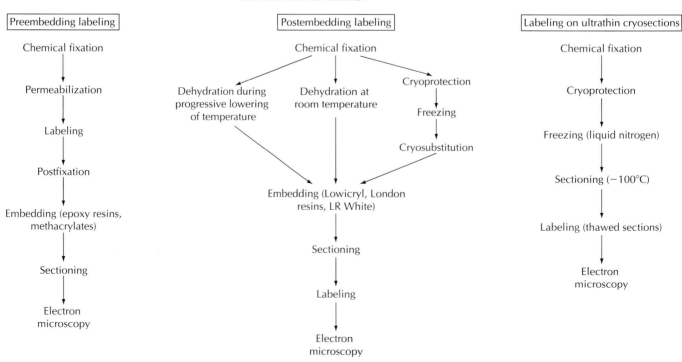

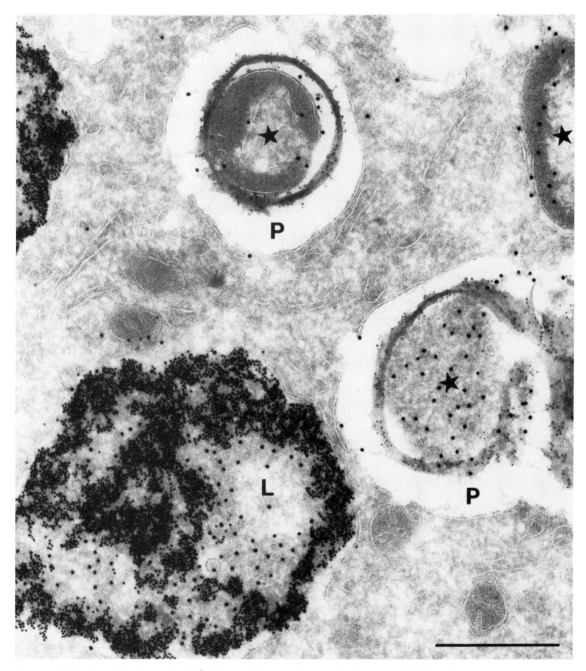

Figure 17.10   Localization of cell and bacterial constituents by immunolabeling of ultrathin cryosections. Mouse peritoneal exudate cells had sequentially internalized (i) 10-nm colloidal gold particles derivatized with BSA (BSA-gold) overnight to label lysosomes and (ii) heat-killed *L. monocytogenes* particles (stars) for 40 min. The cells were then fixed in a mixture of acrolein and paraformaldehyde and processed for cryoimmunogold labeling. Cryosections were doubly immunolabeled for cathepsin D with 15-nm gold particles and for *Listeria* cell wall antigens with 5-nm gold particles. Cathepsin D is present in lysosomes and has penetrated the heat-killed *L. monocytogenes* particles present in the phagosomes (P). The lysosomes (L) are marked with intermediate 10-nm BSA-gold particles, but the phagosomes are not labeled, indicating that they have not yet fused with lysosomes. Bar, 0.5 μm. Kindly provided by Hans J. Geuze, University of Utrecht Medical School, Utrecht, The Netherlands.

Crucial Points

Fixation. Whatever the method used to localize a given molecule, samples must first be fixed. In the absence of fixation, antibodies are internalized by endocytosis, sequestered, and eventually degraded in the lysosomes.

The choice of fixative is crucial because chemical fixatives often introduce artifacts and morphological changes. Chemical fixations are performed with paraformaldehyde or glutaraldehyde (and sometimes acrolein) either alone or in combination. Paraformaldehyde is often preferred to glutaraldehyde. Although the latter permits much better preservation of ultrastructure, it often causes irreversible changes in the antigenic determinants due to efficient cross-linking. Antibodies are then unable to bind to the specific epitopes, especially when monoclonal antibodies are used. Some antibodies are totally unable to recognize chemically fixed antigens. Cryofixation methods must then be used. When deciding on the fixative of choice for a given method and even for an unknown antigen-antibody combination, different fixation procedures should be tested to achieve optimal labeling.

Permeabilization. One of the important obstacles in the detection of intracellular antigens by immunolabeling methods (both fluorescence and electron microscopy) is the low penetration of antibodies through cell membranes, a problem that is enhanced by cell fixation. To improve the permeability of the cell membrane for preembedding labeling, methods include freeze-thawing, mechanical sectioning, and partial permeabilization with detergent. Saponin, which interacts with cholesterol in membranes, is generally preferred to Triton because it is a mild detergent and hence avoids extraction of antigens and can be used after mild fixation with aldehydes without causing damage to the ultrastructure. The necessary concentration must be adjusted to the cell type and fixative and must be added throughout the labeling procedure since permeabilization with saponin is reversible. If labelings are instead performed on ultrathin resin-embedded or frozen sections, the difficulties of antigen accessibility are removed and samples need not be permeabilized.

One must also keep in mind that membrane antigens are more difficult to localize than antigens in the lumen of cell organelles, because the former are less abundant and less accessible to antibodies.

Labeling. With preembedding methods, only one antigen can be labeled at a time. The most common procedure is to first incubate cells with a specific antibody and then with immunoglobulin G (IgG) from another species coupled to HRP. Cells are then stained for HRP by the usual procedure. For postembedding labelings or immunolabeling of cryosections, gold probes can be used instead of HRP. The most commonly used single-labeling procedures are (i) specific antibody followed by protein A coupled to gold particles (PAG), (ii) specific antibody followed by bridging antibody and then by PAG, and (iii) specific antibody followed by IgG of a different species coupled to gold particles. The specific antibody can be a monoclonal or polyclonal antibody; the bridging antibody and the IgG conjugated to the gold probe originate from different species. A bridging antibody can be used either to amplify the signal or in situations where specific antibodies do not react with PAG. The two-step PAG labeling (specific antibody followed by PAG) gives the best result in terms of resolution and low background. Moreover, the use of PAG permits double and even triple labelings with antibodies from the same species, which is not as readily achievable when specific IgG-gold complexes are used.

In performing double and triple labelings with antibodies from the same species, it is necessary to beware of possible pitfalls due to cross-reactions of secondary antibody and/or gold probes. The size of gold particles (5, 10, or 15 nm) and the sequence of antibodies should be chosen carefully. There are no general rules, but it is advisable to do the most critical reactions first and with the largest gold particles to facilitate detection. In multiple-labeling procedures, when antigens are present in the same intracellular location, the yield of subsequent labeling steps is lower than in a single-labeling experiment, probably due to steric hindrance. Another problem is false-positive reactions (colocalizations) due to the reaction of the gold probe or bridging antibody with previously used immunoreagents. To overrule this problem, a glutaraldehyde cross-linking step is used between successive labelings.

Quantitative immunocytochemistry.   The use of particulate markers such as colloidal gold has the advantage that the immunolabeling can be quantitated. However, quantitation in immunocytochemistry is usually not absolute. This method generally cannot reveal the quantities of antigen present, because it is not known to what extent a measured labeling density reflects the actual concentration of antigen. Some of the variables which affect the labeling efficiency, such as fixation, dehydration, and embedding, can be standardized. Variables due to the antigen itself are difficult to control. Differences in the intracellular matrix densities will also introduce variables in the labeling efficiency. This generates differences in accessibility of antigens to the immunoreagents due to differential penetration into the section. This can be avoided by embedding cells in Lowicryl or LR White prior to sectioning. In this way, the surface of the section is labeled and no variations in penetration can occur. However, the labeling efficiency will still decrease since only antigens at the thin-section surface are accessible. Simple methods that consist of counting gold particles and determining surface areas by morphometric methods will give very meaningful results, provided that the gold particles are uniformly distributed over organelles. This method has been successfully used to estimate the intraphagosomal pH of *M. avium*-containing phagosomes.

Conclusion

A wide variety of morphological methods at both the light and electron microscopy levels can be used to analyze the molecular mechanisms involved in the survival of pathogens within the host cell. From the different illustrations and comments, it is clear that electron microscopy techniques are the most suitable methods to obtain maximum resolution and precision when studying interactions between phagosomes in which pathogens reside and the different compartments of the endocytic pathway and defining the specific molecules involved in survival of pathogens in general.

Concerning the immunolocalization of pathogen or cell constituents, there is no standard procedure. Each method has its advantages and limitations, and each antigen-antibody combination may require a different approach to achieve optimal labeling. Before using the more complicated electron microscopy methods, immunofluorescence should be performed as a first assay. Nevertheless, a positive result does not guarantee a satisfactory reaction at the electron microscopy level. All three immunoelectron microscopy methods permit the study of the distribution of antigens with good resolution. However, ultracryomicrotomy and immunogold labeling are the techniques of first choice since they have proven to combine excellent preservation of ultrastructure and antigenicity and a very precise localization of antigens.

Selected Readings

de Chastellier, C., T. Lang, A. Ryter, and L. Thilo. 1987. Exchange kinetics and composition of endocytic membranes in terms of plasma membrane constituents: a morphometric study in macrophages. *Eur. J. Cell Biol.* **44:**112–123.

de Chastellier, C., T. Lang, and L. Thilo. 1995. Phagocytic processing of the macrophage endoparasite *Mycobacterium avium,* in comparison to phagosomes which contain *Bacillus subtilis* or latex beads. *Eur. J. Cell Biol.* **68:**167–182.

de Chastellier, C., and L. Thilo. 1997. Phagosome maturation and fusion with lysosomes in relation to surface property and size of the phagocytic particle. *Eur. J. Cell Biol.* **74:**49–62.

Griffiths, G. (ed.). 1993. *Fine Structure Immunocytochemistry.* Springer-Verlag, KG, Berlin, Germany.

Harding, C. V., and H. J. Geuze. 1992. Class II MHC molecules are present in macrophage lysosomes and phagolysosomes that function in the phagocytic processing of *Listeria monocytogenes* for presentation to T cells. *J. Cell Biol.* **119:**531–542.

Maniak, M., R. Rauchenberger, R. Albrecht, J. Murphy, and G. Gerisch. 1995. Coronin involved in phagocytosis: dynamics of particle-induced relocalization visualized by a green fluorescent protein tag. *Cell* **83:**915–924.

Ojcius, D. M., F. Niedergang, A. Subtil, R. Hellio, and A. Dautry-Varsat. 1996. Immunology and the confocal microscope. *Res. Immunol.* **147:**175–188.

Raposo, G., M. J. Kleijmeer, G. Posthuma, J. W. Slot, and H. G. Geuze. 1997. Immunogold labeling of ultrathin cryosections: application in immunology, p. 208.1–208.11. *In* L. A. Herzenberg and D. M. Weir (ed.), *Weir's Handbook of Experimental Immunology,* vol. 4. Blackwell Science Inc., Oxford, United Kingdom.

Sluder, G. and D. E. Wolf (ed.). 1998. *Video Microscopy.* Academic Press, London, United Kingdom.

Stearns, T. 1995. The green revolution. *Curr. Biol.* **5:**262–264.

Tougard, C., and R. Picart. 1986. Use of pre-embedding ultrastructural immunocytochemistry in the localization of a secretory product and membrane proteins in cultured prolactin cells. *Am. J. Anat.* **175:**161–177.

18

Promising New Tools for Virulence Gene Discovery

Timothy K. McDaniel and Raphael H. Valdivia

As the preceding chapters have demonstrated, the starting point of many cellular microbiology studies is identification of the microbial genes necessary for virulence. The logic underlying the identification of such virulence genes is codified into a set of rules known as molecular Koch's postulates. These rules establish the criteria to prove that a suspected virulence gene is necessary for a given pathogenic phenotype, just as the original Koch's postulates established the criteria to prove that a suspected pathogen causes its associated disease. The basis of these molecular postulates is that the relationship between a gene and a functional phenotype, such as virulence, may be established by investigating the phenotypes of strains differing only in a defined genetic lesion. The molecular Koch's postulates are fulfilled if all three of the following conditions are met:

1. a gene is found in strains with a certain virulence phenotype
2. mutating the gene abolishes the phenotype
3. reintroducing the gene reconstitutes the phenotype in the mutant strain

Obviously, the conditions required for fulfillment of molecular Koch's postulates are attainable only within organisms that can be genetically manipulated. Beyond this requirement, a candidate virulence gene is needed. As more microbial genome sequences become available, it becomes increasingly possible to identify virulence gene candidates in these organisms by DNA sequence comparisons to known virulence genes from other pathogens. However, this approach is useless if the candidate genes are unrelated to previously characterized sequences. As much as ever, investigators need rapid laboratory methods to screen bacterial genomes for candidate virulence genes. Such methods have been developed in recent years and are the subject of this chapter. Although these new methods use a variety of technologies, they all identify virulence gene candidates by one of two general approaches: mutagenesis or analysis of differential expression.

333

Identifying Candidate Virulence Genes by Mutation

Early forays into identifying virulence genes used the genetic tools available at the time: mutation and complementation. In such screens, DNA-damaging agents or transposons are used to mutate a pathogen, generating a collection of bacteria known as a mutant library or a mutant bank, where each member of the collection contains a random mutation.

The most direct way to screen such libraries would be to test individual library members for the ability to infect and cause disease in model animals. However, screening all the genes in a bacterial genome of typical size would require testing thousands of mutants, which would be unrealistic due to the time and expense of animal tests. The development of in vitro infection models involving cultured mammalian cells has permitted more rapid screening of mutant libraries for loss of phenotypes relevant to the host-pathogen interaction, such as adherence, invasion, and intracellular survival. Although more convenient than assays with whole animals, assays based on cultured cell models are still time-consuming and cumbersome, especially when library members must be individually tested to find a desired mutant. Despite some successes, the time and expense associated with such brute-force screens have limited their use.

Enrichment Strategies

The overwhelming difficulty of individually screening members of a mutant library can be avoided if the investigator devises a genetic selection in which the survival of desired virulence mutants is enriched. Where such selections can be devised, whole mutant libraries can be tested simultaneously in pools. For example, a decades-old strategy for isolating non-replicative mutants has been adapted to isolate *Salmonella typhimurium* mutants defective for growth during intracellular residence. This strategy relies on the fact that β-lactam antibiotics, such as penicillin, kill only bacteria that are actively dividing. To select the desired mutants, bacteria residing in cultured mammalian cells are exposed to a β-lactamase antibiotic. Bacteria that replicate within the cell are selectively killed by the antibiotic, thus enriching for nonreplicating mutants. This approach revealed a number of mutants with altered cell envelope components, nutritional requirements, and susceptibility to host cell antibacterial compounds. A conceptually similar approach has been used to isolate *Legionella pneumophila* mutants defective for intracellular growth.

Signature-Tagged Mutagenesis

In most cases, an investigator cannot devise an enrichment strategy for desired mutants and brute-force testing of individual mutants would be impractical. A technique that overcomes these limitations, known as signature-tagged mutagenesis (STM), allows the simultaneous screening of large numbers of distinct mutants for those that fail to survive a challenge of the investigator's choosing. This is achieved by tagging each bacterium with a unique DNA sequence, called a signature tag, which allows any given mutant to be tracked within a large pool of bacteria. The original STM procedure was a transposon-mutagenesis-based screen used to identify *S. typhimurium* genes necessary for survival in the mouse (Figure 18.1). A collection of transposons is engineered such that each contains a unique 40-bp DNA sequence (tag). The delivery of a transposon to the *Salmonella* genome thus provides the means of mutating, as well as tagging, each

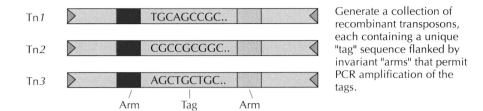

Tn*1*

Tn*2*

Tn*3*

Arm   Tag   Arm

Generate a collection of recombinant transposons, each containing a unique "tag" sequence flanked by invariant "arms" that permit PCR amplification of the tags.

Deliver transposons to genome of pathogen, creating a library of mutants, each with a unique insertion and a unique tag.

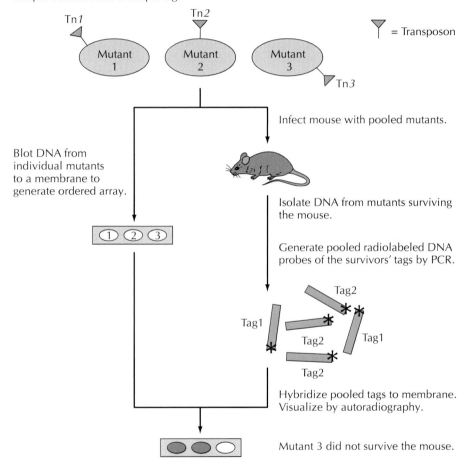

Tn*1*

Tn*2*

Mutant 1   Mutant 2   Mutant 3

Tn*3*

Y = Transposon

Infect mouse with pooled mutants.

Blot DNA from individual mutants to a membrane to generate ordered array.

Isolate DNA from mutants surviving the mouse.

Generate pooled radiolabeled DNA probes of the survivors' tags by PCR.

Tag1

Tag2

Tag2   Tag1

Tag2

Hybridize pooled tags to membrane. Visualize by autoradiography.

Mutant 3 did not survive the mouse.

Figure 18.1   Signature-tagged mutagenesis. This diagram shows the procedure for a pool of three mutants. In practice, pools of up to 96 mutants have been used to infect a single animal.

library member. A pool of tagged mutants is used to infect a mouse, and the bacteria that survive to colonize the spleen are harvested and regrown in vitro. The DNA is isolated from the survivors, and the tags are amplified by PCR to generate a pooled radiolabeled DNA probe (actually a collection of many probes) representing only the tags present in the surviving pool of bacteria. This pooled probe is then hybridized to an ordered array of DNA from all bacteria used in the initial infection, and individuals that failed to colonize the spleen are identified by the failure of their DNA to hybridize to the pooled probe. The STM technique led to the discovery of novel *Salmonella* virulence genes, such as those within the SPI-2 pathogenicity island. Similar STM approaches have been used to identify virulence genes of *Staphylococcus aureus* and *Vibrio cholerae.*

The fundamental biology of some host-pathogen interactions can limit the ability to apply STM. For some bacterial species, only a few individual "founder" bacteria establish an animal infection even when large numbers of bacteria are used for the inoculation. If the number of founders that establish an infection is smaller than the number of distinct mutants in the initial pool used for infection, tags will be randomly missing from the output pool even when their corresponding mutations are not deleterious. In the example of the successful *Salmonella* STM screen described above, 96 individual mutants could be screened simultaneously in a single animal. However, when the same number of mutants is used in STM screens in *V. cholerae* and *Yersinia pseudotuberculosis,* only a minority of the initial pool survives to seed the infection. This low survival rate necessitates using smaller mutant pools for each infection and correspondingly increasing the number of animals needed to screen the same number of mutants.

Another potential complication of STM is that mutants containing disruptions in essential virulence genes can survive selection if their functions can be complemented by other members of the input library. For instance, a secreted bacterial toxin may alter the host environment in a way that is necessary for survival of the bacterium, and a mutant deficient in toxin synthesis, which would be expected to be killed, might survive if its environment is altered by toxin produced by the neighboring bacteria.

Identifying Candidate Virulence Genes by Differential Expression

Bacteria are frugal and, as a rule, express a gene only when doing so benefits survival. Accordingly, the coordinate expression of a subset of genes during residence in vivo is necessary for the organism to colonize, survive, replicate, and cause disease within its host. This principle underlies several strategies for isolating candidate virulence genes, in which genes are isolated based on their preferential expression within animals or cultured cells or under laboratory conditions that mimic the environment within the host. Often the "host-like" conditions studied are merely laboratory media with slight recipe changes, such as the addition of acid to mimic the acidic lysosome of phagocytic cells or the omission of iron to mimic the iron-sequestered environment of host tissues.

Various techniques are available to identify these differentially expressed genes. Some entail isolating mRNA and proteins whose expression is induced upon association with the host. Other approaches involve sophisticated genetic selections that use the host environment as a selective medium to identify promoters with host-specific activity. Once the mRNAs, proteins, or promoters associated with a host-induced gene are identified, the gene itself can be identified by using a short segment of the corresponding amino acid or nucleotide sequence for comparison to genomic DNA sequence databases or as a probe to clone the gene from an unsequenced genome.

Differential RNA Display

Differential RNA display analysis is a technique that allows the isolation of RNA species synthesized preferentially by an organism under certain environmental conditions. Applied to bacterial pathogenesis studies, differential display analysis has been used to identify microbial mRNA synthesized under host-like conditions. Current approaches to the identification of such mRNAs use PCR-based amplifications of cDNA synthesized from RNA extracted from bacteria grown in environments of interest. PCR amplification permits the investigator to work with small starting amounts of RNA, which is particularly useful when isolating bacterial RNA from infected host cells or tissues. In typical applications, total RNA is isolated from bacteria and treated with reverse transcriptase to synthesize cDNA, which is used as substrate for PCR with arbitrary primer sets. Each primer set leads to the amplification of a subset of the total cDNA population. The amplified products are separated by size by using polyacrylamide gel electrophoresis (PAGE). Comparisons between the lengths and abundances of PCR-amplified fragments from cDNA isolated under host-like conditions and cDNA isolated from bacteria grown in normal laboratory media indicate which transcripts are induced (and repressed) in the host environment. Analysis of these differentially expressed genes and their roles in adaptation to the host environment requires their isolation and mutation. A fragment can be excised from the polyacrylamide gel, subjected to further PCR amplification, and used as a probe to isolate full-length genes out of a genomic library. This general approach, for example, has been used to identify *L. pneumophila* genes preferentially expressed in human monocytes.

With the increasing availability of microbial genome sequences, efforts based on differential display are focusing on genome-wide searching of differentially expressed genes rather than relying on subsets of genes amplified by arbitrarily primed PCR reactions. One approach to genome-wide identification of differentially expressed mRNA transcripts uses ordered grids of DNA immobilized on a solid platform, such as a nylon membrane or glass slide, where each spot of the grid contains a DNA sequence corresponding to a different gene from the probed pathogen. By hybridizing labeled cDNA or RNA extracted from bacteria grown under different environments and observing which grid positions hybridize, it is possible to identify genes preferentially expressed under the environments of interest (Figure 18.2).

To fully exploit sequenced microbial genomes, investigators are assembling these ordered DNA arrays into microscopic grids to create thumbnail-sized DNA chips, named for their resemblance (but not functional similarity) to silicon computer chips. Fluorescently labeled cDNA or mRNA probes are synthesized from bacteria grown under different conditions and hybridized to a chip made for the pathogen in question. Automated readers use lasers to scan the array and read the pattern of hybridization. In this way, it is possible to rapidly and simultaneously read the expression levels of every gene of a sequenced genome in response to conditions of interest.

A challenge in all approaches based on differential RNA display is to isolate high-quality bacterial mRNA. In eukaryotes, the long poly(A) tail found uniquely on mRNA presents a convenient handle for its isolation; in contrast, there is no distinguishing characteristic of bacterial mRNA to permit its easy isolation from rRNA, the overwhelmingly predominant component of RNA preparations. It is especially difficult to isolate sufficient quantities of high-quality bacterial mRNA in animal tissues, where one faces the additional challenge of separating bacterial RNA from that of the host.

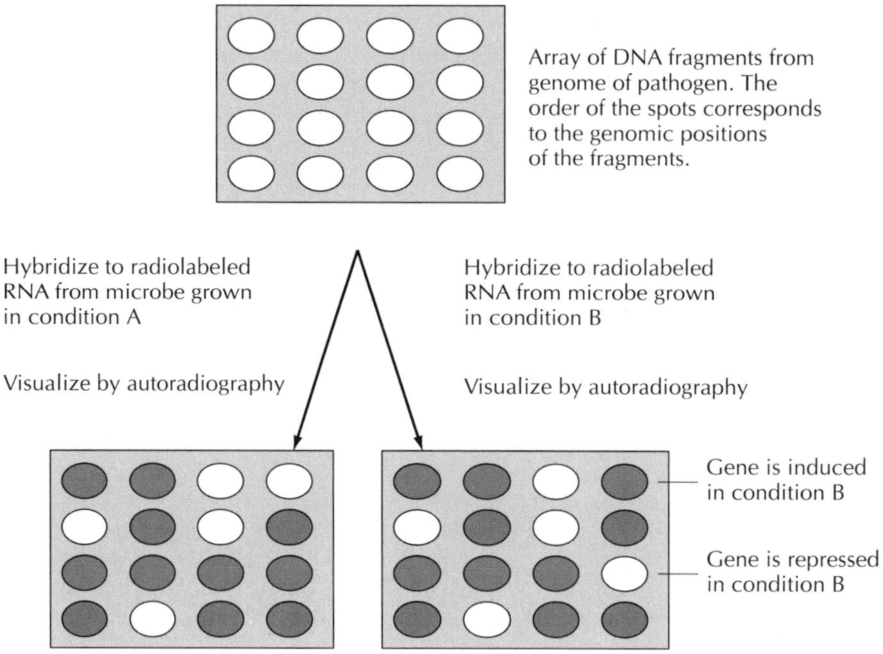

Figure 18.2   Differential RNA display with an ordered genomic array.

Promoter Traps

Because of the difficulties in isolating RNA samples that reflect true expression profiles of the microbe, the most widely used technique for isolating differentially regulated genes is a genetic strategy known as a promoter trap, in which genes are isolated on the basis of the strengths of their promoters rather than the levels of their RNA and protein products. Once conditionally active promoters are isolated, they can be sequenced and the genes or operons under their control can be determined by sequencing downstream regions or by comparison to sequence databases.

In a promoter trap, promoter activity is measured by fusing promoters to a reporter gene, a gene whose product is easily measured and thus permits the investigator to distinguish bacteria that are expressing the gene product from those that are not. For example, the *lacZ* gene is widely used as a reporter because the activity of its product, β-galactosidase, is easily detected by using indicator media that change color in the presence of this enzyme. A promoter trap begins with the construction of a promoter library or promoter bank, a collection of bacteria genetically engineered such that each bacterium contains the same reporter gene inserted downstream of a different promoter. The reporter gene contains no promoter of its own, and so the product is made only if the gene is fused downstream of an active promoter. This allows one to isolate conditionally active promoters by selecting for library members exhibiting reporter activity preferentially under the conditions of interest.

The activities of traditional reporter genes are easily measured only in relatively simple environments such as laboratory media or agar plates. As a consequence, promoter traps originally had limited utility when applied to probing bacterial interactions with host cells or tissue. New selection schemes with different combinations of selectable markers and recently developed fluorescent reporters have expanded the use of promoter traps to environments of interest to the cellular microbiologist, including the surface and interior of cultured mammalian cells and even the interior of organs of living animals.

In Vivo Expression Technology

A variety of promoter traps permit the direct selection of bacteria bearing gene fusions induced during animal infections; this type of trap is termed in vivo expression technology (IVET). The original IVET made use of an *S. typhimurium* library consisting of a promoterless two reporter operon fused randomly in the bacterial chromosome (Figure 18.3). One of the two re-

Figure 18.3   In vivo expression technology with *purA* and *lacZ*.

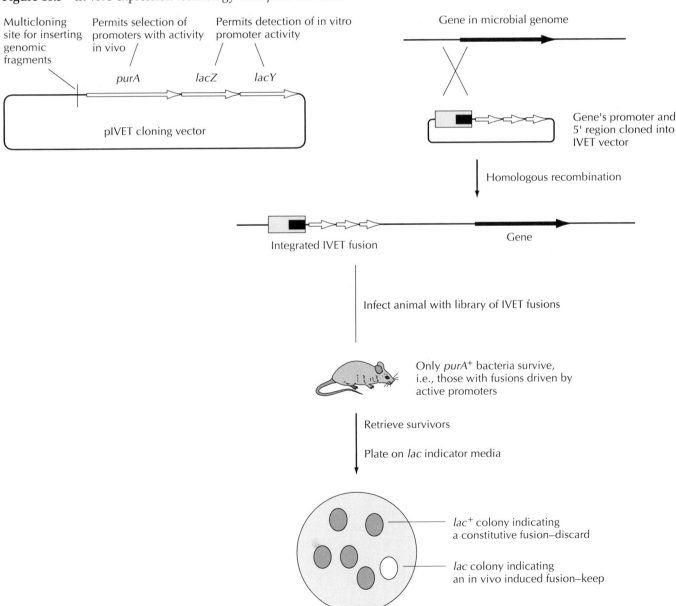

porters, *purA*, encodes an enzyme involved in purine biosynthesis whose expression is necessary for survival in mice. A mouse is infected with the library, and only bacteria containing in vivo-induced fusions survive, since only they express *purA*. Bacteria are then harvested from the animal and grown on laboratory media. It is at this step that the IVET operon second reporter, *lacZ*, comes in. The investigator determines *lacZ* activity of bacteria grown on indicator media, and only those with low *lacZ* activity (i.e., those with promoters active in the mouse but inactive in vitro) are kept for further study.

A bias in this IVET strategy is that the *purA* gene used to assess in vivo promoter activation must be continuously active for survival in the mouse. As a result, the original IVET screen identified promoters that are continuously induced during infection but not those that are only transiently induced. This bias meant that the screen was liable to miss promoters driving genes needed only during restricted times or in restricted sites of infection. This problem has been addressed in more recent IVET strategies involving alternative reporters to *purA*, which allow the investigator to choose the time after infection when selection is applied or to avoid lethal selections within the animal altogether.

One way investigators have avoided the problem of continuous selection within the animal is by using genetic recombination in concert with antibiotic selection as a reporter. In recombination-based IVET screens (Figure 18.4), the promoter library is constructed in a background strain containing an antibiotic resistance gene flanked by copies of a DNA sequence that is the substrate of a DNA resolvase enzyme. The resolvase catalyzes recombination at the repeat sequences, resulting in a deletion of the DNA flanked by these sequences. A library is made within this strain in which the reporter operon contains a promoterless gene for the resolvase. If the fused promoter is even transiently activated, the resolvase is made and catalyzes the deletion of the antibiotic resistance gene. This deletion in effect permanently records the fact that the promoter was activated and thus allows the antibiotic resistance status of the bacterium to be determined, even after the bacteria have been harvested from the animal, if a promoter was active at any time during the infection.

Figure 18.4 Resolvase as an IVET reporter. Promoter fragments are inserted before a promoterless *tnpR* gene, which encodes a resolvase enzyme. If the gene is fused to an active promoter, the resolvase is produced and deletes an antibiotic resistance gene that has been inserted into the bacterial chromosome flanked by two resolvase recognition sites. Loss of antibiotic resistance therefore permanently records that the promoter was activated.

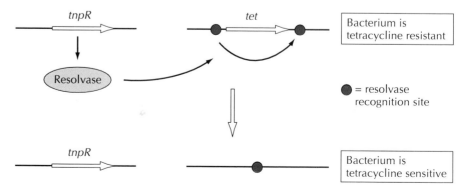

Regardless of the in vivo reporter(s) used, IVET screens have a history of yielding a high proportion of promoters for genes involved in metabolism and nutrient biosynthesis. This trend is probably seen because such genes are needed for survival in host compartments, in which essential nutrients are depleted or sequestered, whereas they are not needed in the nutrient-rich laboratory media often used during the in vitro phase of the IVET procedure. While this teaches us about the ecology of the host environment, it has resulted in a low yield of bacterial genes directly involved in the interactions of microbes and host cells, which are the primary interest of cellular microbiologists. A recent IVET-based screen for *Yersinia* virulence genes avoided this problem by using defined minimal media to grow the bacteria during the counterscreen against genes that are active in vitro. This design ensured that promoters for biosynthetic genes would be induced during both phases of the selection and that these promoters would thus be eliminated from the final pool. Therefore, the choice of growth media can be an important consideration when designing an IVET-based screen.

Differential Fluorescence Induction

Perhaps the most versatile single reporter for probing host-pathogen interactions is the *gfp* gene, which encodes green fluorescent protein (GFP), a jellyfish protein that fluoresces green when excited by blue light. An important feature of GFP is that its fluorescent signal can be detected by a fluorescence-activated cell sorter (FACS), an instrument that sorts living cells into pure populations on the basis of their light-scattering and fluorescence properties. This amenability to FACS analysis means that bacteria harboring libraries of promoters fused to the *gfp* gene can be screened for activated promoters in an automated fashion, and thus thousands of desired promoters can be isolated in minutes. The FACS can also sort eukaryotic cells to which GFP-expressing bacteria have adhered or internalized, making possible the detection of bacterial genes activated by interactions with host cells.

The promise of GFP as a tool for genetic selection has been realized in the development of a promoter trap known as differential fluorescence induction (DFI). In DFI, conditionally active promoters are isolated from a GFP-promoter library by selecting library members whose fluorescence is stimulus dependent (Figure 18.5). In addition to its amenability to auto-

Figure 18.5 Isolation of host cell-induced microbial genes by differential fluorescence induction.

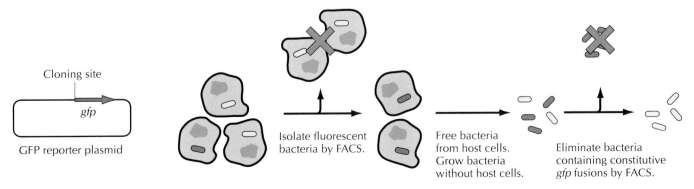

Infect cultured cells with *gfp* library.

mation, other features of the GFP reporter contribute to the versatility of DFI. The fluorescent signal of GFP requires no enzymatic substrate, and so promoter activity can be monitored in environments such as living tissues and subcellular compartments of living cells, which cannot be probed by enzymatic reporter genes that require artificial substrates for their detection. GFP is a stable molecule, and so its signal is maintained even after multiple steps of processing, such as those needed to harvest bacteria from animal tissues. Indeed, GFP-promoter fusions have been used to track the expression of individual bacterial virulence genes directly in mouse urinary tracts and spleens, suggesting that screens of libraries harvested directly from experimentally infected animals should be feasible. Finally, the ability to visualize GFP in situ by fluorescence microscopy means that the promoter fusions retrieved from a DFI screen will provide the researcher with a set of ready-made tools for detailed cellular microbiology studies (Box 18.1 and Figure 18.6).

▶ *For Figure 18.6, see color insert.*

The properties of GFP also place some limitations on the utility of DFI. Unlike enzymatic reporters, which amplify their signal by converting multiple substrate molecules into signal-generating products, the fluorescent signal of GFP comes directly from the protein itself. This lack of enzymatic amplification means that the GFP signal is weak compared to the best enzymatic reporters. Hyperfluorescent variants of GFP that considerably improve the sensitivity of the reporter have been developed, but identification of weak-to-medium-strength promoters still requires the use of libraries based on plasmids, which amplify the fluorescent signal via an increased copy number of the *gfp* gene. When cloned on multicopy plasmids, promoters can behave differently from when they are found on the bacterial chromosome, so that results generated from plasmid-based screens should be viewed cautiously until confirmed by other means.

BOX 18.1

Beyond genetic screens: the many uses of GFP

GFP is a tool with many uses. Because it can be observed in live cells that have not been fixed or stained, it can be used to observe bacterium-specific proteins that interact with host cells and tissues. Examples of specific applications are given below.

Identifying Subsets of Host Cells Infected within Animal Tissues
S. typhimurium targets, among other sites, the spleen, an organ containing a mixture of different immune system cells. To find which of the spleen cells become infected, investigators infected mice with *S. typhimurium* constitutively expressing GFP, excised their spleens, and stained the spleen cells with a mixture of fluorescently tagged antibodies, where each antibody was a different color and was specific for a different type of cell. These stained cells were then analyzed by a FACS machine: the green signal of the bacteria was associated with phagocytes, such as neutrophils and macrophages, but not with B or T lymphocytes.

Tracking Bacterial Gene Expression in Host Cells
Several *S. typhimurium* genes are activated in association with macrophages, but which step of the interaction confers the inducing signal? *S. typhimurium* organisms containing *gfp* driven by the promoter of the macrophage-induced *aas* gene were observed by time-lapse fluorescence microscopy as they infected macrophages. The bacteria fluoresced only after entering the cells and associating with host vacuoles. In this experiment, the gene was never induced in bacteria outside macrophages or in those merely adhering to macrophages.

Monitoring Subcellular Localization of Bacterial Toxins within Host Cells
Genes for bacterial toxins can be fused to the *gfp* gene to create toxin-GFP hybrid proteins. Such hybrids can retain both the activity of the toxin and the fluorescence of GFP. With these hybrids and a fluorescence microscope, an experimenter can record in real time where the toxin accumulates in the host cell and can identify probable sites of action.

Despite this theoretical problem, DFI-based screens have yielded a high proportion of virulence genes relative to housekeeping genes. Part of this success results not from any particular advantage of the DFI technique per se but from a principle that is important when designing any screen based on analysis of differential expression, be it differential RNA display, IVET, DFI, or the proteomic techniques discussed below. This principle is that the investigator should minimize the differences between inducing and non-inducing conditions. The most successful gene hunts have been those designed around narrow changes in the environment of the bacterium. For instance, the successful DFI-based screen for macrophage-induced *Salmonella* genes discussed above compared genetic expression of bacteria grown in the presence and absence of macrophages but otherwise in identical growth medium. When many variables are changed at once, a plethora of genes responding to the various environmental cues will be retrieved, only some of which will be relevant to the response of the pathogen to the host.

Proteomics

The promoter activities and mRNA levels monitored in screens based on promoter trapping and differential display can be invaluable in deciphering the adaptations of a pathogen to the host, but they can also be misleading. The investigator is ultimately interested not in genes and mRNAs but in gene products such as adhesins and toxins. The level of a protein at any time is affected by its rates of synthesis and degradation, either of which can vary independently of mRNA levels. Investigators are approaching this problem by complementing genomic sequence data with proteomic studies, in which the proteins that are expressed during different phases of the life cycle of an organism are tracked from among the total set of possible proteins of the organism (the proteome).

The myriad proteins synthesized by an organism in virtually any environment can be monitored by two-dimensional PAGE (2D-PAGE). Environmentally induced proteins can be isolated from these gels, and their amino termini can be sequenced. The resulting amino-terminal sequence can then be compared to DNA sequence databases to identify the corresponding gene. In the field of bacterial pathogenesis, these studies have been limited to cell-free conditions believed to mimic host conditions or to simple tissue culture models of infection. The total set of bacterial proteins synthesized during residence within a mammalian cell has been resolved for several pathogens by 2D-PAGE. When looking at protein synthesis in a system where two organisms (host and pathogen) are interacting, it is crucial to distinguish bacterial proteins from host cell proteins. This is usually achieved by preincubating the host cells with cycloheximide, an inhibitor of eukaryotic, but not bacterial, protein synthesis, and then adding radioactive amino acids. With host protein synthesis inactivated, these labeled amino acids are incorporated exclusively into bacterial proteins, which can then be visualized by autoradiography after 2D-PAGE. A problem with this technique is that the profile of bacterial proteins synthesized in the presence of host cells is altered by cycloheximide, most probably because the bacterium alters its protein expression in response to host factors synthesized during natural infection.

Recent studies have tried to circumvent this problem by exploiting differences between bacterial and eukaryotic amino acid biosynthetic pathways. In bacteria, but not in eukaryotes, a precursor of lysine synthesis is diaminopimelic acid (DAP). Investigators have used tritiated DAP to specifically label proteins made by *S. typhimurium* residing in epithelial cells without the need for cycloheximide. While DAP labeling is sound in principle, it is expensive and time-consuming in practice. The tritium label requires approximately 2 months for visualization, and the degree of its incorporation into bacterial proteins varies greatly depending on which type of host cell the bacteria are found. Nonetheless, the logic of the technique is sound, and with technical improvements in labeling and protein detection, the general approach could become widely useful.

Promoter traps and RNA display methods cannot reflect changes in phosphorylation or other processing steps of proteins that may occur in response to host environments. The increasing resolving power of commercial 2D-PAGE systems and the development of novel electrophoresis methods coupled with mass spectrometry should allow up to 10^4 proteins to be resolved in a single gel. Such tools should allow investigators to monitor a myriad of posttranslational responses during infection; these include changes in protein turnover and lipid or phosphorylation modifications, and any of them could be modulated by the bacterium in response to the host environment. Techniques for rapid, proteome-wide analysis are only in their infancy and have not achieved the ease of use offered by many of the above-described genetic screens. However, proteomics is an area of rapid development and will be a field to watch in the coming years.

Conclusion

The preceding sections have outlined a variety of approaches used to identify candidate virulence genes. Each technique has its weaknesses and strengths, which may leave the reader wondering which approach is best. The answer is simple: none of them is! All have limitations that bias toward the isolation of certain genes and not others. In cases where more than one technique has been applied to the same pathogen, each approach has yielded genes missed by the others. In many cases, the choice of technique will be dictated by the availability of genetic tools for use with the pathogen in question. For instance, the requirement by DFI for medium- or high-copy-number plasmid vectors will limit the use of this technique to pathogens for which such vectors are available.

It is crucial to remember that the techniques described in this chapter merely identify virulence gene candidates; these cannot be considered true virulence genes until they have passed further tests to satisfy the requirements of the molecular Koch's postulates. Mutational strategies such as STM offer the advantage that when the screen is finished, the investigator can satisfy these requirements simply by complementing the mutations in the isolated mutants. Screens based on differential gene expression (differential RNA display, proteomics, and promoter trapping) necessitate several additional steps: isolating the full gene, mutating the gene, testing the mutant in assays for pathogenicity, and, finally, showing that the defect can be corrected by complementation. However, at the end of a well-designed differential expression screen, the investigator better understands the exact conditions under which the gene is induced, which can be a crucial clue to guide further studies into the role of this gene in disease. In this regard, DFI screens are especially useful, due to the ease with which the isolated *gfp*-promoter fusions can be incorporated into further cell biology studies.

The genome sequence of every major pathogen (and most minor ones) is or will soon be available. These genomic sequences, augmented by the gene screens described in this chapter, should permit the discovery of novel virulence genes at an unprecedented pace. Future editions of this textbook will almost certainly describe presently unknown mechanisms of microbial infection that will have been discovered by the approaches described in this chapter.

Selected Readings

Burns-Keliher, L. L., A. Portteus, and R. Curtiss III. 1997. Specific detection of *Salmonella typhimurium* proteins synthesized intracellularly. *J. Bacteriol.* **179**:3604–3612.

Camilli, A., and J. J. Mekalanos. 1995. Use of recombinase gene fusions to identify *Vibrio cholerae* genes induced during infection. *Mol. Microbiol.* **18**:671–683.

Chuang, S. E., D. L. Daniels, and F. R. Blattner. 1993. Global regulation of gene expression in *Escherichia coli. J. Bacteriol.* **175**:2026–2036.

de Saizieu, A., U. Certa, J. Warrington, C. Gray, W. Keck, and J. Mous. 1998. Bacterial transcript imaging by hybridization of total RNA to oligonucleotide arrays. *Nat. Biotechnol.* **16**:45–48.

Falkow, S. 1988. Molecular Koch's postulates applied to microbial pathogenicity. *Rev. Infect. Dis.* **10**(Suppl. 2):S274–S276.

Hensel, M., J. E. Shea, C. Gleeson, M. D. Jones, E. Dalton, and D. W. Holden. 1995. Simultaneous identification of bacterial virulence genes by negative selection. *Science* **269**:400–403.

Mahan, M. J., J. M. Slauch, and J. J. Mekalanos. 1993. Selection of bacterial virulence genes that are specifically induced in host tissues. *Science* **259**:686–688.

Shevchenko, A., O. N. Jensen, A. V. Podtelejnikov, F. Sagliocco, W. Wilm, O. Vorm, P. Mortensen, A. Shevchenko, H. Boucherie, and M. Mann. 1996. Linking genome and proteome by mass spectrometry: large-scale identification of yeast proteins from two dimensional gels. *Proc. Natl. Acad. Sci. USA* **93**:14440–14445.

Valdivia, R. H., and S. Falkow. 1997. Fluorescence-based isolation of bacterial genes expressed within host cells. *Science* **277**:2007–2011.

About the Contributors

Klaus Aktories is professor and director of the Institute of Pharmacology and Toxicology (Department I) at the University of Freiburg, Freiburg, Germany. He started his scientific career with studies on the regulation of adenylyl cyclase by G proteins, cholera toxin, and pertussis toxin. Since 1986 his main research topics have been bacterial protein toxins acting on actin and small GTPases. Major work focused on actin-ADP-ribosylating toxins, C3-like transferases which ADP-ribosylate Rho proteins, the family of large clostridial toxins, and toxins which deamidate Rho GTPases. He is interested in the structure-function analysis of the toxins and in their actions on signal transduction processes.

Avri Ben-Ze'ev works in the Department of Molecular Cell Biology, Weizmann Institute of Science, Rehovot, Israel, studying different aspects of cell-cell interactions, especially the roles of β-catenin and plakoglobin in signaling and tumorigenesis. He received his Ph.D. at the Hebrew University of Jerusalem.

Alexander D. Bershadsky works in the Department of Molecular Cell Biology, Weizmann Institute of Science, Rehovot, Israel, studying the role of cytoskeleton and cell contractility in formation of focal adhesions and adherens junctions. He received his Ph.D. at the Cancer Research Center of the Russian Academy of Medical Sciences, Moscow.

Patrice Boquet spent several years at the Pasteur Institute in Paris, and is now Professor of Bacteriology and head of an INSERM laboratory in the Faculty of Medicine in Nice on the French Riviera. He is working on the mechanisms of action of bacterial toxins, with particular interests in how diphtheria and tetanus toxins translocate across cell membranes, the endocytotic mechanisms of toxins such as diphtheria toxin and ricin, and the molecular mechanisms and cellular effects, mainly on the actin cytoskeleton, of toxins affecting small GTP-binding proteins of the Rho family.

Michael Caparon is currently an Associate Professor of Molecular Microbiology at the Washington University School of Medicine, St. Louis, Mo.

Following undergraduate training at Michigan State University and graduate training at the University of Iowa, he began his work on the genetics and biology of the streptococci during his postdoctoral studies under the guidance of June Scott at Emory University. His work has focused on the development of streptococcal genetics and its application in understanding the interaction of *Streptococcus pyogenes* with host epithelial cells. His work has contributed to the identification of adhesins, the regulation of their expression, and their role in manipulating the signaling responses of epithelial cells.

Singh Chhatwal is head of the Department of Microbial Pathogenesis and Vaccine Research at the National Research Centre for Biotechnology (GBF), Braunschweig, Germany. He also belongs to the Basic Sciences Faculty of the Technical University, Braunschweig. His research has focused mainly on the interaction of gram-positive bacteria with extracellular matrix and plasma proteins. His group was the first to identify SfbI, a fibronectin-binding protein, which is the major adhesin and invasin of group A streptococci. His current research is on the development of antiadhesive streptococcal vaccines. Before joining GBF, he was at the University of Giessen working on bacterial toxins and was involved in the identification of C3 botulinum toxin.

Peter J. Christie received his Ph.D. in microbiology with Gary Dunny, Cornell University, Ithaca, N.Y., in 1987. He then trained as a postdoctoral fellow with Eugene Nester, University of Washington, Seattle, Wash., and Virginia Walbot, Stanford University, Stanford, Calif. In 1991, he joined the Microbiology and Molecular Genetics Department at The University of Texas Health Sciences Center, Houston, Tex. His laboratory characterizes the structure and function of the T-DNA transport system of *Agrobacterium tumefaciens.*

Pascale Cossart is the head of the Unité des Interactions Bactéries Cellules in the Pasteur Institute, Paris, France. She has been working on the molecular basis of *Listeria monocytogenes* infection since 1986 and has specialized in studying entry into epithelial cells and actin-based motility.

Antonello Covacci was born in Tuscania, Italy, on 14 December 1957. During medical studies, he was a junior fellow of the Sclavo Research Center (now IRIS), Siena, Italy. He graduated summa cum laude from the University of Florence Medical School and was promoted to a practicing physician in internal medicine. His postdoctoral work was at the Hormone Research Institute, University of California, San Francisco. He is a senior scientist at the Immunobiological Research Institute of Siena (IRIS), Chiron Vaccines, and Professor of Molecular Microbiology at the University of Siena. In 1995, he was a visiting scholar at Stanford University.

Chantal de Chastellier is a Research Associate, employed by Institut National de la Santé et de la Recherche Médicale, Paris, France. She worked at the Institut Pasteur, Paris, France, from 1976 to 1989 and since then has worked at the Necker Faculty of Medicine, Paris, France, in INSERM Unit 411. At present, she is group leader of a program on the cell biology of endoparasites. Her research activity first involved work on the ultrastructure of bacteria and then on endocytosis in the amoeba *Dictyostelium dis-*

coideum, with special emphasis on the morphological characterization of organelles of the endocytic pathway, endomembrane trafficking, and exchange of endocytic contents. Since 1983, her work has specialized in endocytic and phagocytic processing in macrophages with the aim of characterizing the intraphagosomal survival strategies of intracellularly growing pathogens (e.g., *Mycobacterium* and *Listeria*).

B. Brett Finlay is a Professor in the Biotechnology Laboratory and the Departments of Microbiology and Immunology and of Biochemistry and Molecular Biology at the University of British Columbia, Vancouver, B.C., Canada. He obtained his B.Sc. (Honors) in biochemistry at the University of Alberta (1981) and obtained his Ph.D. in biochemistry (1986) at the University of Alberta, studying F-like plasmid conjugation and pilus production with William Paranchych. He did his postdoctoral fellowship studies with Stanley Falkow (1986–1989), Department of Microbiology and Immunology, Stanford University, Stanford, Calif., studying *Salmonella* interactions with host cells. He then started his independent academic career in the Biotechnology Laboratory at the University of British Columbia (1989), and continues to study bacterial pathogenesis.

Åke Forsberg is associate professor at the Department of Microbiology at the Defense Research Establishment, Umeå, Sweden. His work has focused on the regulation and function of virulence determinants in pathogenic bacteria. His current work involves characterization of virulence determinants of *Yersinia pestis* and *Pseudomonas aeruginosa.*

Matthew S. Francis is a postdoctoral researcher in the *Yersinia* group at the Department of Cell and Molecular Biology, Umeå University, Umeå, Sweden.

Raluca Gagescu has been a Ph.D. student in Jean Gruenberg's laboratory at the University of Geneva, Geneva, Switzerland, since 1995.

Benjamin Geiger works in the Department of Molecular Cell Biology, Weizmann Institute of Science, Rehovot, Israel, studying different aspects of cell-cell and cell-extracellular matrix interactions and their role in cell regulation and signaling. He received his Ph.D. at the Weizmann Institute of Science.

Jean Gruenberg received his Ph.D. from the University of Geneva, Geneva, Switzerland, in 1980. He was then a postdoctoral student at the University of California, Riverside, and at EMBL from 1980 to 1987. After being a group leader at EMBL from 1987 to 1993, he moved to the University of Geneva, where he has been a professor since 1994.

Ilona Idanpaan-Heikkila received medical training in Helsinki, Finland. She joined the laboratory of Elaine Tuomanen, then at the Rockefeller University. After a postdoctoral stay in the laboratory of Arturo Zychlinsky at the Skirball Insititute, New York University Medical Center, she joined the vaccine division of SmithKline Beecham.

Marc Lecuit is pursuing a doctoral degree in clinical infectious diseases and microbiology. He is interested in developing the interface between clinical infectious diseases and fundamental aspects of microbial pathogenicity. His present work at the Pasteur Institute, Paris, France, in Pascale Cossart's laboratory concerns the study of the interaction between *Listeria monocytogenes* internalin and its eukaryotic receptor E-cadherin both in vitro and in vivo.

Sandra J. McCallum is a postdoctoral fellow in the laboratory of Julie Theriot at Stanford University, Palo Alto, Calif. Her work is focused on the mechanisms of actin-based motility mediated by the IcsA (VirG) protein of *Shigella flexneri*. She obtained her Ph.D. from Cornell University, Ithaca, N.Y., in the field of signal transduction by Rho family GTPases.

Timothy K. McDaniel is a Damon Runyon-Walter Winchell Postdoctoral Fellow in the laboratory of Stanley Falkow at Stanford University, Stanford, Calif. He is a molecular geneticist who studies the virulence of *Helicobacter pylori*. While a graduate student in the laboratory of James Kaper at the University of Maryland, Baltimore, Md., he identified the locus of enterocyte effacement pathogenicity island of enteropathogenic *Escherichia coli*.

Jeremy E. Moss is an M.D.-Ph.D. candidate at the New York University Skirball Institute, New York, N.Y. He is working in the laboratory of Arturo Zychlinsky, studying the pathogenic mechanism of *Shigella flexneri*, the etiologic agent of bacillary dysentery. He received his B.A. degree from the Stanford University Program in Human Biology in 1993. During that time, he worked in the laboratory of Mark Holodniy, analyzing mechanisms of human immunodeficiency virus drug resistance.

Mariagrazia Pizza obtained her Ph.D. in Chemistry and Pharmaceutical Technology at the University of Naples, Naples, Italy. Following stays at EMBL, Heidelberg, Germany, and at the University of Naples, she joined the Laboratory of Molecular Biology at Sclavo Research Center, Siena, Italy. Her research interests have focused on the design and construction of nontoxic mutants of bacterial toxins (pertussis, cholera, and heat-labile toxins). Her main achievement has been the development of the first recombinant bacterial vaccine that contains a genetically detoxified form of pertussis toxin. She is a member of the senior staff at the Research Center of Chiron SpA, Siena, Italy.

Klaus T. Preissner is head of the Institute for Biochemistry, Medical Faculty, University of Giessen, Giessen, Germany. The main focus of his research is dedicated to cellular adhesion mechanisms in the vascular system and in pathogen-host tissue interactions. His group identified novel mechanisms that are relevant for vascular remodeling processes related to angiogenesis and atherogenesis, inflammation, and humoral defense systems. Formerly, he joined the Department of Immunology, Scripps Clinic and Research Foundation, as a postdoctoral fellow and was group leader at the Clinical Research Unit for Blood Coagulation and Thrombosis as well as the Max-Planck-Institut, Kerckhoff-Klinik, Bad Nauheim, Germany.

Daniel L. Purich is Professor of Biochemistry and Molecular Biology, University of Florida College of Medicine, Gainesville, Fla. He is a specialist in enzyme kinetics and protein polymerization. His major research accomplishments include the demonstration that exchangeable-site GTP hydrolysis controls microtubule assembly and stability; determination of the role of repeated sequences in microtubule-associated protein (MAP) interactions with microtubules; discovery of MAP2-polymerization into Alzheimer-like filaments; and identification of ABM-1 and ABM-2 oligoproline docking sites that regulate intracellular actin-based motility of *Listeria*, *Shigella*, and vaccinia virus.

Rino Rappuoli obtained his Ph.D. in Biological Sciences from the University of Siena, Siena, Italy. Following stays at the Rockefeller University and the Harvard Medical School, he returned to the Sclavo Research Center in Siena, first as head of the Laboratory of Bacterial Vaccines and subsequently as head of the R&D Division. His research interests include various aspects of bacterial pathogenesis, focusing on bacterial toxins and vaccines. His major achievements have been the primary structures of diphtheria, pertussis, and *H. pylori* cytotoxins and the development of the first recombinant bacterial vaccine that contained a genetically detoxified form of pertussis toxin. He is Head of Research at Chiron SpA, Siena, Italy.

David G. Russell is a professor in the Department of Molecular Microbiology, Washington University School of Medicine, St. Louis, Mo. He trained at Imperial College, London, United Kingdom, and the University of Kent, Canterbury, United Kingdom, before assuming positions at the Max-Planck-Institut, Tübingen, Germany, and NYU School of Medicine, New York, N.Y. His work has always focused on the interface between the microbe and its host, originally on apicomplexan and kinetoplastid parasites and more recently on *Mycobacterium* spp. His continuing characterization of the mycobacterium-macrophage interplay constitutes his most substantive contribution to the field.

Philippe J. Sansonetti is a professor and principal investigator at the Institut Pasteur, Paris, France. He has been studying the mechanisms involved in the virulence of *Shigella*, the causative agent of bacillary dysentery, for the last 19 years. His laboratory has made key contributions in the identification and characterization of bacterial factors and cellular and tissue responses implied in the colonization of the colonic mucosa by *Shigella*.

Kurt Schesser is a postdoctoral researcher in the *Yersinia* group at the Department of Cell and Molecular Biology, Umeå University, Umeå, Sweden.

Frederick S. Southwick is Professor and Chief of the Division of Infectious Diseases, Department of Medicine, University of Florida College of Medicine, Gainesville, Fla. He is an infectious-diseases specialist with a long-standing interest in actin-based motility. His major research accomplishments include the discovery of CapG, a protein that caps the barbed ends of actin filaments in macrophages; demonstration of the association of defective neutrophil motility with impaired actin filament assembly in vivo; discovery of the central role of actin in the intracellular motility of *Listeria monocytogenes;* and identification of ABM-1 and ABM-2 oligoproline docking sites that regulate intracellular actin-based motility of *Listeria*, *Shigella*, and vaccinia virus.

Julie A. Theriot is an assistant professor of biochemistry at Stanford University, Palo Alto, Calif., with a secondary appointment in Microbiology and Immunology. Her laboratory studies the actin-based motility of the bacterial pathogens *Listeria monocytogenes* and *Shigella flexneri*. She obtained her Ph.D. in the field of actin dynamics at the University of California, San Francisco, and then completed her training as a Whitehead Fellow at the Whitehead Institute, Cambridge, Mass.

Guy Tran Van Nhieu is an INSERM fellow at the Institut Pasteur, Paris, France. He worked as a postdoctoral fellow on bacterial entry into epithelial cells mediated by the *Yersinia pseudotuberculosis* invasin protein in the laboratory of Ralph Isberg (Boston), before joining the Sansonetti group in 1994.

Emil R. Unanue is the Head of the Department of Pathology, Washington University School of Medicine, St. Louis, Mo. His work has centered on the analysis of antigen-presenting cells, particularly macrophages, in the immune response. His laboratory has shown the requirement for antigen processing for T-cell recognition. Their analysis of processing led to the findings that class II MHC molecules were peptide-binding molecules. His laboratory has also defined the role of the innate system in resistance to intracellular pathogens.

Raphael H. Valdivia is a Damon Runyon-Walter Winchell Postdoctoral Fellow at the University of California, Berkeley, Calif., in the laboratory of Randy Schekman. While a graduate student in the laboratory of Stanley Falkow at Stanford University, Stanford, Calif., he developed the differential fluorescence induction (DFI) technique for identifying bacterial virulence factors and codeveloped the hyperfluorescent green fluorescent protein variants that made DFI possible. He currently studies protein trafficking in yeast.

Hans Wolf-Watz is professor and chairman of the Department of Cell and Molecular Biology, Umeå University, Umeå, Sweden. His work during the last 15 years has focused on the molecular mechanisms of bacterial virulence with emphasis on the pathogenic *Yersinia* species. The major contribution to this field by his research team is the finding that *Yersinia* is an extracellular pathogen with the ability to block phagocytosis and the discovery of the type III-mediated translocation of virulence effector proteins into eukaryotic cells.

Eli Zamir works in the Department of Molecular Cell Biology, Weizmann Institute of Science, Rehovot, Israel, studying the structure and dynamics of focal adhesions. He is a Ph.D. student in the laboratory of Benjamin Geiger.

Arturo Zychlinsky received his undergraduate education at the Escuela Nacional de Ciencias Biologicas, Mexico City, Mexico, and received his Ph.D. in immunology from the Rockefeller University, New York, N.Y. After a postdoctoral fellowship in the laboratory of Philippe Sansonetti at the Institut Pasteur, Paris, France, he joined the Skirball Institute and the Department of Microbiology at the New York University Medical Center, New York, N.Y.

Index

353